Archaeological Sciences 1989

Proceedings of a conference on the application of scientific techniques to archaeology Bradford, September 1989

Edited by

P. Budd, B. Chapman, C. Jackson, R. Janaway, B. Ottaway

Oxbow Monograph 9
1991

Published by
Oxbow Books, Park End Place, Oxford OX1 1HN

© The individual authors 1991

ISBN 0 946897 29 8

This book is available direct from
Oxbow Books, Park End Place, Oxford, OX1 1HN
(Phone: 0-865-241249; Fax: 0-865-794449)
Payment may be made by credit card

Printed by
The Short Run Press, Exeter

PREFACE

The Archaeological Sciences '89 conference, held at the University of Bradford in September 1989, continued the tradition established at Glasgow two years before of arranging a small archaeometry conference in Britain in alternate years to the International Symposia on Archaeometry. The theme of the conference, the application of scientific techniques to archaeology, was particularly appropriate to the conference venue; the first undergraduate degree course in Archaeological Sciences having been founded at Bradford from where the concept has spread to a number of other institutions in recent years.

The presentation of so many papers over the three days of the conference reflected the higher profile of scientific techniques in archaeology which has developed in recent years. Papers dealt with ongoing projects and new research initiatives and featured new techniques and refinements to established methods which were of interest both to the specialist and the wider science based archaeology community. The papers are presented thematically in sections on the physical and chemical analysis of inorganic materials, archaeometallurgy and mining, dating, remote sensing, the analysis of animal and plant remains and the study of human remains.

Many people were involved in the organization of the conference and we would like to thank all those members of staff and students of the Department of Archaeological Sciences who gave up their time to help and in particular our colleagues on the organizing committee who did so much to make the conference a success. Furthermore, we would like to thank our colleagues on the conference proceedings editorial committee for their initial enthusiasm and effort in the compilation of this volume.

Paul Budd and
Caroline Jackson.

Department of Archaeological Sciences
University of Bradford.

December 1990.

CONFERENCE ORGANIZERS

Paul BUDD
Barbara CHAPMAN
John HUNTER
Caroline JACKSON
Robert JANNAWAY
Barbara OTTAWAY

Department of Archaeological Sciences
University of Bradford
Bradford, BD7 1DP.

CONFERENCE DELEGATES

M AITKEN, RLAHA, 6 Keble Road, Oxford, OX1 3QH

J AMBERS, 90 Sandwich House, Sandwich Street, London WC1H 9PW

S ANDERSON, Department of Anthropology, 43 Old Elvet, Durham, DH1 3HN

S ANTOINE, School of Chemistry and Applied Chemistry, University of Wales, College of Cardiff, PO Box 912, Cardiff

M ARMOUR CHELU, Dept of Palaeontology, British Museum (Natural History), Cromwell Road, London, SW7 5BD

C J ARNOLD, Bronawel, Green Lane, Abermule, Montgomery, Powys, SY15 6LB

A ASPINALL, Department of Archaeological Sciences, University of Bradford, Bradford, BD7 1DP

D ATKINSON, Geomagnetism Laboratory, University of Liverpool, Oxford Street, Liverpool.

I BAILIFF, TL Laboratory, Archaeology II, Woodside Building, University of Durham, South Road, Durham, DH1 3LE

M G L BAILLIE, Palaeoecology Centre, Queens University, Belfast BT7 1NN

C BATT, C/O Geology Department, Beaumont Building, University of Sheffield, Brookhill, Sheffield, S3 7HF

J BAYLEY, Ancient Monuments Laboratory, 23 Savile Row, London, W1X 2HE

L S BELL, Hard Tissue Unit, Dept of Anatomy and Developmental Biology, University College, Gower Street, London, WC1N 6BT

M BELL, 223 Western Avenue, Dagenham, Essex, RM10 8UL

T J BELLERBY, Department of Geology, Beaumont Building, University of Sheffield, Brookhill, Sheffield, S3 7HF

P H BETHELL, 22 Montpellier Terrace, Swansea, SA1 6JW

N W BLADES, Department of Geology, Royal Holloway and Bedford New College, Egham Hill, Egham, Surrey, TW20 0EX

J BOND, Department of Archaeological Sciences, University of Bradford, Bradford, BD7 1DP

W H M BOUTS, Hinthamerstraat 153, 5211 MK's Hertogenboch, The Netherlands

P BUDD, Department of Archaeological Sciences, University of Bradford, Bradford, BD7 1DP

J BUXEDA, Dpt. Prehistòria, Història Antiga i Arqueologia, Facultat Geografia i Història Universitat de Barcelona, C. Baldiri i Reixac, s/n, 08028 Barcelona, Spain

C CAPLE, Dept of Archaeology, University of Durham, Durham

S P CAUVAIN, 97 Guinions Road, High Wycombe, Bucks, HP3 7NU

B CHAPMAN, Department of Archaeological Sciences, University of Bradford, Bradford, BD7 1DP

A M CHILD, Department of Chemistry, University of Wales, College of Cardiff, PO Box 912, Cardiff

P CLARK, SURRC, East Kilbride, Glasgow, G75 0QU

R J CLARK, SURRC, East Kilbride, Glasgow, G75 0QU

H CLYNE, 11 Fountain Drive, London, SE19 1UW

M COWELL, Research Laboratory, British Museum, Great Russell Street, London, WC1B 3DG

M COX, 7 Port Close, Bearsted, Maidstone, Kent

A CRAWSHAW, Y.A.T., Conservation Laboratory, Galmanhoe Lane, Marygate, York YO3 7DZ.

J CSAPO, University of Agriculture, Faculty of Animal Science, H7400 Kaposvar, Hungary

L R DAY, 16 Whadden Chase, Ingatestone, Essex, CH4 9HF

C DICKSON, C/O Dept of Botany, University of Glasgow, Glasgow, G12 8QQ

S DOCKRILL, Department of Archaeological Sciences, University of Bradford, Bradford, BD7 1DP

A J DUGMORE, Department of Geography, University of Edinburgh, Edinburgh, EH8 9XP

L A DUTTON, Gwynedd Archaeological Trust Ltd, Ffordd y Coleg, Bangor, Gwynedd, LL57 2DG

R EDWARDS, University of Wales, College of Cardiff, Department of Inorganic Chemistry, PO Box 912, Cardiff, CF1 3TB

T ELLIOTT, Department of Anthropology, University of Durham, 43 Old Elvet, Durham, DH1 3HN

R P EVERSHED, Dept of Biochemistry, University of Liverpool, PO Box 147, Liverpool, L69 3BX

M FABRIZI, 42 Tyrwhitt Road, Brockley, London, SE4 1QG

P J FASHAM, Gwynedd Archaeological Trust Ltd, Ffordd y Coleg, Bangor, Gwynedd, LL57 2DG

K FOLEY, Ancient Monuments Laboratory, English Heritage, 23 Saville Row, London, W1X 2HE

N GARLAND, Dept of Pathology, Stopford Building, Oxford Road, Manchester, M13 9PT

G GAUNT, Ascham House, 10 Foxhill Crescent, Weetwood, Leeds, LS16 5PD

H GEAKE, Sutton Hoo Research Project, Sutton Hoo, Woodbridge, Suffolk

G GILMORE, Universities Research Reactor, Risley, Warrington, WA3 6AT

D R GRIFFITHS, Institute of Archaeology, 31-34 Gordon Square, London, WC1H OPY

M F GUERRA, C Ernesi-Babelon, URA 27, CNRS-IRHT, 3 bis, Av. de la Recherche Scientifique, 45071 Orleans Cedex 2, France

J HAIGH, Department of Mathematics, University of Bradford, Bradford, BD7 1DP

D N HALE, Geology Department, University of Sheffield, Brookhill, Sheffield, S3 7HF

S HARDMAN, Dental School, University of Wales, College of Medicine, Cardiff, CF4 4XY.

I HEASMAN, Particle Characterisation Services Ltd, 9 Woodhouse Grove, Manor Park, London, E12 6SR

C HERON, Dept of Biochemistry, University of Liverpool, PO Box 147, Liverpool, L69 3BX

M HEYWORTH, Ancient Monuments Laboratory, 23 Savile Row, London, W1X 2HE

T HOLDEN, 50 Grovelands Road, Palmers Green, London N13 4RH

M J HUGHES, British Museum, Research Laboratory, Great Russell Street, London, WC1B 3DG

M S HUMPHREY, Sedgwick Museum, Department of Earth Sciences, Downing Street, Cambridge

J R HUNTER, Department of Archaeological Sciences, University of Bradford, Bradford, BD7 1DP

C M JACKSON, Department of Archaeological Sciences, University of Bradford, Bradford, BD7 1DP

R JANAWAY, Department of Archaeological Sciences, University of Bradford, Bradford, BD7 1DP

D JENKINS, c/o SAFS - Soil Science, UCNW Bangor, Gwynedd, LL57 2UW

A K G JONES, Department of Archaeology, University of York, Heslington, York

D JORDAN

P KENT, West Bury Farm, Whaddon Road, Shenley Brook End, Milton Keynes

A G LATHAM, Institute of Prehistoric Sciences and Archaeology, Liverpool University, Liverpool, L69 3BX

F MACALISTER

M MACQUEEN, 7 Furzefield Road, Blackheath, London, SE3 8TU

K MANCHESTER, c/o Dept of Archaeological Sciences, University of Bradford, Bradford BD7 1DP

B MARGERISON, Department of Archaeological Sciences, University of Bradford, Bradford, BD7 1DP

M MARLOW, Department of Anthropology, University of Durham, 43 Old Elvet, Durham DH1 3HN

G MARTIN, Conservation Department, Victoria and Albert Museum, South Kensington, London, SW7 2RL

C McELROY, RLAHA, 6 Keble Road, Oxford

C MILLS, AOC, SDD, 23-25 Chamber Street, Edinburgh

B MOFFAT, 36 Hawthornvale, Edinburgh, EH6 4JN

C MORTIMER, Research Laboratory, Archaeology, 6 Keble Road, Oxford

D de MOULINS, Museum of London, London Wall, London, EC2

J MURRAY, Dept of Environmental Archaeology, Museum of London, London Wall, EC2Y 5HN

N NAYLING, The Museum of London, London Wall, London, EC2Y 5HN

P T NICHOLSON, 111 Heavygate Road, Crookes, Sheffield, S10 1PE

R A NICHOLSON, 4 Abbey Street, Clifton, York, YO3 6BG

Y NISHIMURA, 9-1, 2 Chom Nijyocho, Japan

M NOEL, Department of Geology, Geology Department, Beaumont Building, University of Sheffield, Brookhill, Sheffield, S3 7HF

K OSBORNE, 32 Burghley Road, St Andrews, Bristol, BS6 5BN

R L OTLET, UK Atomic Energy Authority, Harwell, Oxon

B OTTAWAY, Department of Archaeological Sciences, University of Bradford, Bradford, BD7 1DP

S OVENDEN, Department of Archaeological Sciences, University of Bradford, Bradford, BD7 1DP

M PARKER PEARSON, Room 313, Fortress House, 23 Savile Row, London, W1X 2HE

C PATHY-BARKER, 3 Britain Street, Dunstable, Bedfordshire, LU5 4JA

P M PAYTON, 130 Parkway, Welwyn Garden City, Herts, AL8 6HP

J PILC, Scientific Department, National Gallery, Trafalgar Square, London, WC2N 5DN

M PIKE, c/o Applied Geology Ltd, Techbase 3, Newtech Square, Deeside Industrial Park, Deeside, Clwyd, CH5 2NT

C PLANAS, Dpt. Prehistòria, Història Antiga i Arqueologia, Facultat Geografia i Història Universitat de Barcelona, C. Baldiri i Reixac, s/n, 08028 Barcelona, Spain

M POLLARD, School of Chemistry and Applied Chemistry, University of Wales, College of Cardiff, PO Box 912, Cardiff, CF1 3TB

A H POWERS, Dept of Archaeology and Prehistory, Sheffield, South Yorks, S10 2TN

C PRICE, Ancient Monuments Laboratory, English Heritage, 23 Saville Row, Londom, W1X 2HE

A QUYE, Royal Museum of Scotland, Chambers Street, Edinburgh

J REEVE, Glebe Farm Bungalow, Carleton Formoe, Norwich, Norfolk, NR9 4AP

F REHMAN, Chemistry Department, The University, Manchester, M13 9PL.

P R RITCHIE, 35 Castle Avenue, Edinburgh, EH12 7LB

C ROBERTS, Dept of Archaeological Sciences, Univ of Bradford, Bradford, BD7 1DP

D ROBINSON, National Museum 8th Dept., NY Vestergade II, 1471 Copenhagen K, Denmark

V J ROBINSON, Chemistry Department, The University, Manchester, M13 9PL.

P ROTTLANDER, Wacholderweg 2, D 7407 Rottenburg 6, Germany.

R C A ROTTLANDER, Inst. f. Urgeschichte Schloss, Archaeochem. Lab., D 7400 Tubingen, Germany

D SANDERSON, SURRC, East Kilbride, Glasgow, G75 0QU

M SAX, Research Laboratory, British Museum, London, WC1B 3DG

E SLATER, Department of Archaeology, University of Glasgow, Glasgow, G12 8QQ

F SMITH, c/o SERC, Polaris House, North Star Avenue, Swindon, SN1 2ET

R SMITH, The Royal Armouries, Tower of London, London, EC3N 4AB

L SOMERS, Box 363, Sea Ranch, California, 95497 USA

J SPENCER, SURRC, East Kilbride, Glasgow, G75 0QU

S STALLIBRASS, 6 Grays Terrace, Durham City, DH1 4AU

Z STANCIC, Filozfska fakulteta Arheolog, Askerceva 12, 6100 Ljubljana, Yugoslavia

P SUTTON, School of Chemistry and Applied Chemistry, University of Wales, College of Cardiff, PO Box 912, Cardiff

R SWAN, Deynes Farm, Debden, Saffron Walden, Essex, CB11 3LG

J SZYMANSKI, Dept of Electronics, University of York, Heslington, York, YO1 5DD

J O TATE, Royal Museum of Scotland, Chambers Street, Edinburgh.

R G THOMAS, Department of Chemistry, University of Wales, College of Cardiff, PO Box 912, Cardiff

R S THORPE, Department of Earth Sciences, Open University, Milton Keynes, MK7 6AA

S A TIMBERLAKE, 12 York Street, Cambridge, CB1 2PY

M S TITE, British Museum Research Laboratory, Great Russell Street, London WC1B 3DG

J E TOMLINSON, Chemistry Department, The University, Manchester, M13 9PL.

B TUGRUL, Instanbul Technical University, Institute for Nuclear Energy, Ayazaga Kampusu, 80626 Maslak, Instanbul, Turkey

R F TYLECOTE, Yew Tree House, East Hannay, Nr Wantage, Oxford, OX12 0HT

J WAKELY, Anatomy Department, University of Leicester, PO Box 138, University Road, Leicester, LE1 9NH

S WALES, 14 Chapel Hill, Soulbury, Leighton Buzzard, Beds, LU7 0BZ

A R WALKER, Geoscan Research, Heather Brae, Chrisharben Park, Clayton, Bradford, BD14 6AE.

J N WALSH, Department of Geology, Royal Holloway and Bedford New College, Egham Hill, Egham, Surrey, TW20 0EX.

S E WARREN, Department of Archaeological Sciences, University of Bradford, Bradford, BD7 1DP

L C WINTER, Dept of Anthropology, 43 Old Elvet, Durham, DH1 3HN

A G WOODCOCK, County Archaologist, East Sussex County Council, County Planning Department, Southover House, Lewes, East Sussex BN7 1YA

J D WILCOCK, Dept. of Computing, Staffordshire Polytechnic, Blackheath Lane, Stafford, ST18 0AD

P A WILLIAMS, School of Chemistry, University of Wales, Cardiff.

R WILLIAMS, School of Chemistry and Applied Chemistry, University of Wales, College of Cardiff, PO Box 912, Cardiff, CF1 3TB

O WILLIAMS-THORPE, Department of Earth Sciences, Open University, Milton Keynes, MK7 6AA

L WILLIES, Peak District Mining Museum, Matlock Bath, Derbyshire

P T WILTHEW, Royal Museum of Scotland, Chambers Street, Edinburgh

H WOUTERS, University of Antwerp, Dept of Chemistry, Universiteitsplein 1, B-2610 Wilrijk, (Antwerp), Belgium

CONTENTS

ARCHAEOMETALLURGY AND MINING

DATING

REMOTE SENSING

THE ANALYSIS OF ANIMAL AND PLANT PRODUCTS AND RESIDUES

THE STUDY OF HUMAN REMAINS

COORDINATOR'S REVIEW 1989

A.M. Pollard

School of Chemistry and Applied Chemistry, University of Wales, Cardiff, CF1 3TB.

When the idea of a Coordinator's review was first discussed, it was intended to be a presentation of the contents of the Coordinator's Report, which was to be the final act of a two year term of office. That is not now the case. The Coordinator's Report exists in draft form, but has not yet been considered by the three bodies jointly supporting the position (the Science-based Archaeology Committee of the Science and Engineering Council, English Heritage and the British Academy). Nor has the term of office come to an end - it has been extended for a further year. I would therefore like to take this opportunity to review the progress made in the support of archaeological science since the Hart report, published in 1985 (Hart, 1985).

Few who were present at the meeting organised by SERC in Manchester in October 1985 to discuss the findings of the Hart report will ever forget the experience. At the time, it seemed as if the entire structure of science-based archaeological research in the UK was about to be demolished, along with archaeology in general. Professor Hart made a lasting impression; he said that he was only trying to be helpful, and that we had to get our act together. With hindsight, both were correct.

In 1987, the three bodies named above appointed a National Coordinator for Science-based Archaeology, essentially with a remit to improve communications between funders and funded, and to liaise between funding bodies, as well as between the various groups of workers interested in archaeological science. Much has happened since then, although in fairness the Coordinator can only claim a catalytic action in some of the proceedings. The following is a brief survey of some of the developments.

The Forum for Coordination in the Funding of Archaeology

This is a discussion body composed of representatives of the major funding bodies in British Archaeology - not just archaeological science (the British Academy, English Heritage, SBAC, SDD, Cadw, DoENI, Local Authorities, Museums and Galleries Commission, the British Museum, the British Property Federation, the Society of Antiquaries of London and the Royal Commissions). The Coordinator attends the meetings by invitation, and the secretary is Miss Justine Bayley of the Ancient Monuments Laboratory, English Heritage.

The Forum discusses issues of joint concern, which transcend the individual remits of each constituent body. Of interest to archaeological scientists is the work of the Sub-Committee on the Funding of Applied Archaeological Science (discussed below), and the consideration of the provision of radiocarbon dates for UK researchers. The establishment of the Forum directly counters one of the major criticisms in the Hart report, ie. that the structure of UK archaeology is 'complex and fragmented'.

The Forum is deliberately designed to be a discussion vehicle for the major funders of UK archaeology. A parallel initiative has resulted in the establishment of the British Archaeological Research Forum, comprising, in addition to others, of the Council for British Archaeology, the Institute of Field Archaeologists, and the Standing Committee of University Professors and Heads of Archaeology. This met for the first time in summer 1989, and it is to be hoped that a fruitful dialogue will ensue between the two Fora, and that this does not contribute further to the 'fragmentation'.

The Funding of Applied Archaeological Science

A Sub-Committee was established by the Forum to report on the funding of applied archaeological science, under the chairmanship of Professor Renfrew. Because of the historical development of the subject, archaeology has inherited an unfortunate funding structure, with scientific developments being funded by SERC, but archaeological research using those techniques being, in the main, the responsibility of the British Academy. English Heritage and the equivalent bodies in Scotland, Wales and Northern Ireland are responsible for Rescue archaeology, although the last few years have witnessed a rapid rise in 'Developer funding'.

It has become apparent, and was highlighted in the Hart report, that applied archaeological science - the application of established scientific techniques to archaeological research - is not currently being funded from any of the major bodies at a sufficient level to meet the demand. This is the result of a combination of factors: the remit of some bodies effectively limits the support which can be given to this category of work; in other cases, the funds are insufficient to meet the demand.

The Sub-Committee recommended that:

i)   all archaeological science related to the production of the site report and archive is, in the main, the responsibility of the body (or bodies) funding the work, and should be included as far as possible in the research design;

ii)   University and Polytechnic staff should make more use of existing Research Council facilities (listed in an Appendix to the report);

iii)   there should be a Fund for Applied Archaeological Science, initially intended to pay for access to existing scientific services, and made available to a wide range of archaeological users;

iv)   in the longer term, there is a need for a number of regional archaeological science laboratories, which would probably grow out of existing nuclei.

The British Academy has taken the lead in establishing a fund, and has made available £20,000 for this financial year (1989/90). Discussions are in hand between the Academy, English Heritage and SBAC about the establishment of a larger fund next year. Although this is a relatively small sum, it is a major step in addressing a problem which has bedevilled archaeological science for a number of years.

The Science-based Archaeology Committee

SBAC has prospered since 1985; in the last year, there has been a 90% increase in research funding, and also increases in the number of studentships available. Despite this, there continues to be a vast over-application for both research grant funds and studentships. It has maintained a high profile, both in the archaeological community and on the Science Board of SERC. The current awareness of environmental issues has brought some of SBAC's research support into the limelight: the construction of long-term tree-ring chronologies and the subsequent calibration of the radiocarbon timescale, and studies of mankind's impact on the landscape are now appreciated as substantial contributions to the 'green' debate. It is noteworthy that the most effective legislation for the protection of the archaeological heritage has been achieved by linking archaeology with the general theme of environmental protection.

The current chairman, Professor Rosemary Cramp, retires from the Committee this month (9/89), to be replaced by Professor David Harris. It is doubtful whether the archaeological science community will ever be fully aware of the debt it owes to Professor Cramp. The high regard in which she is held, both on the Science Board and, equally importantly, within the archaeological community, has been of vital importance over the last three years. The Committee changes every year, with three or four of the members being replaced at any one time. This year, the new appointees are Drs. Martin Jones (Durham), Stephen Shennan (Southampton) and Colin Shell (Cambridge).

There continues to be a problem with the funding of environmental archaeological research, and this is to be addressed at a day meeting on the topic at the British Museum (Natural History) on the 16th January 1990. The current atmosphere of collaboration between the Research Councils is helping this matter, and SBAC continues to receive full cooperation in this matter from the assessor representing NERC on the Committee, as well as from those representing English Heritage and the British Academy.

English Heritage

The Ancient Monuments Laboratory continues to provide scientific support to rescue archaeologists in England, either through the in-house AML staff, or via university-based contract staff. The rapid rise in developer funding is causing a re-appraisal of the way in which post-excavation support is provided through English Heritage, as summarised in the latest policy document (English Heritage, 1989). This recommends an assessment stage at the end of the excavation and after the production of the site archive and narrative account, during which the potential for analytical work on the data is examined. If further work is merited, a new research design is submitted to English Heritage, together with costings. If approved, this work will constitute a research archive, which will contain all the records of the objects and ecofacts, together with reports on the analyses carried out on them. Throughout the document, there is a constant demand for the relevant scientists to be involved at all stages of the planning and execution of the project, and as such the proposals should meet the demands of archaeological scientists involved with rescue excavations.

One of the tasks of the Coordinator over the past two years has been to visit archaeological trust and units, in an attempt to assess the awareness of field archaeologists about scientific techniques; this remains extremely variable. The larger operations (eg. London, York, Wessex) now employ a large number of scientifically-trained archaeologists, and maintain close links with a number of university departments. This is unfortunately not so in the case of smaller organisations, and any initiatives to improve this (eg. through IFA or the CBA) should be encouraged. A positive sign of field archaeologists' interest in archaeological science has been the growing number of applications to SBAC for CASE studentships sponsored by archaeological trusts, as well as by English Heritage itself - these are an extremely effective way of sponsoring research which is of direct interest to field archaeology.

The British Academy

The Academy is the primary funding body for archaeological research in the UK, and it is increasingly being expected to act as the Research Council for the humanities - a role for which it lacks infrastructure. Section 10 (the archaeology committee), under the chairmanship of Dr. Roger Moorey, has been extremely active in attempting to streamline the decision-making processes within the Academy, and has taken the lead in important initiatives such as the Forum and the Fund for Applied Archaeological Science.

The University Sector

The long-awaited UGC review of archaeology has placed archaeological science at the heart of the universities' provision for archaeology, with a proposed funding structure based on the scientific content of the teaching and research. 'Fully science-based' departments are recommended to receive funding on a par with the physical sciences, with other departments being funded at the geography level or better. In the long term, this is probably the most significant development over the last five years, although we must await the response of the Universities Funding Council to the proposal.

The retirements of Professors Hall and Aitken at Oxford, and Stanley Warren at Bradford mark the end of an era as far as British archaeometry is concerned. Fortunately, all are being replaced, and Cambridge is making a further spectacular expansion in its scientific provision - archaeological science has never been healthier in British Universities and Polytechnics. At the other end, as it were, of the career spectrum, the quantity, quality and enthusiasm of research students has never been higher. The 'angry young men (and women) of archaeological science' (Thomas, 1989) are determined to demonstrate that archaeological science is not a 'second class science' (Jones, 1988). Increasingly, it is becoming clear that archaeological science is setting its own research goals, in addition to being a service to archaeology.

Summary

The overall picture is now considerably better than that perceived in Manchester in 1985. There is, however, still a need for vigilance. We must ensure that the proposed re-structuring of the Research Councils (the Morris Report) does not damage the position of SBAC. Nor should we rely too heavily on developers to fund archaeology - property development is a notoriously cyclical business. There are still sources of funding within SBAC which are relatively untapped - eg. fellowships and overseas travel grants. We must re-double our efforts to bring together 'field' and 'university' archaeology, as well as 'archaeology' and 'archaeological science'; the distinction between archaeology and archaeological science is largely a bureaucratic one, arising from the historical development of the funding bodies and the maturation of the discipline of archaeology. The impending restructuring of English Heritage and the relocation of the Ancient Monuments Laboratory will no doubt cause short-term problems, but should be turned to the long-term advantage of archaeology as a whole.

The Coordinators report will be a detailed discussion of these and other matters. It will be considered by the three funding bodies during the autumn, and will hopefully be published in early 1990. It is likely that SBAC will present it to the science board, together with a reminder that in 1985 Professor Hart recommended that science-based archaeology should be reviewed again in 1990. It would be a measure of the success of all involved over the last five years if the board were to decide that archaeological science had put its own house in order, and that further review would not be necessary.

References

English Heritage (1989). *The Management of Archaeology Projects*, English Heritage.

Hart, M. (1985). *Review of Science-based Archaeology 1985*, SERC.

Jones, R.F.J. (1988). Questions, answers and the consumer in archaeological science. In: *Science and Archaeology*, Glasgow 87, BAR British Series 196, eds. E.A. Slater and J.O. Tate, pp 1-7.

Thomas, R.G. (1989). Pers. comm.

THE TECHNOLOGY OF LATE LA TÈNE 'PAINTED POTTERY' DECORATION.

K. Andrews

Department of Archaeology and Prehistory, University of Sheffield, Sheffield, S10 2TN, England.

Summary

Thin decorative coatings on Late La Tène 'painted pottery' from the site of Aulnat, France have been analysed using scanning electron microscopy coupled with energy dispersive x-ray microanalysis (SEM/EDS). The analyses indicate that the decorative coatings are clay based slips rather than organically based pigment paints. The results have important implications in modelling the technological aspects of 'painted pottery' production.

Introduction

The following study represents the results of research into the nature of various coloured decorative coatings characteristic of the late Iron Age 'painted pottery' fine ware ceramic class from the Auvergne, France. The central research aim was to define the nature of the coatings and the implications results could hold for their production technology. Such technological aspects have important implications for theories of the economics of production. The question as to the exact nature of the decorative coatings is also central to understanding, and effectively modelling, the production stages and degree of specialisation involved in their manufacture. It is hoped then that this analytical programme will provide information which contributes to an understanding of how the 'painted' pottery was made; the levels of investment in materials, equipment and labour required to produce it and reveal any significance results may hold for understanding the firing technology employed.

The analysis of decorative coatings on La Tène 'Painted pottery' described here represents the largest set of analyses of this type. Previously only tentative conclusions have been formed as to the nature of such coatings based on limited analysis of very similar La Tène 'painted pottery' from Czechoslovakia and Poland (e.g. Cumberpatch and Pawlikowski, 1988; Wirska-Parachoniak, 1980). The present analytical programme has allowed more robust conclusions to be drawn which broadly agree with East European results.

Materials and methods

'Painted pottery' is a widely recognized element of the homogeneous Late La Tène material culture of Europe (sometimes referred to as 'red and white painted ware'). As a ceramic type it is noted extensively in European archaeological literature. All languages refer to this type as 'painted' pottery: German-'Bemalte keramik'; Czech-'Molvana keramika'; Polish-'Malowana keramika; and French-'Ceramique peinte'(Cumberpatch pers. comm.). It was made on a fast wheel, of a fine paste body fabric. Most, if not all, of the external (and rarely some internal) surfaces are finished with decorative coatings of one or more colours (Watson, 1988). The most common colours are reds and whites (red-oranges, dark reds/browns and white/creams, and beiges) and rarely mauve/greys. The coatings can be divided into a variety of styles and motifs ranging from simple banding, to zoomorphic and geometric designs. Some of the coatings have been burnished to give a lustrous finish and in many cases, selective burnishing was employed to create grid or curvilinear patterns from reflected light, whereby the burnished areas appear deeper coloured and gloss, whilst unburnished areas appear matte.

Previous studies have concentrated on motif design and art-historical aspects. There has been comparatively little archaeometric work carried out to define the composition of the decorative coatings. Are they organically based paints applied prior to or after firing; are they clay based slips - if so are they highly refined, are the coatings related to the body fabrics (self-slipped) or were separate clays prepared for the decoration; what materials were used to colour the coatings?

Examination of the coatings under a binocular microscope revealed that they vary in thickness from sample to sample from extremely thin (<50 micro metres) to comparatively thick (up to 0.4 mm) although the majority were thin (<0.1mm) and could not be seen with the unaided eye in fresh fracture cross sections. Most, however thin, gave a uniform and densely coloured surface finish appearance. Measurement under binocular microscope using an eye piece graticule showed that the majority of coatings were in the order of 0.06-0.08 mm thick.

In order to study the decorative coatings effectively, it is necessary to examine their structural characteristics as well as their chemical composition. The structural characteristics may include for example: particle size; nature of the interface between the coating and body fabric; microstratigraphy of the coating layer(s); and presence, distribution, shape and size of inclusions. The scanning electron microscope (SEM) offers exceptional opportunities of observing such features, most importantly the interface between the body fabric and coating, whilst analysis of each can be selectively carried out using an energy dispersive spectrometer (EDS) attached to the SEM.

Samples undergo uniform sample preparation therefore there are no problems of differing sample-detector geometries. A sample of the sherd cross section is presented to the analysing beam (which also produces an image of the area under analysis) which can be focused on any part of the cross section. When analysing the coating, it may be regarded as effectively of infinite depth or thickness (since it is presented in cross section) so there are no problems of unknown depth effects.

Two Link systems EDS analysis programmes were used, one a qualitative programme which analyses for most major and minor elements (above atomic number 11) and a semi quantitative ZAF 4 flat surface programme.

Results

In total 446 unknown spectra were recorded and analysed from 97 sample sherd cross sections. Several spectra from the coating layers of each sherd were analysed in order to obtain a mean reading which would overcome the problems of EDS microanalysis, namely sample heterogeniety causing unrepresentative results if only one atypical spectra were recorded (for example an analysis recorded over an area that was not recognized as an inclusion). Spectra were also recorded and analysed of the body fabric clay phases, i.e. avoiding large inclusions, in order to examine the possibilities of relationships between coating clay phases and body fabric clay phases. The element suites for both the coatings and body fabric clay phases were found to be very similar comprising magnesium, aluminium, silicon, phosphorous, sulphur, potassium, calcium, titanium and iron. This range of elements is typical of ceramic analyses (Tite and Maniatis, 1975:222; Magetti and Galetti, 1980; Freestone, 1982). Not all of the nine elements revealed by EDS displayed differences in amount between coating colour groups. Of the nine, four (Ca, Fe, Si, Al) were selected as displaying significant variation between coating colour groups. Table 1 illustrates the simplified results of the analytical programme, the median (M) and mean (x) values are shown. Values represent the simplified results of all analyses which were by percentage weight normalised to 100%.

With respect to body fabrics, there is no significant difference between the body fabrics of all decorative coating colour groups in their content of all characteristic elements indicating perhaps a universal source of body fabric clay, although there are obvious

2

Table 1. Group Means and Medians for Selected Elements

Group	Ca		Fe		Si		Al	
	x	M	x	M	x	M	x	M
1 C	7.73	7.45	14.8	15	48.6	47.2	20.8	20.2
1 B	3.76	3.60	13.3	13.3	54	53.1	19.3	19.2
2 C	11	8.4	13.8	15.7	47.6	48.4	19.4	19.1
2 B	5.13	3.5	13.0	12.4	54.5	54.8	18	18.1
3 C	11.7	4.5	6.3	5.3	52.5	57.1	21.8	22.5
3 B	2.4	2.2	12.9	12.6	56.2	55.6	19.8	19.0
4 C	32.8	32.3	6.13	5.6	41.3	42.2	12.9	12.3
4 B	4.55	3.4	12.7	12.2	54.5	55.3	18.5	18.3
5 C	30.2	29.5	5.65	5.70	41.0	40.6	15.7	14.9
5 B	5.5	4.2	14.0	14.4	51.4	52.2	17.9	18.3

Groups are as follows: 1 = dark reds; 2 = reds; 3 = whites; 4 = pale browns; 5 = greys. C denotes values for coatings, whilst B denotes values for body fabrics.

differences in added inclusions. The body clays are low in calcium (generally less than 5%) indicative of non calcareous clay. The body clays are also ferruginous bearing at least 12% Fe.

With respect to calcium content of the coatings, there is a marked difference between the content of groups 4 and 5 (33% and 30% mean values respectively) and the contents of groups 1-3 which contain less than 12%.

With respect to iron content of the coatings, there is a marked difference between the content of group 3, 4 and 5 (non reds) and groups 1 and 2 (reds). The mean content of iron for the reds is 14% contrasting with the non red mean content of less than 6%.

With respect to silicon content, there is less in the coatings of groups 1 and 2 when compared to their respective body fabrics.

For group 3 (whites) there is no significant difference between coating and respective body fabric indicating comparatively high coating content of silicon for group 3. In the pale browns and greys the difference is obvious, the coatings containing over 10% less silicon.

With respect to the aluminium content of coatings, there is no significant difference between groups 1, 2 and 3 (each containing c. 20%). Groups 4 and 5 show no significant difference, the average content being 12%. Statistical testing has shown this dichotomy between groups 1-3 and 4-5 with respect to aluminium content to be significant.

The significance of these various differences were supported by statistical testing including the rank sum test and analysis of variance which show that they are not random effects.

Definitions of a slip and a paint

A slip is better defined in the archaeological literature as an aqueous slurry of clay particles formed from washing out such fine fractions from a source clay (a process known as elutrification, where a clay is purified and fine fractions of it separated by the actions of straining and/or decanting in water). Slips may also be processed by levigation, whereby fine clay particles are suspended by grinding the source clay in water. In both cases fine clay particles are collected and mixed with water to form slips. Slips are applied onto leather hard ceramics prior to firing (Rice, 1987).

Paints are made by the addition of water to concentrated extracts of organic matter and admixtures of various minerals, especially metal oxides such as copper oxide (Gillies and Urch, 1983). A paint may be applied before or after firing of the ceramic although post firing painting is rare if not absent from European prehistory (Noll, 1977).

There may not be a sharp compositional distinction between a slip and a paint if the first has a large mineral admixture or the source clay of a slip has a large organic component. The presence of organic matter used in production of paints (for example carbohydrates such as starches) as binding media would be impossible to detect using most spectrometric analytical methods which do not detect elements such as hydrogen and carbon. A better distinction between a paint and a slip is a structural one. Structurally a paint contains a wider range of particle sizes and consists of various assorted flakes which can be detected under the SEM. A slip however, being water sorted, contains particles of a more uniform size and has a more isotropic structure when compared to a paint. Also a slip invariably shows a distinct line of demarcation from the body fabric (Tite and Maniatis, 1975).

In the case of the Aulnat coatings, the structural characteristics of a slip were noted under the SEM. The textural demarcation between coarser body fabric and slip coatings could be more easily observed in SE mode pick-up owing to the topographic contrast between the higher polish achieved consistently on the coatings when compared to that of the body fabric. Such demarcation indicative of a slip was also observed by colour contrast under the binocular microscope.

The EDS analysis results show uniform similarity between the elemental suites of the body fabric clay fractions and the decorative coatings. It may therefore be concluded that the Aulnat 'painted' pottery are decorated with clay based slip rather than organically based paint. In modelling production processes, the preparation of slips rather than other types of pigmented coatings has to be considered. This principle conclusion agrees with other limited analyses of surface painted wares from Iron Age Europe.

With regard to the elemental suites revealed by EDS analysis, it can be said that they represent the total components of fired clays (Baille et al., 1978) namely alumina (Al_2O_3), silica (SiO_2) and the fluxes Fe_2O_3, calcium oxide (CaO), magnesium (Mg) and titanium (Ti) with small amounts of sulphur (S) and phosphorous (P). No major differences in elemental suite revealed by EDS analysis were found between body fabric and coating suggesting similar bulk composition, and that the last represents a refined clay based slip.

The contrasting chemical content between coating and body fabric, especially in Ca, Fe, and Al content of group 4-5 when compared to groups 1-2 point to different source clays for different coatings.

Al:Si ratios can be used as an index of clay mineral content (Tite et al., 1982). For the Aulnat material such ratios are similar between analyses of the fine fraction of the body fabric and coatings. These uniform ratios reinforce the conclusion that the coatings are clay based slips. Higher clay content of groups 1, 2 and 3 is also illustrated perhaps by higher aluminium content which can be used as a guide to clay content (Gillies and Urch, 1983:37). Lower silicon content of the slips when compared to body fabrics is common to all groups (less pronounced in

4

group 3 - the whites) representing removal of course particulate grains such as quartz from the source clays. For group 3 the comparatively higher silicon is probably the result of the presence of kaolinite.

Body fabric and coating clays - possible firing conditions

It has been shown that the body fabric clays are low refractory (invariably containing over 10% fluxes) and non calcarious having calcium contents of less than 5%. They are also ferruginous containing c. 14% iron. The body fabric structure as shown under the SEM in preliminary examination of fresh fracture cross section (unpolished) did not display the characteristic open structure associated with vitrified calcareous clays. In contrast the calcium content of the slips is higher suggesting perhaps, use of a different calcareous source slip clay with higher flux content. For the red slips however, higher aluminium and similarity between the body clay and slip suggest refinement of the body clay fabric to produce the iron bearing slips. It is likely therefore that calcium was added to basic non-calcareous slips in various amounts to modify the colours produced in firing. The use of non-calcareous clays for both the body and coatings has a technological advantage in that the thermal expansion coefficient for body and slip would be similar (for non-calcareous clays 2-3.5 x 10-16/°C, Tite et al., 1982: 119). This explains the general very good adherence of the slips. The disadvantage of using non-calcareous clays for production of the 'painted' pottery is that the control of the firing temperature is more critical for calcareous clays, meaning a greater degree of control and sophistication of firing technology.

Group 3 (white) slips stand out as different from the remainder of the non-red groups in that they have markedly higher Si and Al content and are relatively free of iron. In this case it is likely that a different kaolinitic clay source was used, kaolin ($AlO_2.2SiO_3$) being indicated by the Si and Al content.

In the case of red slips, slurries of ferruginous clays were prepared and fired in short oxidising firings, Haematite (aFe_2O_3) a red pigment is formed and fixed in oxidising firing conditions at a temperature of c.820 - 880°C (Longworth and Tite, 1979).

The grey cores of the body fabrics of sherds involving red slip decoration (such grey cores were noted in practically the whole of the slip decorated pottery sample from Aulnat) indicate that the oxidising firing must have been short and at the end of firing since residual carbon has not been burned out of the cores which would have occurred during a prolonged oxidising firing. The oxidation of the iron contained in the slips to aFe_2O_3 probably occurred at the end of a semi-reducing firing when brief oxidation conditions were induced by the introduction of air into the kiln, or produced during cooling of the fired ceramic in air.

That the iron contents of groups 1 and 2 are not significantly different points to the fact that variation in red colour is resultant of different firing conditions, not different concentrations of iron, whereby a less oxidising firing produced a less concentrated colour.

For group 4 (pale browns) the addition of considerable amounts of calcium (perhaps in the form of lime from burnt/crushed limestone or chalk) is indicated by the analyses. Calcium acts to suppress formation of haematite since any iron contained in the slip forms calcium iron silicates which are not red in colour. This suggestion is supported by variation in the colour of the pale browns which, although not reflected in the Münsell measurements undertaken to differentiate major colour groups, were (rarely) observed to have orange taints, but an overwhelming impression of a beige/white appearance common to all members of group 4.

Discussion - the technology of decoration

Vestiges of brush marks were observed macroscopically on the red banding overlying pale brown coatings indicating the use of a brush to apply the red designs. The pale brown coatings which commonly formed the background to the red designs seem to have been applied by dipping the vessel into the slip as no brush marks were observed.

Decorative schemes adopted for the 'painted' pottery from Aulnat rely quite heavily on burnishing whereby leather hard body fabrics or slips are polished smooth before firing. Darker coloured effects as well as producing a lustre were attained by burnishing the slip surface to form grid, curvilinear and geometric designs. The effects of burnishing are perhaps aesthetically pleasing but place technological restrictions on the firing process if their effect is to be retained. If the burnish is highly fired, shrinkage disrupts the burnish effect making it appear dull (Shepard, 1980:123). This often means that burnished pottery is non-utilitarian in nature since this class is low fired. That the burnishing of the Aulnat pottery is often intense supports the idea of a relatively low temperature firing of a luxury fineware decorated pot not intended for a utilitarian role. That the coatings are matte or semi-matte (except where burnished) and are permeable (indicative of a lesser degree of vitrification) supports the conclusion of a non-utilitarian role for 'painted' pottery (this does not rule out uses such as storage of dried food stuffs or use as containers, but general heavy domestic roles such as cooking and water storage are excluded).

That the burnished motifs require a semi-matte background for effect may have implications for the degree to which the slips were refined, since very fine particle slips attain a high lustre in firing (Gillies and Urch, 1983:37) which would reduce the effect of the burnishing decorative technique. Contrast between the matte of the coating and lustre of the burnished areas would be reduced if very highly refined clay slips were used. That many of the slips contain small inclusions (e.g. of quartz) observed macroscopically and also via the SEM, would support the idea that the slips were not refined to a very high degree.

Conclusions

The analyses indicate that the decorative coatings are elutriated or levigated clay based slips rather than organically based pigment paints or stains. The slip coatings contain little in the way of added pigments, rather the colours (dark reds, red/oranges, whites, pale browns and light greys) were developed and fixed during firing of the thin clay slurries, whereby haematite is formed by oxidation, resulting in a variation of reds. Calcium was added to suppress the formation of haematite in order to modify the colours produced in the firing process.

The characterization of the decorative coatings on 'painted' pottery undertaken here, has allowed limited conclusions to be made concerning the technology involved in their production. Uniformity of composition and quality, as well as the established uniformity of design motif, argues for a sophisticated production technology which is perhaps beyond the scope of the so called 'house-hold' mode of production. The production technology involved selection and refinement of raw materials, application of decorative slips, treatment such as selective burnishing, and their controlled firing to achieve the desired variation in colour and lustre.

Further work

Further research is called for to examine a wider range of Late Iron Age and early Gaulish/Roman coated wares, including for example: terra nigra, grey reduced/black burnished wares and Campanian imports. A regional sample of slip decorated red and white 'painted' wares also requires analysis in order to put the results contained here into greater perspective. More intensive study of the technology of such quality finewares will make an important contribution to understanding the development of specialised ceramic industries in Late Iron Age/Roman Gaul.

Acknowledgements

I am grateful to Dr. Barbara Ottaway for her supervision of this research programme and to Dr. Margeret Rebbeck of the Materials Science Department at Bradford University for her instruction in use of the SEM/EDS. Thanks are also offered to Dr. John Collis and Chris Cumberpatch of Sheffield University for helpful discussion and for providing material from the Auvergne Archaeological Survey, and to Frank Jolley and Stuart Mullins for their help in sampling.

References

Baille, P.J., and Stern, W.B. (1984). Non-destructive surface analysis of Roman terra sigillata: a possible tool in provenance studies. *Archaeometry* 26(1), 62-68.

Cumberpatch, C., and Pawlikowski, M. (1988). Preliminary results of mineralogical analysis of Late La Tène painted pottery from Czechoslovakia. *Archeologicke Rozhledy* 40, 184-193.

Freestone, I. (1982). Current research in ceramic cross-section studies. *Archaeometry* 24(2), 99-116.

Gillies, K.J.S., and Urch, D.S., (1983). Spectroscopic studies of iron and carbon black surface wares. *Archaeometry* 25(1), 29-44.

Longworth. G. and Tite, M.S. (1979). Mossbauer studies on the nature of red or black glazes on Greek and Indian painted wares. *J. Phys. Colloq.* 40(2), 460-461.

Maggetti, M., and Galetti, G. (1980). Composition of Iron Age fine ceramics from Chatillon-s. Glane (Kt. ribourg, Switzerland) and the Heuneburg (Kr. Sigmaringen). *West Germany Journal of Archaeological Science* 7, 87-91.

Noll, W. (1977). Kaltbemalung Antiker gefasskeramik. *Archaeoligie und Naturwissenschaften* 1, 1-19.

Rice, P. M. (1987). *Pottery Analysis*, University of Chicago Press.

Shepard, A. (1980). *Ceramics for the Archaeologist*, Ann Arbor, Washington.

Tite, M.S., Maniatis, Y., Meeks, N.D., Bimson, M., Hughes, M.J., and Leppard, S.C. (1982). Technological studies of ancient ceramics from the Near East Aegean and Southwest Europe. In: *Early Pyrotechnology*, eds. Wertime and Wertime, Washington.

Tite, M.S. and Maniatis, Y. (1975). Examination of ancient pottery using the scanning electron microscope, *Nature* 251, 222-223.

Watson, C. (1988). *The Painted Pottery from Aulnat*. Unpublished B.A. Thesis, La Trobe University.

Wirska-Parachoniak, M. (1980). Produkcja ceramiczna Celtow na terenach Polski Poludniowej. *Materialy Archeologiczne Nowej Huty* 6. Krakow.

THE ICPS ANALYSIS OF ANCIENT COPPER ALLOYS

N.W. Blades[1], J. Bayley[2] and J.N. Walsh[1]

[1]Royal Holloway and Bedford New College, Egham Hill, Egham, Surrey, TW20 0EX.

[2]Ancient Monuments Laboratory, English Heritage, 23 Savile Row, London, W1X 2HE.

Introduction

Copper alloy artefacts have been subject to much archaeological research in the past by typological study, and in more recent years using metallography and chemical analysis, particularly the techniques of X-ray fluorescence (XRF), Atomic absorption spectrometry (AAS), and now Inductively-coupled plasma spectrometry (ICPS).

In the present project ICPS is being used to analyse a wide range of copper alloy finds from sites in England dating from late Roman to late medieval times. Advantage will be taken of the multi-element analysis ability of the method to study the major elements, with a view to discovering what alloys were in use at what time and location; and also the trace elements which may enable us to assist in the provenancing of copper alloy finds by identifying different trace element patterns which could be tied in with different geographical locations and periods in history.

Description of the ICP Spectrometer

The inductively-coupled plasma spectrometer (ICPS) is an emission spectrometer using an argon plasma, at a temperature of approximately 10 000 K to ionise the sample, and bring about emission. The sample is first dissolved in acid, and then pumped into the ICP flame where it is ionised and emits light at wavelengths characteristic of the elements present. This light passes through a diffraction grating which splits into its component wavelengths, and then on to the spectrometer itself. Here there is fitted a photomultiplier tube set to detect light at each particular wavelength used to determine an element; this converts the light it receives into a quantifiable electrical signal that is proportional to the amount of that element in the sample. By comparing this signal with that obtained from a calibration standard the weight percent of the element in the sample can be calculated.

An ICPS can theoretically analyse for as many elements (with the exception or argon) as there are photomultipliers fitted. In this project the following elements are being sought: copper, zinc, tin, lead, iron, arsenic, antimony, cadmium, bismuth, sulphur, phosphorus, silver, gold, nickel, cobalt, chromium, manganese and vanadium. The major elements, copper, zinc, tin and lead, were generally, though not always, deliberate additions in antiquity whereas the remaining minor and trace elements, usually adding up to less than 2% of the total, are present by accident. Because ICPS is able to analyse for many elements simultaneously, it is an ideal technique for a project of this type as the 18 elements given above can be analysed for it a very short space of time, using a small volume of solution, and hence a small metal sample of about 5-10 mg.

The metal samples are prepared for analysis by weighing out c. 5 mg accurately on a 5 decimal place balance, dissolving this in 0.5 ml reverse aqua regia (4 vols nitric acid: 1 vol hydrochloric acid, used in this case) and 0.5 ml distilled water. This solution is then made up to a final volume of 5 ml in a volumetric flask using distilled water.

Precision of the method

The analytical method, tested using commercial alloy standards, has a precision of 2-3% for major elements and about 10% for minor and trace elements. This lower precision, although not ideal, is acceptable on an absolute concentration of say 0.15%. It was found that precision, and analytical total varied with sample weight. Solutions prepared from 5 mg samples gave similar, if slightly worse precision to those made up from 100 mg of metal. Solutions prepared from 2-3 mg samples gave analytical totals in the range 94-106%. At this sample size it is thought that weighing errors start to become significant.

Sampling of artefacts.

This is mostly carried out using an engineer's pillar drill. Although rather heavy, this has the advantage of being able to run a low speed without losing power, and is fitted with a stop so the depth of the sample hole can be controlled - this is particularly important when drilling thin or delicate objects. Artefacts are usually sampled using a drill bit in the range 0.5-1.5 mm diameter. The surface corrosion products are first drilled through, and discarded, to expose bright metal from which the sample is collected. After drilling it is recommended that the sample hole is left unfilled - this leaves visible evidence that an artefact has been sampled, so that any future researchers need not duplicate the analysis. This should present no problems for artefact conservation - the Ancient Monuments Laboratory has many objects that were sampled in 1973 and show no signs of decay at the present time. In the case of sheet and wire metalwork drilling may not be feasible. If it is permissible a sample can be taken by cutting a small piece away. However, this method should be used with some caution: it has been found that many pieces of sheet metal are severely, or totally corroded, making it impossible to obtain a wholly metallic sample, and hence a reliable analysis of the alloy.

Project aims

In the course of this study mainly material from Pagan Anglo-Saxon up to late medieval times is being analysed. A small number of Roman artefacts will also be analysed, principally to see if we can use the data in assessing the level of Roman scrap used in later times. Generally Roman and prehistoric alloys, have received quite extensive analytical study. By comparison copper alloys of the Anglo-Saxon period have received little attention; though examples of work in this field include Mortimer et al. (1986) and Mortimer (1988). A similar situation exists with medieval copper alloys - few workers have carried out a systematic programme of quantitative analysis of material from sites in different parts of the country. An example of work from the medieval period is that by Brownsword & Pitt (1983).

Types of artefact that will be analysed include brooches, rings, pins, buckles, wristclasps, strap-ends, and other general everyday items. At present the policy of analysing whatever artefacts become available is being practised. In this way the artefacts that offer the most fruitful lines of research will be discovered.

The aims in carrying out chemical analyses are to address the following questions:

1. What alloys were in use at what time and what location?

2. What was the usage of scrap over the period 400-1600 AD? Can we identify alloys containing recycled scrap metal?

3. Were different alloys used for different object types/manufacturing methods? What does this tell us about the technical knowledge of the ancient metalworker?

4. Brass was extensively used in Roman times, but there is then little evidence for its use before the later Saxon period. Did the alloy go out of use between these times?

The use of trace element patterns to discriminate between objects found at sites in different parts of the country and from different periods in history is a real possibility. However, this aim is complicated by the re-use of scrap metal in antiquity. Thus it is unlikely that a metal alloy can be traced back to its original mineral source, but there is a strong possibility that different sources of raw metal can be identified according to their trace element patterns.

Results

To date two set of archaeological samples have been analysed - 65 late Roman finds from the site of Richborough in Kent, and 29 artefacts from the Anglian cemetery at West Heslerton, North Yorkshire. The late Roman finds are thought from typological evidence to include items brought to Britain from the continent by Germanic peoples who were in Roman military service during the 4th century. These can be considered as alloys of the very early Migration period, predating those from West Heslerton.

Only a preliminary interpretation of the data is attempted here, because, as more analyses are carried out the explanations presented will, no doubt, be modified and extended. it was found that the alloys from the sites show distinct differences, both in major and trace element composition. The results were plotted on ternary diagrams, which show the relative amounts of the alloying elements zinc, lead and tin. Minor and trace elements are ignored. The elements zinc, lead and tin are normalised to 100% and their relative proportions plotted. This way of displaying data gives an immediate, visual representation of the alloy types present - for instance, a point close to the zinc apex would indicate an alloy containing mostly zinc, with very little tin or lead - hence this alloy would be a brass. A point plotting in the centre of the diagram would correspond to an alloy containing equivalent amounts of zinc, lead and tin - a gunmetal. However, points that plot close together will have similar compositions only when the copper contents all fall within a limited range. To ensure this is the case, only analyses having a copper content between approximately 75 and 85% have been plotted in this way.

Note on alloy nomenclature

The following alloy nomenclature has been used:

Brass - an alloy of mainly copper and zinc.

Bronze - an alloy of mainly copper and tin.

Gunmetal - an alloy of copper with significant amounts of zinc and tin.

The term 'leaded' has only been applied to alloys containing deliberately added lead - at a level of perhaps 6-8% or more.

The alloys from Richborough and West Heslerton

The ternary diagram for the finds from Richborough (Figure 1) shows that they fall into the alloy categories of brasses, bronzes, gunmetals and leaded bronzes. However the West Heslerton finds (Figure 2) are almost all gunmetals or leaded gunmetals, containing significant amounts of zinc, lead and tin but in varying proportions: none of the alloys could be considered as brasses, and only one was a bronze. An attempt was made to correlate artefact type with composition for the Richborough finds, by plotting several more ternary diagrams. The bangles (or armlets) are thought to be of the late 3rd or early 4th century origin, and are of the same types as those depicted in Crummy (1983, 37-45). The buckles are late 4th, early 5th century types, which are thought to have a continental origin, and include examples of Hawkes and Dunning's types IA, IIA, IIB, and IIIA (Hawkes & Dunning, 1961). Figure 3 is a ternary plot for the strip bangles (labelled 'solid bangles') from Richborough, revealing these to be made from brasses, bronzes, and gunmetals, with generally a low lead content. Figure 4 is the plot for cable bangles (labelled 'interwoven wire bangles'). These follow a similar pattern to strip bangles, but with a slightly lower lead content. Figure 5 is a plot for the late Roman buckles found at Richborough, and these are largely leaded bronzes, with in some cases over 20% lead.

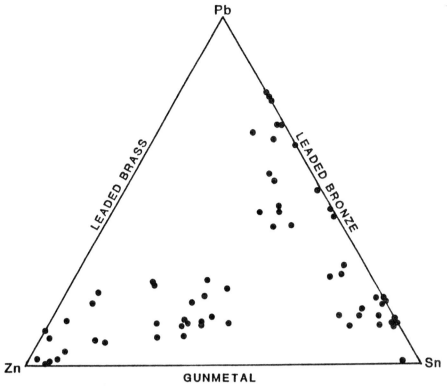

Figure 1. Ternary diagram for Richborough copper alloys, showing relative Zn, Pb and Sn concentrations.

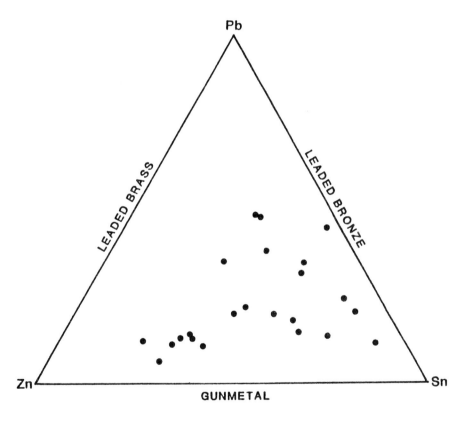

Figure 2. Ternary diagram for West Heslerton copper alloys (1977-1978 dig.), showing relative Zn, Pb and Sn concentrations.

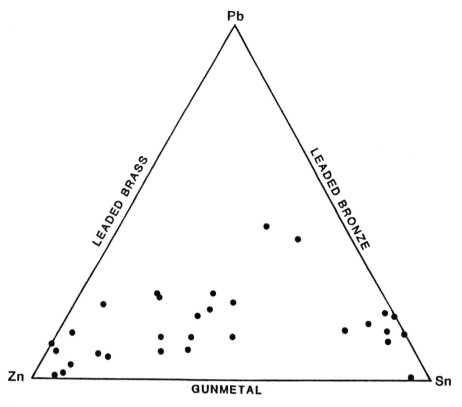

Figure 3. Ternary diagram for solid bangles, Richborough.

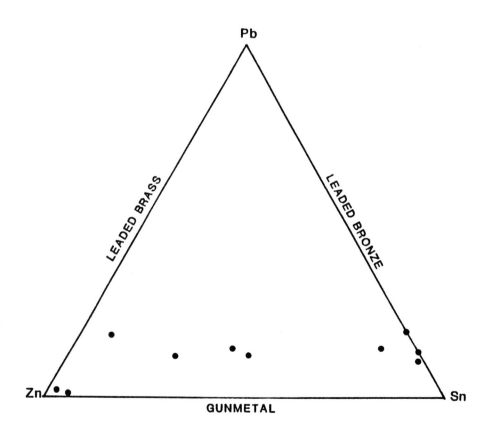

Figure 4. Ternary diagram for interwoven wire bangles, Richborough.

12

The alloys used for each object type would appear to follow good metallurgical practice - the bangles would have been worked in some way, either during manufacture, or by stamping or carving for decoration. For this purpose the alloy used would have to be low in lead because over a few percent of lead would cause it to crack on working. However with the buckles, which are largely cast items a higher lead content would have no detrimental effects, and would probably make for a cheaper alloy. The finds analysed from West Heslerton fall mostly into the categories of gunmetal and leaded gunmetal, with one find, an annular brooch fragment of heavily leaded (24% lead) bronze. This level of lead would yield an alloy of poor mechanical strength, and may account for the brooch's failure (the remaining brooches were complete, apart from their pins).

The minor and trace elements show up clear differences between the two sites. For West Heslerton there is a clear co-variation between nickel and arsenic (Figure 6), whereas with the Richborough finds (Figure 7) these elements are more randomly distributed.

The absolute nickel and arsenic levels from West Heslerton are higher than those from Richborough. From the above it would seem reasonable to conclude that some of the metal used to make the finds at West Heslerton showing this co-variation came from a specific source which was high in nickel and arsenic. This would probably have been diluted with metal from another source, possibly scrap, which would appear to be low in both elements, so preserving the co-variation. The presence of all the major elements, copper, zinc, lead and tin in significant amounts in most of the West Heslerton alloys points towards their containing recycled metal.

Future work

It is planned to improve the precision of the analytical method by introducing an internal standard element, and to search for several of the trace and minor elements down to the lower level of 0.001% (at present the spectrometer is set to detect all elements down to a level of 0.01%); this should still be

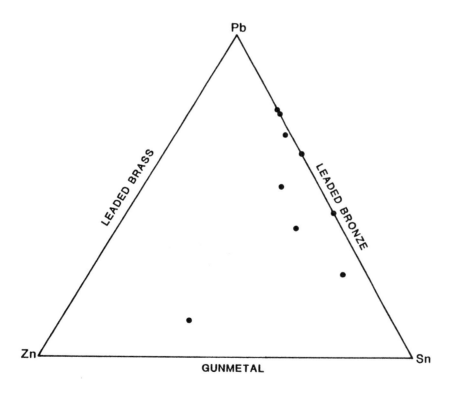

Figure 5. Ternary diagram for all buckle types, Richborough.

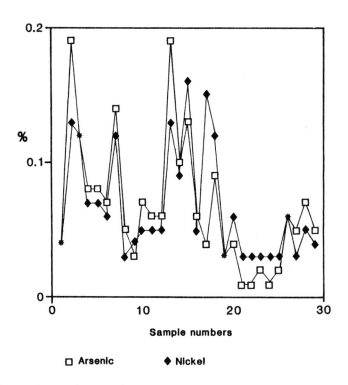

Figure 6. Graph of sample number against % nickel and arsenic for West Heslerton results.

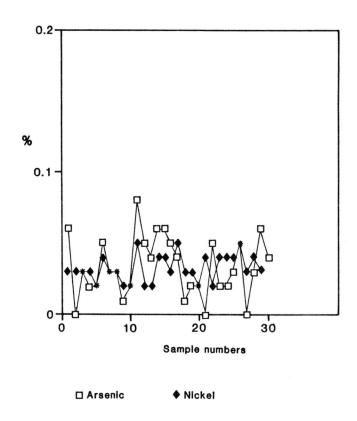

Figure 7. Graph of sample number against % nickel and arsenic for Richborough results.

14

well within the detection limits of the spectrometer for most of these elements. In the sample preparation method there is scope for improvement of the weighing technique to reduce errors due to such factors as air currents, static electricity, and temperature differences between balance and weighing vessel. Archaeologically, we shall be going onto analyse finds from many other sites covering the period 400-1600 AD.

References

Brownsword, R. & Pitt, E.E.H. (1983). Alloy composition of some cast 'latten' objects of the 15th and 16th centuries. *Historical Metallurgy* 17, 44-49.

Crummy, N. (1983). *Colchester Archaeological Report* 2: The Roman small finds from excavations in Colchester 1971-9, Colchester Archaeological Trust, Colchester.

Hawkes, S.C. & Dunning, G.C. (1961). Soldiers and settlers in Britain, fourth to fifth century: with a catalogue of animal ornamented buckles and related belt fittings. *Medieval Archaeology* 5, 1-70.

Mortimer, C., Pollard, A.M. & Scull, C. (1986). XRF analyses of some copper alloy finds from Watchfield, Oxfordshire. *Historical Metallurgy* 20, 36-42.

Mortimer, C. (1988). Anglo-Saxon copper alloys from Lechlade, Gloucestershire. *Oxford Journal of Archaeology* 7, 227-233.

NEUTRON ACTIVATION ANALYSIS OF LATE ROMAN AND DARK AGE GLASS FROM KENT

L.R. Day and D.R.J. Perkins

Division of Environmental Sciences, Polytechnic of East London, Stratford, London E15 4LZ, Great Britain.

Summary

Neutron Activation Analysis was employed in the examination of thirteen archaeological glass samples. Chronologically, these ranged from the late third century to the ninth, with ten samples being derived from the fifth-seventh century Jutish burial grounds of East Kent. Claw and cone beakers were represented, as were palm cups and unclassified vessels. This work is intended to form part of a study in which Atomic Absorption Spectrometry will be used to quantify the oxides forming the major components. The results of analysis to date are discussed and contrasted with similarly obtained data derived from recent studies of Anglo-Saxon glass.

Introduction

The Pagan Jutish cemeteries of East Kent were discovered by antiquaries in the last quarter of the eighteenth century. Their investigation during the next hundred years provided a plethora of artefacts, whose published descriptions loomed large in the early annals of British archaeology. Among the more impressive items were a variety of glass vessels, the beakers, bottles and cups subsequently classified by D.B. Harden (1956).

Little if any of this material has been subjected to modern analytical techniques. Vessels were often excavated intact, or in reconstructable condition. When found hopelessly smashed, the fragments do not seem to have been retained. Modern archaeologists have exercised a moratorium with regard to the few surviving cemeteries, only occasional rescue excavations taking place. It will be understood therefore, that the ten 'Jutish' glass samples dealt with in this study were assembled with some difficulty. Although this sample population is small, it does include six vessel classes, as well as unidentified fragments, one from an as yet unclassified type. The full colour range of Jutish glass is also represented. Two Late Roman samples, and a 'cane bead', (a seventh century import), have been included as comparisons.

It is hoped that, pending the availability of a more comprehensive sample population, (or the development of a totally non-destructive analytical technique), this study will at least provide a general picture of the composition of these samples for comparison with the generally later and much more abundant material from the Saxon town sites.

The glass samples and their provenance

In Table 1, over, the basic sample data is arranged by groups for ease of consultation. The classifications from Harden (1956) are used throughout.

Samples within the foregoing groups are described in detail below and their provenance is given so far as it is known at the time of writing.

The Jutish Group

Beakers: Sample 9 is a fragment from a claw beaker, (class II,c), in clear 'beer-bottle brown' glass from the Jutish cemetery at Bifrons, Kent. This is from one of the graves excavated in 1867 by Lord Conyngham's gamekeeper, of which no grave numbers or other data has survived. The beaker is in the keeping of maidstone Museum as part of the Tomlinson loan, the sample fragment being one of several that could not be utilised in its

Table 1. Roman and Dark Age glass samples, the basic data.

Sample	Vessel-Type	Class	Colour	Estimated Date (AD)	Origin
Roman Glass:					
7	Flagon	-	Clear blue	Third century	Romano-British
8	Flagon	-	Clear blue	Third century	Villa sites
'Jutish' Glass:					
9	Claw beaker	II,c	Clear brown	Mid-sixth century	Grave goods
10	Cone beaker	III,c	Clear white	Late-fifth century	Grave goods
11	Bag beaker	VI,a	Clear white	Seventh century	Grave goods
1	Palm cup	X,b	Clear blue	Seventh century	Grave goods
13	Palm cup	X,b	Clear amber	Seventh century	Grave goods
2	Unknown	-	Clear white	Seventh century	Grave goods
3	Unknown	-	Clear white	Seventh century	Grave goods
5	Bottle?	-	Clear white	Seventh century	Grave goods
Imported Dark Age Glass:					
4	'Base' Cup	-	Clear green	late-seventh century	Grave goods
6	Cane Bead	-	Blue-white-red	Seventh century	Grave goods
Later Anglo-Saxon Glass:					
12	Bottle?	-	Blue-white-clear	Tenth century	Farmstead site

restoration. Harden places it among the 'degenerate types' of the mid- sixth century (Harden 1956, p.159).

Sample 10 is a fragment from a cone beaker (class III,c), in clear white glass. This is from the Bifrons cemetery, Grave 1, and is illustrated by Harden (1956, p.140, Pl. XVI,f), who suggests a fifth or early sixth century date. It is in the keeping of Maidstone Museum, again a fragment un-usable in restoration.

Sample 11 is a fragment from a bag beaker (class VI,a), in clear white glass. It was a nineteenth century find from Faversham, Kent, and nothing further is known. It is ascribed by Harden (1956, p.141) to the seventh century.

Cups: Sample 1 is a rim and body sherd from a palm cup, (class X,b), in a clear blue glass. This was found during exploratory trenching in an unexcavated sector of the Jutish cemetery at Sarre, Thanet, Kent, in 1983, (Perkins, 1988, p.293). It accompanied the burial of a young female with late sixth-seventh century grave goods, the sherd itself probably being worn as an item of adornment, as also samples 2 and 3, see below.

Sample 13 is a fragment of a palm cup (class X,b), in clear amber glass, from an excavation in 1974 at Kingston Down, Kent, Site IX, 123, layer 2. This was a rescue excavation on the border of Kingston Down Jutish cemetery. The context is believed to be the plough-damaged mound of a Jutish barrow (MacPherson-Grant, pers. comm.).

Unidentified material from Jutish Graves: Sample 2 is a rim fragment with rounded hollow lip in clear white glass, found in association with Sample 1, in Grave 2, (1983 series), in the Sarre Jutish cemetery, Thanet, Kent. Sample 3 was found with Sample 2, and is very similar but from another vessel. Sample 5 is a body sherd in clear white glass believed to belong to an unclassified vessel of bottle type. This is a find from Dr. A. Rowe's excavations in the Jutish cemetery at Half Mile Ride, Margate, in 1924, (Perkins, 1987, p.228).

Later Anglo-Saxon Glass

Sample 12 is a sherd from the narrow neck-column of a vessel in clear white glass, decorated with alternative opaque blue and white spiral trails. This is from an Early Medieval site at Netherhale Farm, Birchington, Kent (Perkins, 1980). Associated finds indicate a tenth century date.

Imported Glass

The two samples in this group belong to types whose European origins are not contested. Sample 4 is a fragment from a nearly complete 'base cup' in cloudy green glass, (which could not be used in reconstruction). This was re-discovered in 1986 during the cataloguing of museum material at Margate. It was in 'as-excavated' condition, and was accompanied by a buckle of late seventh century type. It almost certainly derives from the unrecorded rescue excavations in the Jutish cemetery at Half Mile Ride, Margate, in 1924 (Perkins, 1987). Harden excluded these vessels from his classification of British Dark Age glass, as he believed them to be the broken-off bases of Tudor goblets, (Harden, 1956, p.166). They are more generally accepted however, as of Dark Age date. Only five such vessels appear in British archaeology, and of these only three can now be traced.

Sample 6 is a fragment from a 'cane bead' of millefiori construction in opaque black, red and white glass. This was a surface find in

the area of an unexcavated Jutish cemetery at Minster in Thanet (Hawkes, 1984). These beads can be firmly dated to the seventh century. A sample fragment was selected so that all three colours were about evenly represented.

Roman Material

Samples 7 and 8, are both fragments of flagons in clear blue glass, both probably third century. Sample 7 which has pincered decoration is from a Romano-British settlement site at Netherhale, Thanet. Sample 8 is from a villa site at the Mount, Maidstone.

Neutron Activation Analysis

After preparatory work of cleaning and weighing, the samples were taken for irradiation in the research reactor at Imperial College's Reactor Centre, Silwood Park, Ascot. The samples were exposed for 7.5 hours at an estimated flux of 1×10^{12} $ncm^{-2}s^{-1}$. In addition to the samples, standards and monitors were irradiated as follows.

Standards: British Museum ceramic standard, (BMSP), British Museum glass standard (76-C-150), Corning Museum glass standard, 'B'. Monitors: Three zinc flux monitors were included. From a comparison of the weights and normalised counts from these, a flux variation over the length of the sample canister of 5.77% was calculated, this being more or less in keeping with the Reactor Centre's estimate of 5%.

A heat sealed plastic packet as used to contain the samples was also included. Measurements on this indicated that the plastic sample packets had an appreciable antimony content. After re-weighing the samples to discover packet weights, it became possible to estimate the portion of the antimony count from each sample that had originated from the packet.

Counts on the samples using a Canberra Pulse Height Analyser commenced on the fifth day after irradiation, and were completed next day. Live times for these counts were in the order of 500 seconds, and samples were presented at 4 cm above the detector head. Short-lived isotopes, in particular sodium were detected. Subsequently, all samples were counted at least twice more for live times of between 1000 and 80 000 seconds, these times being progressively increased with the number of days since irradiation. A new geometry was adopted for these long counts, with the sample platform being lowered to its minimum height from the detector head.

In all, 64 photopeaks were observed, in the range 56.6 kev to 2095 Kev. Of these 13 have not been identified at the time of writing. Most of the isotopes detected were represented by one or more of the standards so that they could be quantified by direct comparison. An exception being tin, quantitative estimates of which are based on a calibration curve constructed by plotting counting efficiency against γ-ray energy. This method may involve errors of $\pm50\%$. Estimates of the concentrations of oxides and trace elements determined from the standards are given in Table 2.

A Discussion of the Results

The analytical work described here is part of a large study. This will embody the use of Atomic Absorption Spectrometry and similar techniques, so that oxides forming major components of the glass samples can be quantified. It is hoped that the combined data will throw some light on the origins, (or at least the affinities), of the Jutish glass, when compared with that obtained in recent studies of similar material. Without possession of the final results, what can be said of the samples on the basis of five oxides and twenty-one trace elements detected is rather limited. Some fairly general and interim comments follow.

The Roman glass, samples 7 and 8: These fragments seem mostly remarkable for the austerity of their composition. Both had comparatively high antimony contents, detectable amounts of rhodium and rhenium.

The cane bead, sample 6: This obviously differs from the vessels in that its manufacture employed opacifiers and colourants. A tin content greater than 10% was estimated, this probably representing both a white colourant and opacifier. The red and blue components being produced by the cuprous oxide and cobalt.

The base cup, sample 4: This is uniquely different in composition to the other vessels, most obviously in its very low sodium content, the latter being confirmed by recently analysis of another example of this rare type. During conservation work on a base cup in the keeping of Liverpool Museum, it was examined by electron microscopy and energy dispersive spectroscopy, this indicating a 4.8% sodium content, (F.A. Philpott, pers. comm.). Alone among the samples, sample 4 contains lanthanum, thorium, and dysprosium. The cerium and tin contents are high. Another fragment of this vessel was among those chosen in an exploration of the viability of Fission Track Dating with respect to these glasses. On irradiation with a standard of known uranium content, (Corning CN2), it exhibited induced tracks suggesting a uranium content greater than 10 $\mu g/g$. It will be interesting to see what potassium content is indicated by subsequent analysis.

The Late Saxon Fragment, sample 12: This has a low sodium content, (6.7%). High values for cobalt and tin are explained as the colourants for blue and white trail decoration.

The Jutish group, samples 1, 2, 3, 5, 9, 10, 11 and 12: The unidentified flask of bottle, sample 5 is set apart only by a low sodium content, (7.6%), and the presence of arsenic. In Table 3 a comparison is made between some means from the eight 'Jutish' samples in this study, and means from seven groups examined by Neutron Activation Analysis and Inductively Coupled Plasma Spectrometry. Groups B and C, (ICPS), are from Saxon Southampton and Winchester (Heyworth et al., 1986). Groups A, and D, E, F, G (NAA), are from: A: Saxon Southampton; D and F: Helgo, Sweden; E: Spong Hill; G: Dorestadt, Holland (Sanderson and Hunter, 1982).

Table 2: Estimated concentrations of oxides and trace elements. Errors for all oxides and elements apart from tin and antimony are in the order of $\pm$7%. Sample No. 6 gave a very high count for ^{117}Sn, probably enhanced by the presence of uncombined tin oxide. Its tin content is therefore estimated as approaching the maximum (saturation) level.

Ele-ment	Sample (% detected)												
	No. 1	No. 2	No. 3	No. 4	No. 5	No. 6	No. 7	No. 8	No. 9	No. 10	No. 11	No. 12	No. 13
Na$_2$O	13.5	15.6	14.6	2.3	7.5	8.81	15.4	18.1	19.9	18.77	15.6	9.06	16.11
Fe$_2$O$_3$	n.d.	n.d.	n.d.	n.d.	n.d.	6.51	n.d.	n.d.	n.d.	n.d.	2.41	2.15	n.d.
CuO	0.38	0.49	0.62	n.d.	0.38	0.58	n.d.	n.d.	n.d.	nd.d	n.d.	0.57	0.61
Sb$_2$O$_3$	0	0.006	0.016	0	0	0.02	0.3	0.56	0.16	0.13	0	0.004	0.0030.
CoO	0.001	0.003	0.003	0.003	0.002	0.03	0.001	n.d.	0.003	0.002	0.011	0.011	001
SnO$_2$	0.63	2.20	1.57	3.80	1.20	≥10.0	1.2	n.d.	1.0	1.0	0.84	5.31	0.16
	(A)												
Sc	1.17	0.70	0.63	1.56	1.10	0.8	1.2	0.5	1.6	1.5	1.74	1.76	1.07
Cs	n.d.	n.d.	n.d.	n.d.	n.d.	n.d.	*	*	*	*	*	n.d.	n.d.
Hf	1.3	1.0	*	5.4	*	n.d.	n.d.	n.d.	3.0	n.d.	*	4.6	0.3
Cr	n.d.	53.3	n.d.	n.d.	n.d.	n.d.	n.d.	n.d.	n.d.	n.d.	n.d.	n.d.	n.d.
Ce	11.2	8.6	10.7	32.8	7.5	n.d.	n.d.	n.d.	n.d.	n.d.	2.2	*	2.5
Eu	0.69	0.8	0.45	0.56	0.13	n.d.	n.d.	n.d.	0.85	0.36	0.39	0.36	0.3
La	n.d.	n.d.	n.d.	*	n.d.	n.d.	n.d.	n.d.	n.d.	n.d.	n.d.	n.d.	n.d.
Sm	3.0	n.d.	*	n.d.	*	45.5	n.d.	n.d.	8.2	n.d.	n.d.	*	*
Th	n.d.	n.d.	n.d.	1.25	n.d.	n.d.	n.d.	n.d.	n.d.	n.d.	n.d.	n.d.	n.d.
Tb	n.d.	n.d.	n.d.	n.d.	n.d.	*	n.d.	*	*	*	n.d.	*	n.d.
Tm	n.d.	n.d.	n.d.	n.d.	n.d.	*	n.d.	n.d.	n.d.	n.d.	n.d.	n.d.	n.d.
Au	n.d.	*	n.d.	n.d.	n.d.	*	n.d.	n.d.	n.d.	n.d.	n.d.	n.d.	n.d.
Mn	n.d.	n.d.	n.d.	n.d.	n.d.	n.d.	n.d.	n.d.	n.d.	n.d.	*	n.d.	n.d.
As	n.d.	n.d.	n.d.	n.d.	*	*	n.d.	n.d.	n.d.	n.d.	n.d.	n.d.	n.d.
Rh	n.d.	n.d.	n.d.	n.d.	n.d.	n.d.	*	*	n.d.	n.d.	*	n.d.	n.d.
Br	n.d.	n.d.	n.d.	n.d.	n.d.	n.d.	n.d.	*	*	n.d.	n.d.	n.d.	n.d.
Ta	n.d.	n.d.	n.d.	0.30	n.d.	n.d.	0.02	n.d.	n.d.	n.d.	n.d.	n.d.	0.09
Dy	n.d.	n.d.	n.d.	^	n.d.	n.d.	n.d.	n.d.	n.d.	n.d.	n.d.	n.d.	n.d.
Re	n.d.	n.d.	n.d.	n.d.	n.d.	n.d.	*	*	*	n.d.	n.d.	n.d.	n.d.

(A) All in μg/g.

Table 3: A comparison between means from the eight 'Jutish' samples in this study, and means from seven groups examined by NAA and ICPS.

	This Study	A	B	C	D	E	F	G
All oxides as %								
Na$_2$O	14.3	12.43	15.6	17.7	19.6	18.9	16.2	15.1
CuO	0.28	0.11	0.1575	0.018	0.01	0.01	0.22	0.19
Sb$_2$O$_5$	0.092	0.0241	0.30	0.08	0.0008	0.0008	0.0157	0.266
CoO	0.005	0.00152	0.0028	0.0014	0.0014	0.0014	0.0182	0.0026
All in μg/g								
Sc	1.2	1.7	2	2	4.1	2.8	2.3	2.1
Hf	1.0	2.9	-	-	5.6	3.9	1.9	2.5
Ce	5.8	12.2	20	16	3.4	12.5	3.2	14
Eu	0.4	0.6	-	-	0.5	0.5	0.7	1.4

day. Live times for these counts were in the order of 500 seconds, and samples were presented at 4 cm above the detector head. Short-lived isotopes, in particular sodium were detected. Subsequently, all samples were counted at least twice more for live times of between 1000 and 80 000 seconds, these times being progressively increased with the number of days since irradiation. A new geometry was adopted for these long counts, with the sample platform being lowered to its minimum height from the detector head.

In all, 64 photopeaks were observed, in the range 56.6 kev to 2095 Kev. Of these 13 have not been identified at the time of writing. Most of the isotopes detected were represented by one or more of the standards so that they could be quantified by direct comparison. An exception being tin, quantitative estimates of which are based on a calibration curve constructed by plotting counting efficiency against γ-ray energy. This method may involve errors of $\pm 50\%$. Estimates of the concentrations of oxides and trace elements determined from the standards are given in Table 2.

A Discussion of the Results

The analytical work described here is part of a large study. This will embody the use of Atomic Absorption Spectrometry and similar techniques, so that oxides forming major components of the glass samples can be quantified. It is hoped that the combined data will throw some light on the origins, (or at least the affinities), of the Jutish glass, when compared with that obtained in recent studies of similar material. Without possession of the final results, what can be said of the samples on the basis of five oxides and twenty-one trace elements detected is rather limited. Some fairly general and interim comments follow.

The Roman glass, samples 7 and 8: These fragments seem mostly remarkable for the austerity of their composition. Both had comparatively high antimony contents, detectable amounts of rhodium and rhenium.

The cane bead, sample 6: This obviously differs from the vessels in that its manufacture employed opacifiers and colourants. A tin content greater than 10% was estimated, this probably representing both a white colourant and opacifier. The red and blue components being produced by the cuprous oxide and cobalt.

The base cup, sample 4: This is uniquely different in composition to the other vessels, most obviously in its very low sodium content, the latter being confirmed by recently analysis of another example of this rare type. During conservation work on a base cup in the keeping of Liverpool Museum, it was examined by electron microscopy and energy dispersive spectroscopy, this indicating a 4.8% sodium content, (F.A. Philpott, pers. comm.). Alone among the samples, sample 4 contains lanthanum, thorium, and dysprosium. The cerium and tin contents are high. Another fragment of this vessel was among those chosen in an exploration of the viability of Fission Track Dating with respect to these glasses. On irradiation with a standard of known uranium content, (Corning CN2), it exhibited induced tracks suggesting a uranium content greater than 10 $\mu g/g$. It will be interesting to see what potassium content is indicated by subsequent analysis.

The Late Saxon Fragment, sample 12: This has a low sodium content, (6.7%). High values for cobalt and tin are explained as the colourants for blue and white trail decoration.

The Jutish group, samples 1, 2, 3, 5, 9, 10, 11 and 12: The unidentified flask of bottle, sample 5 is set apart only by a low sodium content, (7.6%), and the presence of arsenic. In Table 3 a comparison is made between some means from the eight 'Jutish' samples in this study, and means from seven groups examined by Neutron Activation Analysis and Inductively Coupled Plasma Spectrometry. Groups B and C, (ICPS), are from Saxon Southampton and Winchester (Heyworth et al., 1986). Groups A, and D, E, F, G (NAA), are from: A: Saxon Southampton; D and F: Helgo, Sweden; E: Spong Hill; G: Dorestadt, Holland (Sanderson and Hunter, 1982).

As previously stated, this work is part of a study employing several analytical techniques. Its aim being to provide a compositional profile of East Kent's Jutish glass. Table 3 shows what may be an 'identity' for the Jutish glass starting to appear, marked principally by a similar sodium content. It will be most interesting if an affinity between the study group and group A is observed when concentrations of the major oxides in the study group are determined, since this would indicate group A as being a direct continuation of an industry. This following the progression of Jutish settlement in Kent and the Isle of Wight through to Hamwic. Whether or not the wares were indigenous or imported, or made from imported cullet, would of course still be open to question.

References

Harden, D.B. (1956). Glass vessels in Britain and Ireland. In: *Dark Age Britain: Studies presented to E.T. Leeds* (London, 1956), 132-167.

Hawkes, S.C. (1984). The Amerst Brooch. *Archaeologia Cantiana*, C, 129- 151.

Heyworth, M.P., Hunter, J.R. & Warren, S.E. (1986). The Role of Inductively Coupled Plasma Spectrometry in Glass Provenance Studies. *Proc. Archaeometry Symposium* (Athens, 1986), 1-10.

MacPherson-Grant, N., from unpublished fieldwork notes in the keeping of Canterbury Archaeological Trust.

Perkins, D.R.J. (1980). A Medieval enclosure at Netherhale Farm, Birchington. *Thanet Archaeological Unit Brief Interim Reports I*, 25-29.

Perkins, D.R.J. (1987). The Jutish Cemetery at Half Mile Ride, Margate. *Archaeologia Cantiana* CIV, 219-297.

Fiona A. Philpott, Liverpool Museum unpublished conservation notes.

Sanderson, D.C.W. & Hunter, J.R. (1982). The Neutron Activation Analysis of archaeological glasses from Scandinavia and Britain. *PACT* part II, 401-411.

CORROSION OF LEAD IN SEAWATER AND SALINE GROUNDWATERS: FORMATION OF PHASES CONTAINING CHLORIDE

R. Edwards, R.D. Gillard, A.M. Pollard and P.A. Williams

School of Chemistry and Applied Chemistry, University of Wales College of Cardiff, P.O. Box 912, Cardiff CF1 3TB

Introduction

A series of chloride-containing phases are known to be formed on or near the surface of corroding lead in contact with sea water or saline groundwaters. Of these, the most frequently reported are cotunnite, $PbCl_2$, laurionite and paralaurionite, $Pb(OH)Cl$, blixite, $Pb_2(OH)_3Cl$, and phosgenite, $Pb_2CO_3Cl_2$. Other common species found are anglesite, $PbSO_4$, cerussite, $PbCO_3$, hydrocerussite, $Pb_3(CO_3)_2(OH)_2$ and litharge, PbO.

Occasionally, other rarer chloride-containing secondary lead(II) phases such as penfieldite, $Pb_2(OH)Cl_3$, fiedlerite, $Pb_3(OH)_2Cl_4$, mendipite, $Pb_3O_2Cl_2$ and 'lorettoite', $Pb_7O_6Cl_2$, are found. These occur in conjunction with the above-mentioned phases in the oxidized zones of base metal ore bodies, as well as in corrosion products of archaeological lead and lead alloys. The conditions which could give rise to a number of these compounds have been investigated by earlier workers (Kiyama et al., 1976; Wilkinson et al., 1937; Yadava et al., 1968; Humphreys et al., 1980). However, the relative stabilities of some of the phases remain unknown and relationships between these and others undetermined. Thus it seems worthwhile to re-explore the chemistry of the $PbO-HCl-H_2O$ system.

We have determined for the first time a stability constant for $Pb_7O_6Cl_2.2H_2O$, a hydrated form of the discredited mineral species lorettoite. Furthermore, the stability constant for blixite has been redetermined as previous studies (Näsänen & Lindell, 1978) lead to inconsistencies in phase relationships. A summary of the data used and derived experimentally is given in Table 1.

Table 1. Standard Gibbs Free Energy of Formation of Some Secondary Lead(II) Minerals.

Mineral	$\Delta fG^o(298.2K)/kJ\ mol^{-1}$	Reference
Cotunnite	-625.3	Robie et al. (1978)
Laurionite	-389.1	Näsänen & Lindell (1976)
Mendipite	-757.5	this work
Blixite	-832.2	this work
$Pb_7O_6Cl_2.2H_2O$	-2000.0	this work
Litharge	-189.2	Robie et al. (1978)

Discussion

Studies concerning the lead(II) chloride compounds have shown that their formation involves a complex interplay of kinetic and thermodynamic influences. In the $PbO-HCl-H_2O$ system at 298.2K (25°C) the only phases present are cotunnite, laurionite, blixite, $Pb_7O_6Cl_2.2H_2O$ and lead(II) oxide. Figure 1 illustrates this fact; it shows the course of the titration of $PbCl_2(aq)$ (0.754g, 2.71mmol) with NaOH (0.11M) at 25°C. The

23

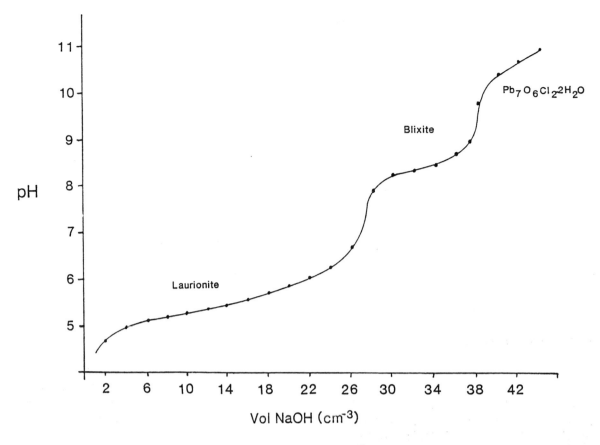

Figure 1. Course of titration of $PbCl_2$ (2.71 mmol in 100 cm^3 H_2O) with NaOH(aq) (0.11 mol dm^{-3}) at 298.2K.

initial lead hydroxychloride to crystallize is laurionite, rather than its dimorph paralaurionite, this being metastable at 298.2K. Lead oxide is also dimorphous, massicot and litharge, the latter being the thermodynamically stable species at ambient temperatures.

Relative stabilities of these solid phases are illustrated in Figure 2. All of these phases have been observed either as corrosion products on archaeological artefacts or as secondary minerals in oxidized base metal sulphide ore bodies. With diagrams such as these, paragenetic sequences for the formation of these minerals can be explained.

One notable absence from the phase diagram is that of mendipite. Its occurrence in oxidized lead ores has been reported a number of times. The type locality of mendipite is in the Mendip Hills, Somerset, England (Symes & Embrey, 1977), where it is found associated with paralaurionite, blixite, hydrocerussite and cerussite. Furthermore, it has also been reported from Laurium (Kohlberger, 1976) along with a host of other secondary lead(II) minerals. Mendipite has not however been found at the Tiger Mine, Arizona, (Bideaux, 1980; Abdul-Samad et al., 1982) where a number of these secondary lead chloride species are known to occur.

The conditions for the formation of mendipite have been investigated, and it has been discovered that mendipite can be synthesised in aqueous solution from the stoichiometric ratio of lead oxide and lead chloride at temperatures greater than 29°C. Below this transition temperature the products of the reaction are a mixture of laurionite and blixite. From these observations it can be concluded that mendipite is a metastable phase at 298.2K, but small increases in temperature can cause it to crystallize as a thermodynamically stable species.

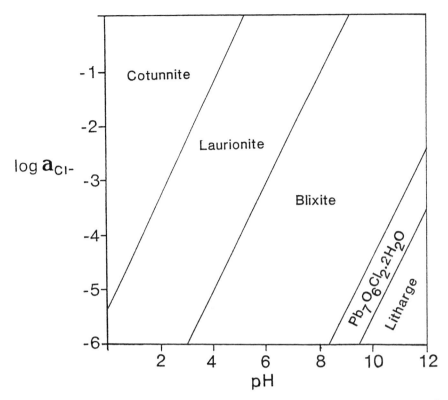

Figure 2. Stability field diagram for the PbO-HCl-H$_2$O system at 298.2K, p = 10^5 Pa.

The presence of mendipite as a corrosion product may, in some circumstances, provide an insight into the conditions under which reaction may have taken place. An example of such is in the Mendip Hills deposits. It is generally thought that the minerals in the Mendips were formed from hydrothermal saline groundwaters. Thus it is unlikely that mendipite would crystallize on ancient lead objects in seawater, unless it were formed metastably. Other kinetic factors are involved in the formation of mendipite. For example, mendipite is not found to be crystallized at 30°C during 2-3 days using the titration method of synthesis. The phases laurionite and blixite persist metastably for considerable periods of time under these conditions.

The phase diagram for the system at 298.2K, including mendipite, ignoring the small temperature increment from 25 to 29°C, is shown in Figure 3. These studies are helpful in explaining the products of corrosion of lead objects in seawater. Of course other compounds can be formed, by virtue of the fact that other chemical species are present. Of these the most important are carbonate and sulphate ions.

Two further basic chlorides of lead(II), penfieldite and fiedlerite, are also known as corrosion products. Fiedlerite is only known from Laurium (Kohlberger, 1976), whereas penfieldite, as well as from Laurium, has been identified in a mine on the western side of the Sierra Gorda, Chile (Palache et al., 1951). It has also been identified as a corrosion product from a few other archaeological contexts.

Goni et al. (1954), have studied the corrosion of ancient lead anchor stocks in seawater from two locations, namely Mahdia, Tunisia, and near the foot of the Isle of Fourmigues, France. The two objects are described as having a corrosion crust of between 1 and 2 cm thick with different minerals mixed with plant remains and shells. Of the phases found, penfieldite was one of the most abundant, with laurionite and paralaurionite being absent or unidentified. Both penfieldite and fiedlerite appear to be metastable with respect to other species, at temperatures from 0 to 100°C. They have not been identified in any titration experiment, and other efforts to synthesize them over a wide range of chloride activity and pH have proved unsuccessful.

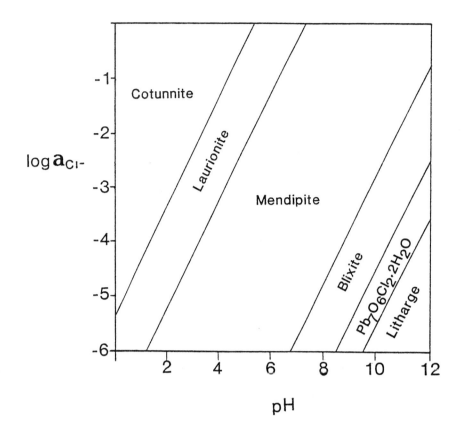

Figure 3. Stability field diagram for the PbO-HCl-H$_2$O system at temperatures $\geq$ 300K, p = 10^5 Pa (see text).

They have not been reported in the past in the PbO-HCl-H$_2$O system.

Lead in the marine environment commonly forms a soft thin corrosion layer, which is greyish white to buff in colour. This corrosion layer forms a protective surface thus preventing further corrosion of the metal. For this reason the majority of ancient lead objects that have been immersed in sea water are usually in good repair. Often the corrosion products are crystalline and can be readily identified using infrared and X-ray diffraction techniques. The predominant phases identified as corrosion products include anglesite, cotunnite, (Borelli, 1975), cerussite, hydrocerussite, (Weier, 1973), phosgenite and laurionite, (Campbell & Mills, 1977). For example, a lead Graeco-Roman anchor stock, found in North Wales, (Robinson, 1980), was covered with a soft buff corrosion layer, with anglesite, cerussite, cotunnite and phosgenite identified as the phases present. Furthermore, laurionite and hydrocerussite have been observed as corrosion products on lead bullets salvaged from the wrecked ships of the British Mediterranean Fleet, west of the Scilly Isles (Campbell & Mills, 1977).

As outlined above, these observations too can be readily interpreted in terms of the equilibrium model which has been developed. However, certain kinetic influences, such as are responsible for the crystallization of penfieldite and fiedlerite, remain to be resolved.

Acknowledgements

The Science-based Archaeology Committee of the SERC is thanked for providing a Studentship to R.E.

References

Abdul-Samad, F., Thomas, J.H., Williams, P.A., Bideaux, R.A. & Symes, R.F. (1982). Mode of formation of some rare copper(II) and lead(II) minerals from aqueous solution, with particular reference to deposits at Tiger, Arizona. *Transition Metal Chemistry* 7, 32-37.

Bideaux, R.A. (1980). Tiger, Arizona. *Mineralogical Record* 11, 155-181.

Borelli, L.V. (1975). Les alterations des bronzes antiques en milieu marin. In: *ICOM Committee for Conservation* 75/13/1, 1-8.

Campbell, H.S. & Mills, D.J. (1977). Marine treasure trove - a metallurgical examination. *The Metallurgist and Materials Technologist*, 551-557.

Goni, J., Guillemin, C. & Perrimond-Tronchet, R. (1954). Description d'espèces minerales neogenes formées sur des jas d'anchres Romaines immergées. *Bulletin de la Société française de Minéralogie (et de Cristallographie)* 77, 474-478.

Humphreys, D.A., Thomas, J.H., Williams, P.A. & Symes, R.F. (1980). The chemical stability of mendipite, diaboleite, chloroxiphite and cumengeite, and their relationships to other secondary lead(II) minerals. *Mineralogical Magazine* 43, 901-904.

Kiyama, M., Murakami, K., Takada, T., Sugano, I. & Tsuji, T. (1976). Formation and solubility of basic lead chlorides at different pH values. *Chemistry Letters* 23-28.

Kohlberger, W. (1976). Minerals of the Laurium Mines, Attica, Greece. *Mineralogical Record* 7, 114-125.

Näsänen, R. & Lindell, E. (1976). Studies of lead(II) hydroxide salts. Part 1. The solubility product of Pb(OH)Cl. *Finnish Chemistry Letters* 95-98.

Näsänen, R. & Lindell, E. (1978). Studies of lead(II) hydroxide salts. Part III. The solubility product of the dilead(II) trihydroxide chloride. *Finnish Chemistry Letters* 227-230.

Palache, C., Berman, H. & Frondel, C. (1951). Penfieldite. In: *The System of Mineralogy*, Vol II, pp 66-67. Wiley and Sons, New York.

Robie, R.A., Hemingway, B.S. & Fisher, J.R. (1978). Thermodynamic properties of minerals and related substances at 298.2K and 1 bar pressure and at high temperature. *United States Geological Survey Bulletin*, No. 1452.

Robinson, W.S. (1980). The preservation and corrosion process of metals from marine archaeological deposits. MSc dissertation, University of Wales, Cardiff.

Symes, R.F. & Embrey, P.G. (1977). Mendipite and other rare minerals from the Mendip Hills, Somerset, England. *Mineralogical Record* 8, 298-303.

Weier, L.W. (1973). The deterioration of inorganic materials under the sea. *Bulletin of the Institute of Archaeology* 131-163.

Wilkinson, L., Bathhurst, N.O. & Parton, H.N. (1937). Equilibria in aqueous lead chloride solutions. *Transactions of the Faraday Society* 33, 623-638.

Yadava, K.L., Padney, V.S. & Krishna, M.L. (1968). Studies in monohydroxy lead chloride. *Journal of Inorganic and Nuclear Chemistry* 30, 2915-2920.

THE DIFFUSION OF THE SILVER FROM POTOSI IN THE 16TH CENTURY SPANISH COINAGE: A NEW ANALYTICAL METHOD

M.F. Guerra[1], J.N. Barrandon[1], B. Collin[2], E. Le Roy Ladurie[3,4] and C. Morrison[4].

[1]Centre E. Babelon, CNRS, 45071 Orléans, France.
[2]Monnaie de Paris, France.
[3]Collége de France, Paris, France.
[4]Bibliothèque Nationale, Paris, France.

Among all the new metal types to arrive in Europe none has empassioned historians and economists as much as the arrival of the precious metals from America during the 16th century - gold from Mexico and silver from Peru (at this time the Potosi mountain were within the territory of Spanish Peru). On arrival in Spain, these metals entered the European and thus World monetary circulation.

In 1971, Gordus et al., tried to date the arrival of the Potosi silver in Spain and neighbouring countries (eg. France) using neutron activation analysis on a micro sample, to determine the concentration of gold in silver coinage. Gold was expected to differentiate between American and European silver.

In the Potosi silver, gold is found in low concentrations, while both European and Mexican silver ores have a high gold concentration. It was hoped that Potosi silver could be traced using this method, however it only allowed for the identification of those Spanish coins struck with pure South American silver.

The economic and historical importance of the Potosi metal during the 16th century was such that a fresh attempts, making use of geochemical data, was necessary. A study of the most recent geochemical data, much of it unpublished, on the silver ores of the Andean region in general, reveals that they contain quite rare elements such as germanium and indium, which do not occur in Mexican and European ores.

Unfortunately, germanium cannot be determined in a non-destructive global way when present at ppm level, so our attention turned to indium. A new analytical technique based on a type of mini-reactor associated with a cyclotron in order to produce moderated neutrons was developed. This technique allows a non-destructive global irradiation of silver coins, with a very high sensitivity for both indium and gold.

The irradiation device

The experimental device is shown in Figure 1 (Guerra & Barrandon, 1988). Charged particles colliding with a beryllium target produce a fast neutron beam which is moderated by a graphite block. The sample is placed inside the graphite block by a rotating system which allows the placing of samples with a maximum diameter of 5 cm. The beam reaches the sample after passing through 15 cm of graphite.

Different particles can be used to produce a neutron beam on the beryllium target. In order to choose the most suitable one, we studied a certain activation coefficient a function of the incident particle type and energy. The maximum value of this coefficient corresponds to the optimum experimental situation, ie. the maximum sensitivity of detection for the minimal interference.

The one meeting our conditions is a HH^+ beam with an energy of 20 MeV and a current of 30 μA. At these experimental conditions, the rotating system allows an homogeneous irradiation, with an 8% precision, on a 80 cm^3 volume. Detection limits for some of the elements studied are shown in Figure 2, for 1 gram Cu matrix. These are not of the same order of magnitude as for activation analysis

28

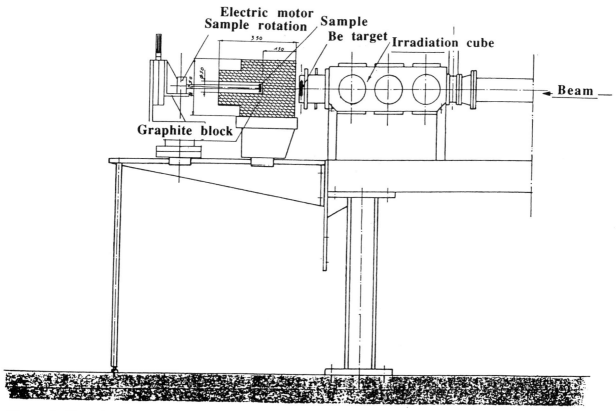

Figure 1: Experimental device used for non-destructive global analysis with moderated neutrons to measure silver, copper, indium and gold concentrations in silver coins.

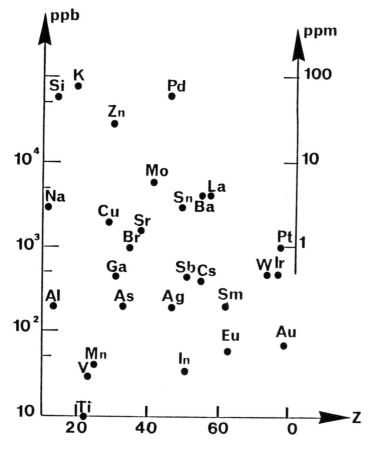

Figure 2: Detection limits (in ppb and ppm) for 1 g sample irradiated with a 20 MeV HH⁺ beam of 30 μA current and counted for 24 hours.

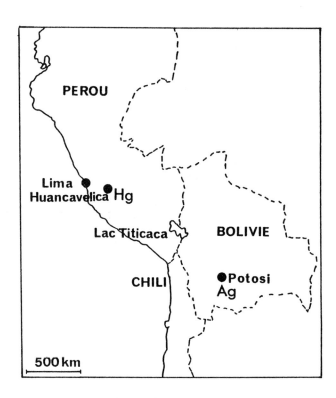

Figure 3: Map showing the Potosi silver mines and the Huancavelica mercury mines in the Andean regions.

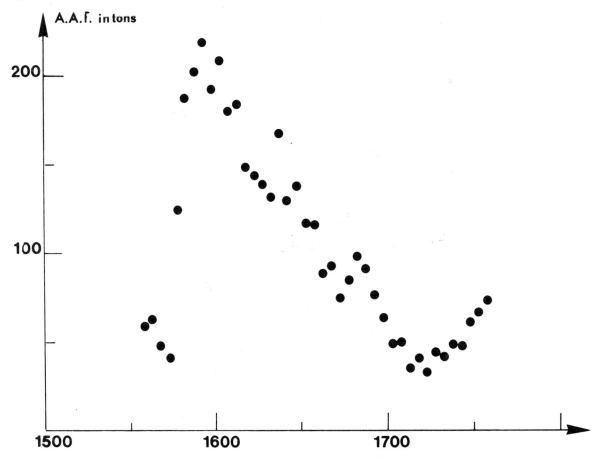

Figure 4: Representation of the average annual silver production (A.A. in tons) at Potosi from 1556-1760; based on Cross (1983) data.

with a reactor, but it must be remembered that these can be improved by varying the mass under irradiation.

In order to consider the shape of the object and the neutrons auto-absorption during irradiation, an internal standard was used. The concentration calculations are based on simulated objects with the same shape as the real objects.

Historical problem

The Potosi mines were found by Francisco Pizarro in 1545. Its first exploitation, which reached only the most superficial veins, took place between 1545 and 1564. These mines are situated in the tin-silver belt region of the Andes, in Cerro Rico, at a height of approximately 4790 m.

The amalgamation method of silver extraction was introduced to the region by Pedro Fernandez Velasco during the visit of the Vice-Roy Francisco de Toledo to Peru in 1571-1572 (Braudel, 1966). Following on from this, the mercury mines found in the Huancavelica region near Lima (Figure 3), as well as the cheap indian labour in the 'mita', started a new era of exploitation of the Potosi mines. Figure 4 shows the evolution of the exploitation of the Potosi silver mine, taken from data published by Cross (1983).

There are some historical documents which refer to the arrival of the South American silver in Europe, and also to inflation observed during the 16th century, and it is hoped these may be reflected in the composition, and may help us date its arrival.

Analytical results

Coins were analysed from the following collections: Cabinet des Medailles de la Bibliotheque Nationale de Paris, Musee Puig de Perpignan and Bourgeay private collection. These coins have the following geographical origins: Mexico, Potosi, Lima and Spain (not including Barcelona). The determination of the gold concentration in the coins allows the comparison between the work of Gordus *et al.* (1972) and this programme. Figure 5 demonstrates that this method is more precise for low gold concentrations than that of Gordus *et al.*

Taking the gold and indium concentrations found in the Mexican and Potosi coins, and plotting the results on a log scale, the graph shows the ratio of Au/Ag as a function of the In/Ag ratio (Figure 6). It is evident that silver ores from Mexico are characterised by a high Au/Ag ratio and a low In/Ag ratio, while silver ores from Potosi have a low Au/Ag ratio and a high In/Ag ratio.

Furthermore, taking the results from the Spanish coins minted before Filipe II and comparing them with the results in Figure 6, it is evident that Spanish and Mexican coins appear to correlate, suggesting that the Mexican/European silver and silver from Potosi can be accurately distinguished (Figure 7).

If the results from all the Spanish coins analysed are treated in the same way, it can be seen that they do not form a group, implying that they were made of a silver alloy which differed in the amount of Potosi silver present (Figure 8).

It is difficult to make precise conclusions with coins which do not present in their legends the date of minting. From Figure 8 we can say that:

- Some coins from Filipe II were struck using a pure Spanish silver, suggesting they were minted before the arrival of silver from Potosi.

- The only coin showing a date of production (1597) is minted with a pure Potosi silver. Other coins showing the same composition, which were also minted with Potosi silver, have no date of issue.

- Some coins thought to have been minted before this date have compositions suggesting they were

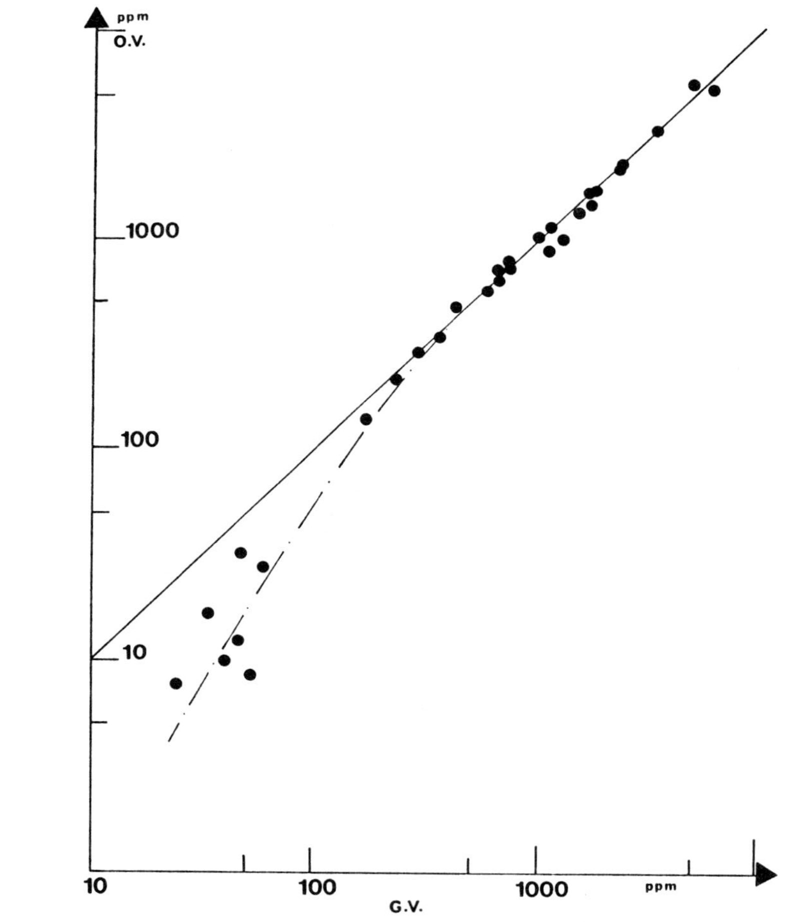

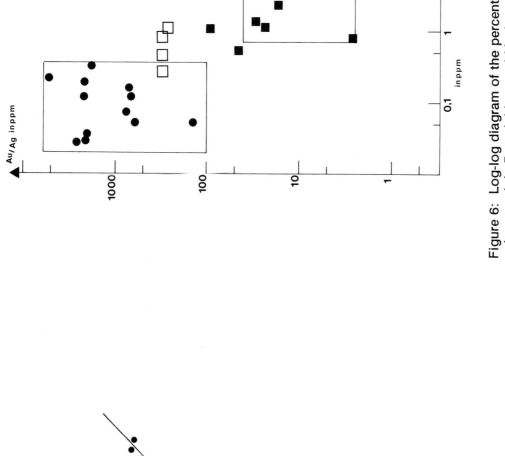

Figure 6: Log-log diagram of the percentage of In and Au to Ag in coins struck in Potosi, Lima and Mexico. Note that the Lima coins struck after 1775 are a mixture of local and Mexican ores.

- black square = Potosi
- black triangle = Lima
- black point = Mexico under Charles V and Philip II.
- black square = Lima from the 18th century onwards.

Figure 5: Comparison between ours (O.V.) and Gordus et al. (1971) values (G.V.) for the concentration of gold in the silver coins.

32

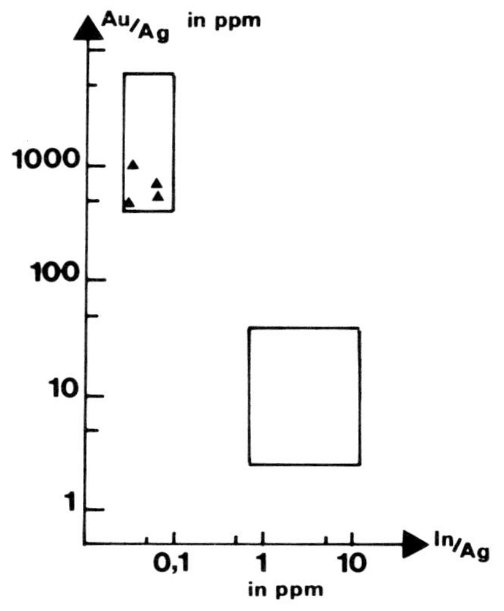

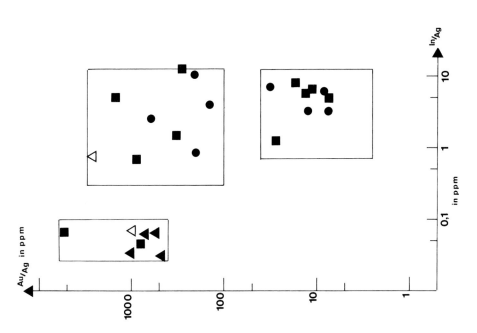

Figure 8: Log-log diagram of the percentage of In and Au to Ag in Spanish coins. The chart makes is possible to classify the coins according to ore used in their production.

- black triangle = prior to Philip II.
- black square = Philip II.
- black point = Philip III and Philip IV.
- black triangle = Charles II.

Figure 7: Log-log diagram of the percentage of In and Au to Ag in Spanish coins struck before Philip II.

33

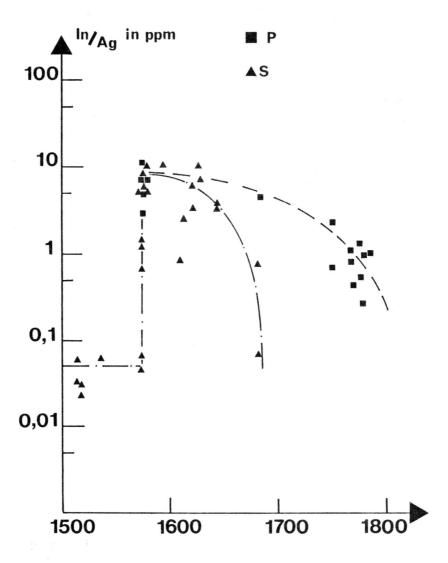

Figure 9: Log scale graph of the percentage of In to Ag by issue date of the coins. The graph shows (i) the scale and relatively short duration of the inflow of the ore from Potosi, and (ii) the dilution of the Potosi ore and Mexican ore in the 18th century.

- P = Potosi coins
- S = Spanish coins.

also made with a European-Potosi silver alloy.

The proportion of Filipe II coins minted with Potosi silver suggest the arrival of Potosi silver during the second half of his reign. This agrees with the date of introduction of the amalgamation extraction method, and suggests that Potosi silver had arrived in Spain by 1570.

If the evolution of the In/Ag ratio through time is plotted (Figure 9), a massive arrival of Potosi silver during the reign of Filipe II is indicated. Furthermore, the remelting of ancient coins to mint new coins has not been evidenced, but that new coins were stuck with pure Potosi silver which is unusual.

However, it is clear that the arrival of the Potosi silver did not last long. In fact, the coins from the reign of Filipe IV (1621-1665) are the only European-Potosi silver alloys found, and those from Carlos II (1665- 1700) show a complete dilution of the silver from Potosi.

Conclusions

A new analytical method for a global non-destructive determination of gold and indium concentrations in silver coins has been developed, which has considerable advantages over traditional methods.

The composition of Mexican, Peruvian and Spanish coins differed in these two elements, which allowed the differentiation between Mexican-European silver and Potosi silver.

This therefore enables the date of 1570 for the arrival of Potosi silver in Spain to be postulated. Although on a very large scale, this arrival was short lived, and in fact seems to vanish during the first quarter of the 17th century. This is certainly due to the low production of the Potosi mines at the time, and to the beginning of the exploitation of the Mexican mines.

Acknowledgements

We would like to thank the CERI-CNRS cyclotron facilities and the financial support of Fundacao Calouste Gulbenkian, Lisboa, Portugal.

References

Gordus, A., Gordus, J., Le Roy Ladurie, E. & Richet, D. (1972). Le Potosi et la Physique Nucleaire. *Annales Econ. Soc. Civilis* 1235-56.

Guerra, M.F. & Barrandon, J.N. (1988). Thermal neutron activation analysis of archaeological artefacts using a cyclotron. In: *Proceedings of the 26th International Archaeometry Symposium*. Ed. R.M. Farquhar, R.G.V. Hancock & L.A. Pavlish, pp. 262-8. The Archaeometry Laboratory, Department of Physics, University of Toronto.

F. Braudel (1966). *La Mediterranee et le Monde Mediterraneen a l'Epoque de Philippe II*, ed. A. Colin, Paris.

Cross, H.E. (1983). South American bullion production and export, 1550- 1750. In: *Precious Metals in the Later Medieval and Early Modern World*, ed. F.J. Richards, pp.397-423. Carolina Academic Press, Durham, N.C.

THE STUDY OF HISPANIC TERRA SIGILLATA FROM THE KILNS OF PLA D'ABELLA (CATALONIA): A PRELIMINARY ANALYSIS

J.M. Gurt[1], F. Tuset[1], J. Buxeda[1], C. Planas[1] & X. Alcobé[2]

[1]Department of Prehistòria, Història Antiga i Arqueologia, Facultat de Geografia i Història, Universitat de Barcelona, C. Baldiri i Reixac, s/n, 08028, Barcelona.

[2]Serveis Científico-Tècnics de la Universitat de Barcelona, C. Martí i Franquès, s/n 08028, Barcelona.

Introduction

The characterisation of the Hispanic Terra Sigillata from Abella has its origin in the rediscovery of its kilns in the 1986 archaeological survey and its further excavation, together with the adjacent 'la Rectoria' settlement. The new kiln excavation allowed us to begin characterisation of the ceramic output, with a wide archaeometric project that is the direct successor of the study initiated by Mayet and Picon (Picon, 1984).

The site

The kilns of the Abella workshop (Figure 1) were discovered by Serra Vilaró in 1912, who excavated three kilns up to 1925, and also walked the area without positive results (Serra Vilaró, 1924; Serra Vilaró, 1925). No publication was made of the stratigraphic data or the specific location of the workshop, which has remained unknown until recently. The archaeological works at Pla d'Abella and La Rectoria were initiated in 1985 with the purpose of distinguishing the features of the

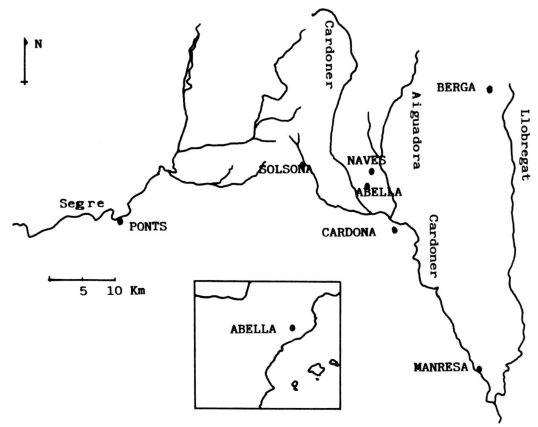

Figure 1. Location of the Abella workshop.

36

ancient population in this area, with the workshop as the main focus.

In 1987, a geo-magnetic survey followed the 1986 field survey at Pla d'Abella, which finally rediscovered the kilns. The 1987-88 excavation confirmed the existence of three kilns cited by Serra Vilaró plus a new smaller one, which is still unpublished.

The scarcity of the remains does not let us establish secure workshop chronologies. But the study of La Rectoria offers chronological data of part of the Abella production (second century to the first half of the third century AD).

Analyses

The first phase of the characterisation was done under a random sampling of 52 potsherds from the re-excavation of the kilns, and one sherd from La Rectoria. Most of the samples are sherds of Hispanic Terra Sigillata, but there are also seven sherds of moulds and four sherds of gas extraction tubes. Such samples were analysed by X-ray fluorescence (XRF), using a glassy pill for the major and minor elements, and a powder pill for the trace elements and sodium oxide. The data collected was firstly corrected for loss on ignition (Dixon, 1983), and then analysed statistically using the BMDP program. For 24 samples X-ray diffraction (XRD) was used, powdering the samples and then using CuKα radiation.

Results of the elemental analyses, when plotted as histograms show quite homogeneous compositions, although with a certain bimodal distribution. This bimodal tendency is emphasises in the resulting dendrogram from a hierarchical cluster analysis, produced using euclidian distance with the average linkage method on the standardized values of the following elements: Fe_2O_3, Al_2O_3, MnO, P_2O_5, TiO_2, MgO, CaO, K_2O, SiO_2, Ba, Rb, Zr, Sr, Zn, Cu, Ni and Ga (Figure 2). NaO was not included in the treatment because of its excessive variability.

The resulting dendrogram, even though the data appears homogeneous, confirms the tendency towards a division into two groups, with some outliers. A closer observation allows us to distinguish between four homogeneous groups, namely GA, GB, GC and GD.

In relation to the abundances of the diverse groups, the percentages of SiO_2, Fe_2O_3 and Al_2O_3 (Figure 3), a similar position of GA, GC and GB are observed, with low medium and high values respectively. Against that, in spite of low values of Fe_2O_3 and Al_2O_3, group GD has high values of SiO_2. The difference within the group GD is seen on the SiO_2/Al_2O_3 diagram (Figure 4). ROB this diagram again makes no mention of group GD series of numbers from 1 to 8 or something. The diagram shows a direct correlation among the GA, GC and GB groups, while the GD turns aside from the regression line. The GC group have a great homogeneity with values between the GA and GB groups, whilst the GD group, with three samples, in spite of its coherent values in accordance with Abella abundances, has a marginal position and singular features. The gradation on CaO abundances from GA, GC to GB, within the group of calcareous pottery is also seen. The SiO_2/CaO scattered plot has an optimum line regression but with the GD turned aside from it (Figure 5). Another problem are the abundances of K_2O for all the samples. We can clearly see the great variability of this element on the diagram (Figure 6). Moreover, by study of the correlation matrix we verified that K_2O did not have a significant correlation with other elements. This is as yet an unsolved problem, but is still under study. The means and standard deviations from the diverse groups are shown in Table 1.

The XRD results suggest the following: Firstly, there is a large variability in all samples with respect to the presence or absence of high temperature phases. Secondly, a correlation appears to exist between the groups from XRF and the relative amounts of such high temperature phases.

On the analysed samples belonging to Groups GA (samples 6, 8, 11, 16, 34) and GC

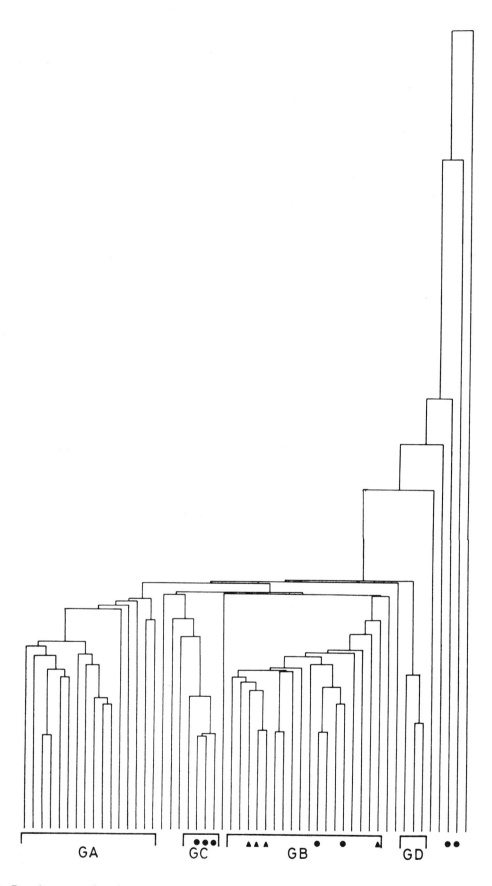

Figure 2. Dendrogram showing four intra-group agroupations (in circles the moulds, in triangles the gas extraction tubes).

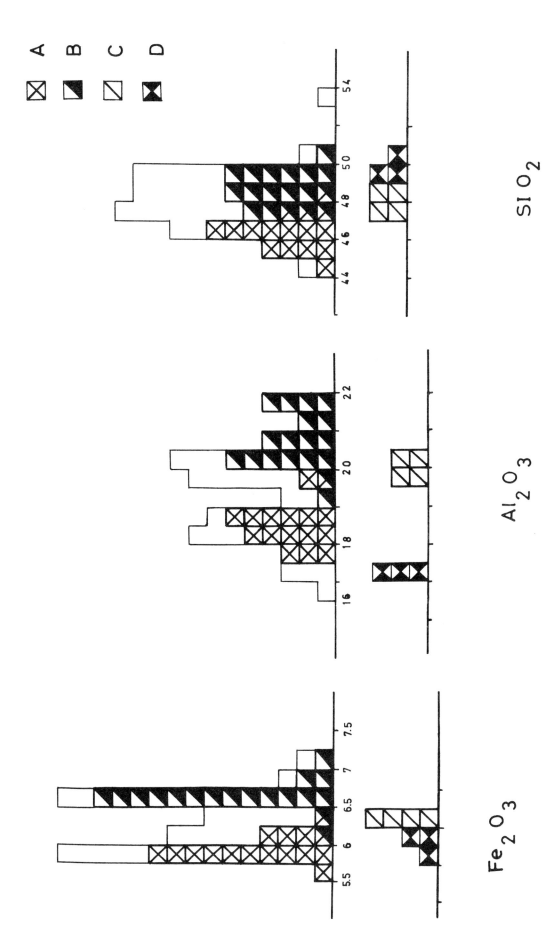

Figure 3. Histograms of Fe_2O_3, Al_2O_3 and SiO_2 values (in %).

elements groups		Fe_2O_3	Al_2O_3	MnO	TiO_2	MgO	CaO	K_2O	SiO_2	Ba	Rb	Zr	Sr
GA	$\bar{m}$	5.95	18.57	0.08	0.69	2.82	21.31	3.58	46.41	913	132	104	1188
	s	0.16	0.57	0.008	0.02	0.17	1.69	0.52	0.95	157	20	7	288
GB	$\bar{m}$	6.65	20.76	0.08	0.76	3.37	15.02	3.73	48.70	599	184	110	449
	s	0.19	0.72	0.005	0.03	0.15	1.74	0.81	0.90	90	20	4	50
GC	$\bar{m}$	6.35	19.97	0.10	0.72	2.97	17.23	4.29	47.90	749	160	95	516
	s	0.05	0.30	0.008	0.01	0.09	0.25	0.05	0.55	47	5	4	39
GD	$\bar{m}$	5.98	17.27	0.09	0.74	2.67	18.99	3.45	50.09	674	123	118	613
	s	0.09	0.15	0.007	0.03	0.15	0.22	0.09	0.32	88	1	3	48

Table 1. Concentration values of grouped samples (m is the mean value of cach group and s is the standard deviation). All numbers are in % except Ba, Rb, Zr and Sr (in ppm).

(samples 20, 25, 55) there is a major phase of quartz alongside significant calcareous phases (calcite and/or gehlenite). Other minerals identified are illite, haematite and K-feldspar, constant in all samples, and diopside and anorthite in some of them. By means of XRD, groups GA and GC are not clearly differentiated, showing variations on the amounts of high temperature phases (mainly gehlenite) and, consequently, on calcite content (Figure 7).

In group GB, diopside and anorthite are significantly more common than in groups GA and GC. In some cases (27, 29, 32 and 47) such minerals are the major phases, appearing as leucite and analcime. In other cases (28, 41, 42 and 45), quartz is predominant, although gehlenite and calcite are still important, and illite, haematite and K-feldspar occur. Calcite is well recorded in all samples of this group, clearly as a second phase (Figure 8). Finally, the three samples of group GD are more homogeneous according

to XRD. Quartz and calcite are the major phases, gehlenite, diopside, illite, K-feldspar and haematite also being present.

We observe that, in samples where analcime was recorded, the NaO content lies between the 1% and 2.5% and that percentages were significantly higher than the mean value of all grouped samples (0.54%). Such fact implies a clear ratio between this mineral phase and the NaO content. Moreover, we must not forget the cited presence of leucite in the same samples. All these point to analcime as a secondary mineral, related to the highest temperature phases.

Archaeological contribution

The analysed samples were, from an archaeological point of view, an homogeneous ensemble. Such homogeneity has been confirmed by the results of the first

40

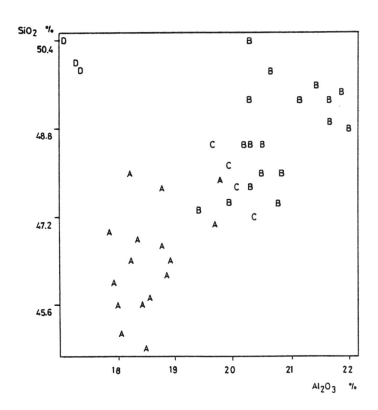

Figure 4. Scatter plot of SiO_2/Al_2O_3 (in %).

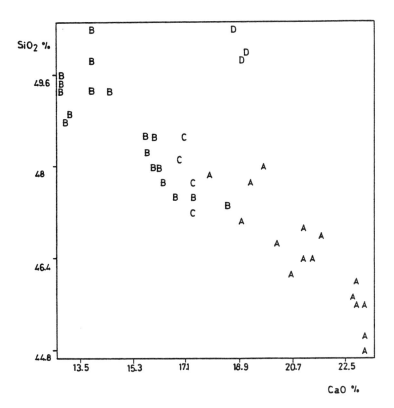

Figure 5. Scatter plot of SiO_2/CaO (in %).

41

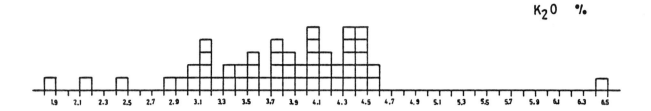

K₂0   %

Figure 6. Histogram of K₂O values (in %).

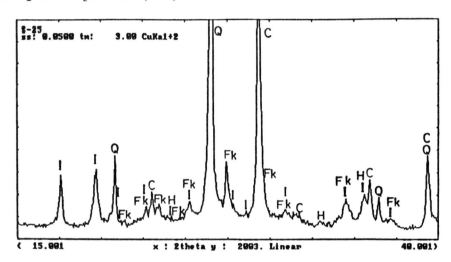

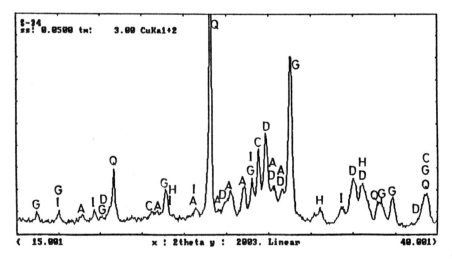

Figure 7. XRD diffractograms of samples 25 and 34 (Q = quartz; C = calcite; G = gehlenite; H = haematite; I = illite; Fk = K-feldspar; D = diopside; A = anorthite).

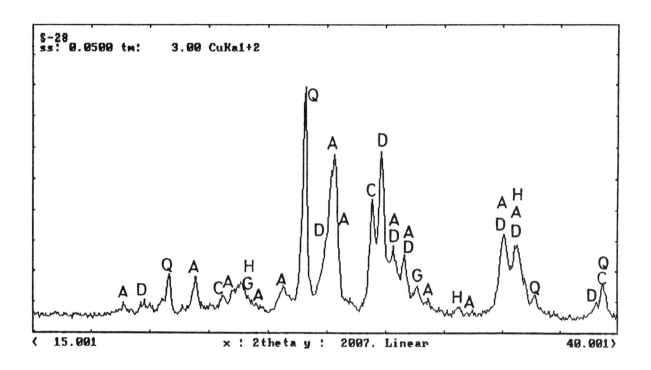

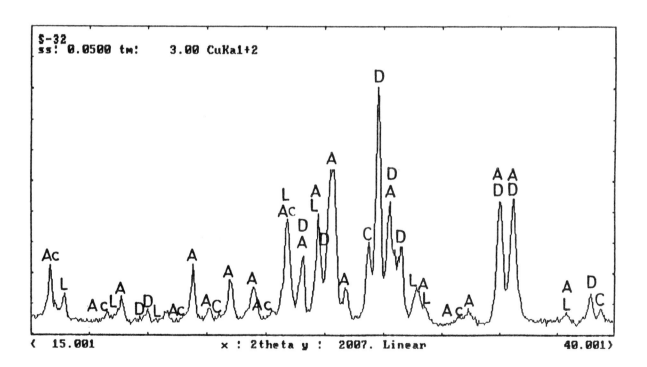

Figure 8. XRF spectrum of samples 28 and 32 (Q = quartz; C = calcite; G = gehlenite; H = haematite; D = diopside; A = anorthite; Ac = analcime; L = leucite).

43

analyses, which have shown small intra-group variations within a coherent production.

The resulting ensembles therefore do not offer considerable differences from the archaeological point of view, but we must emphasise that neither the moulds, the gas extraction tubes nor the over-fired sherds belong to the GA group. Nor do they belong to the GD group, although allowance must be made for the few samples in this group. All these fragments are concentrated in groups GB and GC.

All the Abella production is within group GA, within a wide typological range:- Drag. 37, a possible undecorated one, several with high and straight rims typical of the workshop; Ritt. 8, undecorated and of medium size; Drag. 44, a base of a plate; Herm. 23, a lid of Hispanica 7 and finally, a Drag. 47, a black sigillata with stamped decoration. There were also all the characteristic decorative styles of the workshop, with lineal motives, rosettes, or a combination of both, with stamped decoration, and with an arc over circles on black sigillata, already cited.

The GD group includes two undecorated shapeless fragments and another decorated with rough lines, also typical of the workshop.

Finally the GC group presents three moulds, all of them with lineal decoration, and a shapeless fragment of hispanic terra sigillata without decoration.

The conclusions we can draw are still preliminary. Despite that, taking both the XRF and XRD results, they seem to point to a difference between group GB on one hand and group GA and GC on the other. Nevertheless, taking into consideration the probable random variations in CaO content of the original clays, it appears to demonstrate an inverse ratio between the firing temperature and the CaO content, which would express the internal differences observed at the workshop of Pla d'Abella.

With regard to the samples roughly grouped, two types can be differentiated according to their position on the dendrogram; firstly, there are those among the differentiated groups, and secondly those at the right side of the GD group, with a greater lineal distance from the main block (Figure 2).

The first type of sherds corresponds to the typical workshop shapes: two sherds of decorated Drag. 37, one with lineal motifs and the other combining rosettes with arcs over circles, a motif already seen on the cited black sigillata; and several undecorated shapeless sherds, one with a grey coating and another corresponding to the overfired Herm. 23 sherd with the typical green colouration of overfired calcareous clays.

The second type consists of two moulds, one with lineal decoration, and the other with rosettes on a double frieze; a shapeless fragment of hisp. terra sigillata; a plate of Drag. 18/31 found at La Rectoria (situated on the right part of the dendrogram); and finally a fragment of Drag. 47, possibly a black sigillata with stamped decoration of an arc over circles. All these sherds correspond with the workshop typology and seem to be affected by contamination of several elements: P_2O_5, K_2O and MnO. In this sense, the XRD analysis has not provided sufficiently illustrative data.

Conclusions

After these first series of analyses, it has been possible to make a clear definition of the Abella production, recognising, within a general homogeneity, four minor groups.

The problem of possible contamination in some of our samples the problem established for K_2O has to be solved. Because the samples were recovered from re-excavation, it is not possible to know the depositional environment of our pottery, which again complicates the interpretation of the results.

Certainly, the realisation of further analyses and the termination of our current studies will help us solve our present problems.

Acknowledgements

Financial support of the Comisión Interministerial de Ciencia y Tecnología (investigation project PB85-0086) is gratefully acknowledged. We are indebted to Servei d'Espectroscòpia dels Serveis Científico-Tècnics de la Universitat de Barcelona, especially to the Direct. Dra. Baucells and Dra. Roure, where the samples were analysed. We thank Dra. G. Rauret, Facultat de Químiques, Universitat de Barcelona, and Dr. M. Picon, Laboratoire de Céramologie, C.N.R.S., Lyon, for helpful comments.

References

Dixon, W. (ed). (1983). *BMDP Statistical Software, 1983*, Printing with Additions, University of California Los Angeles Press, Los Angeles.

Picon, M. (1984). Recherches sur les compositions des sigillées hispaniques. Techniques de fabrication et groupes de production. In: *Les céramiques sigillées hispaniques*, F. Mayet, pp. 303-317. Centre Pierre Paris, C.N.R.S., Paris.

Serra Vilaró, J. (1924). Estación ibérica, termas romanas y taller de 'Terra Sigillata'. *Solsona, Memoria de la Junta Superior de Excavaciones y Antigüedades* 69, Madrid.

Serra Vilaró, J. (1925). Cerámica en Abella. Primer taller de 'Terra Sigillata', descubierto en España, *Memoria de la Junta Superior de Excavaciones y Antigüedades* 73, Madrid.

45

ANALYSIS OF ROMAN GLASS-WORKING EVIDENCE FROM LONDON

M.P. Heyworth

Ancient Monuments Laboratory, English Heritage, 23 Savile Row, London W1X 2HE

Introduction

Excavations undertaken in recent years by archaeologists from the Museum of London in the area of the Upper Walbrook Valley have revealed a number of sites with evidence for glass-working and glass-blowing. In particular the sites of Copthall Avenue and Moorgate revealed evidence for an industrial quarter dating to the late first and second centuries A.D. situated on what would have been the northern fringes of the contemporary Roman city. This activity pre-dated the construction of the city wall (to the area's immediate north) but was contemporary with the fort and amphitheatre to the immediate west.

The material from these sites displays the most complete glass-working assemblage so far recovered from Roman London. Included are numerous tank furnace fragments, moils (the waste glass left on a blow-pipe or pontil after the removal of the vessel or object), pot or tank metal fragments (metal is used in this context as the technical term for bulk molten glass), droplets and amorphous blobs of glass and small fragments from possible waster vessels. In collaboration with John Shepherd of the Museum of London a set of objectives was established for an analytical programme to investigate the composition of the different groups of material. The main objectives were related to investigating the composition of the glass associated with the furnace and of the tank metal fragments, and comparing this glass with moils and other waste fragments, and also with vessel fragments.

Analytical Method

A small number of samples from the different groups of material were initially selected for analysis. The analyses were undertaken using inductively coupled plasma atomic emission spectrometry (ICPS). The ICPS technique is becoming increasingly widely used in the analysis of archaeological materials (Heyworth et al., 1988). It gives compositional data for a wide range of elements at the major, minor and trace levels (Thompson & Walsh, 1983). This is particularly important for the analysis of glass where major and minor elements determine the general type of glass and minor and trace elements have an important influence on its colour. In the present programme data was obtained for 32 oxides and elements: Al_2O_3, Fe_2O_3, MgO, CaO, Na_2O, K_2O, TiO_2, P_2O_5, MnO, Pb, Sb, Ba, Co, Cr, Cu, Li, Nb, Ni, Sc, Sr, V, Y, Zn, Zr, La, Ce, Nd, Sm, Eu, Dy, Yb and SiO_2. The figure for silica was obtained by difference as the silica is removed in the sample preparation procedure. Some minor oxides, such as sulphur and tin, which may be present in the glass, were not included in the analytical programme (which was primarily established for silicate rock analysis). The analysis of tin and sulphur is possible using ICPS, though a different sample preparation procedure would be required for the analysis of sulphur.

In this programme, samples of glass for analysis were cut from the glass fragments using a low speed diamond blade saw and milled to a fine powder. A powdered sample of 100 mg was then evaporated to dryness with perchloric and hydrofluoric acid, and the residue dissolved in hydrochloric acid and distilled water before dilution to a standard solution strength.

The analyses were undertaken by Dr J.N. Walsh in the Department of Geology, Royal Holloway & Bedford New College, University of London, using a Philips polychromator ICPS system calibrated for quantitative analysis with multi-element rock standards. The glass solutions were run through the system twice, the first time the majority of major, minor and trace elements were measured, and the second time the solution was diluted to 10% of its original strength to obtain the soda figures. The dilution was necessary to maintain a linear calibration for the soda signal. Multi-element rock standards were analysed at regular intervals during the analytical run to allow for correction of any short-term fluctuations in the system. Three glass substandards were also analysed to check the ICPS calibration.

Results and Discussion

Forty-four samples were analysed by ICPS and were shown to be of basically similar composition, all durable soda-lime-silica glasses which previous work (e.g. Sanderson *et al.*, 1984) has shown to be the standard glass composition in the Roman period in north-west Europe. However there were other features of the analytical data which showed significant differences between the samples.

Most noticeable was the wide variation in the range of values for some of the minor oxides, particularly iron oxide which ranged over an order of magnitude from about 0.2% to over 2.0% (see Figure 1). The variation in iron oxide content correlated with similar variation in the levels of aluminium and titanium oxides, all elements likely to be present at significant levels in the clay of the furnace wall. All the samples with high levels of iron oxide were of glass taken from the surface of the tank furnace and it is therefore likely that there has been some movement of these elements from the clay of the furnace into the glass via some form of exchange mechanism. Great care had been taken when removing the samples for analysis to ensure that no particles of furnace material were left adhering to the glass before the sample was put into solution and any possibility of sample contamination can therefore be ruled out.

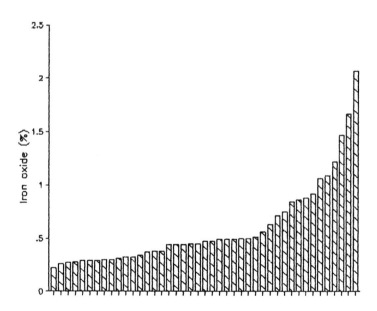

Figure 1. Bar chart showing iron oxide levels in all analyses ranked in ascending order

A cross section was cut through one of the furnace fragments and this shows a number of distinct layers between the clay of the furnace on the outside and the translucent light blue-green glass on the inside (see Figure 2). Further investigation of the movement of elements from the clay to the glass is now planned by looking at the level of particular elements in each layer from the clay through to the glass. A scanning electron microscope (SEM) used in the backscattered mode will be used to look at the distribution of individual elements through the layers shown in Figure 2. It is also planned to investigate the mineralogical composition of the layers using X-ray diffraction.

Figure 2. Photograph showing cross section through a fragment of tank furnace and the glass adhering to the inner surface (from the top of the clay to the bottom of the glass is approximately 1 cm).

Similar work has been reported on glassmaking crucibles from the Wadi el-Natrun in Egypt which were also dated to the Roman period (Saleh et al., 1972). Three layers were distinguished through a vertical section of a fritting crucible which were labelled A, B and C. These layers were distinguished on the basis of their colour and appearance, though they lacked sharply defined boundaries. Layer A is the outside, i.e. that subjected to firing, layer C represents the inner surface of the crucible wall which was in direct contact with the glass layer, and layer B is the middle part of the wall. Analysis of the three layers showed that whilst layers A and B were of similar composition, they were quite different to layer C. In particular the aluminium, iron and titanium oxide levels were reduced in layer C, and the silicon, calcium and sodium oxide levels were increased, in comparison to the composition of layers A and B. This is likely to be a result of the reaction between the crucible material and the glass material. Comparison of the Egyptian data with that obtained from the analysis of different samples of the Moorgate tank furnace showed that a similar pattern could be identified in the latter group (see Table 1).

This clearly has important implications for any attempts to link material from furnaces or crucibles to any waste or finished products from glass production sites. Excavations of archaeological sites do occasionally uncover evidence for glassworking in the form of crucible fragments containing melted glass. As there is only a thin layer of glass usually left adhering to the crucible (often only 1-2 mm) it will clearly be important to investigate the variations in composition through the glass before any assertions can be made as to the exact composition of any glass vessels or objects produced using glass melted in the crucible. It may be that trace elements will be more important in establishing links between glass from crucibles and glass from vessels or objects and the use of multi-element techniques such as ICPS, NAA or electron microprobe will be important to obtain the full glass composition, down to trace element level, for this purpose.

Another interesting aspect of ancient glass technology is the variations in colour within what are variously called colourless, lightly tinted or blue-green glasses. These are glasses which are not deliberately coloured by the addition of specific colourants, but nevertheless have a light blue or green tint. This tint is due to the presence of iron in the glass, usually as a result of the use of raw materials like sand which contain iron as an unrecognised impurity.

The glass adhering to the tank furnace from London showed a considerable variety of blue-green tints and this is probably due to the variation in redox conditions and temperature throughout the furnace. It is not clear if the Roman glassmakers would have had sufficient control over the production method to produce a specific tint on demand, though there is documentary evidence from the medieval period for some limited knowledge of colour changes simply based on extended heating. Theophilus, writing in the twelth century, describes how a colourless glass could, with further heating, turn to either saffron yellow becoming redder with time or a tawny, flesh colour which deepens to a reddish purple with prolonged heating (Hawthorne & Smith, 1963). However there is no evidence to suggest that Theophilus knew how to control whether the glass turned yellow or tawny in the first place, which would have depended on the levels of the transition metals in the glass batch, particularly iron and manganese, and their state of oxidation.

It is known that the use of additives to decolourise glass was widely practised in the Roman period and much work has been undertaken on the pattern of use for specific decolourants (e.g. Sayre, 1963). The principal decolourants used in the Roman period were manganese and antimony oxides, and the analytical data from the London glassworking sites suggests that both were in use there, though it is not clear whether both were deliberately, and separately, added to the glass batch or whether the use of cullet containing either manganese or antimony can be suggested. It is often difficult to judge the level at which something becomes a deliberate addition to the glass. The glass

Table 1. Compositional evidence for chemical exchange of elements between the glass and the clay in glass-melting crucibles

a) Data from Egyptian crucible (from Saleh et al., 1972)

	Crucible rim Layer A	Crucible rim Layer B	Crucible rim Layer C
SiO_2	57.8 %	57.3 %	62.1 %
Na_2O	6.2 %	5.9 %	11.0 %
CaO	6.9 %	6.8 %	7.4 %
Al_2O_3	14.7 %	15.3 %	9.5 %
Fe_2O_3	6.7 %	6.9 %	4.4 %
TiO_2	1.4 %	1.4 %	0.8 %

Layer C, the layer in contact with the glass, shows reduced levels of aluminium, iron and titanium oxides suggesting they have migrated from the clay into the glass, and increased levels of silica, sodium and calcium oxides suggesting they have migrated from the glass into the clay.

b) Data from Roman tank furnace found at Moorgate, London

	Glass fragment adhering to furnace wall	Glass fragment likely to come from the furnace
SiO_2	68.5 %	71.1 %
Na_2O	14.6 %	17.2 %
CaO	4.9 %	6.7 %
Al_2O_3	6.7 %	2.3 %
Fe_2O_3	2.1 %	0.4 %
TiO_2	0.4 %	0.1 %

The glass adhering to the furnace wall shows increased levels of aluminium, iron and titanium oxide suggesting these have migrated from the clay into the glass, and reduced levels of silica, sodium and calcium oxide suggesting these have migrated into the clay from the glass.

NB.   However it is possible in the case of some oxides, particularly silica, that the apparent differences are due to the nature of percentage data where a real fall in one component leads to an apparent rise in others.

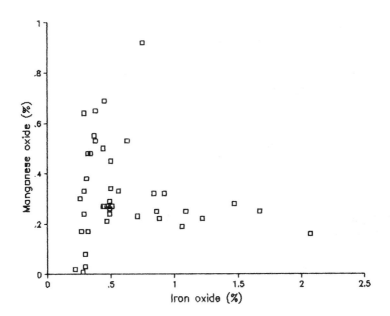

Figure 3. Plot of iron oxide against manganese oxide for all analyses

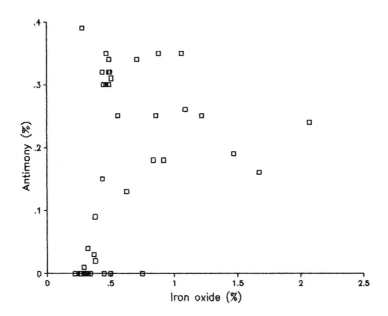

Figure 4. Plot of iron oxide against antimony for all analyses

from the London sites frequently had a manganese oxide level of about 0.25% however there were one group of fragments analysed which had a higher manganese level than most and these fragments also tended to have lower iron oxide contents (see Figure 3). This suggests that the glassmakers readily understood the properties of manganese as a decolourant and also preselected the raw materials containing less iron oxide in an attempt to produce a more colourless glass. The levels of manganese added would not have been sufficient to have had a significant decolourising effect on glass with a high iron content.

Antimony is a more effective decolourant than manganese and tends to produce a more brilliant colourless glass. It was therefore of particular interest to find that the glasses containing higher levels of iron usually contained a significant level of antimony (see Figure 4), though in some cases the measured iron oxide level is affected by the diffusion from the furnace wall.

From the glass fragments found associated with the tank furnace at Moorgate, and vessel fragments of similar date found in other contexts on the site, it is likely that the production of glass tableware was carried out in the area in the second century A.D. The glasses are not completely decolourised so, if antimony and manganese oxides were deliberately added, it appears that their use as decolourisers in the glass was not intended to produce a pure colourless glass but may have been an attempt to vary the tint of the glass to a paler shade. It is, however, possible that the differences result from the use of cullet in the vessel production. Further work is now needed to see if it is possible to link specific decolourants or glass "recipes" to the production of specific glass vessel types which may indicate a more sophisticated technological knowledge. This more detailed analytical study will hopefully lead to a greater understanding of the compositional variation within the Roman glass from London.

Acknowledgements

The work was undertaken as part of a project in the Ancient Monuments Laboratory of English Heritage. I would particularly like to thank Dr J.N. Walsh of the Dept of Geology, Royal Holloway & Bedford New College, for undertaking the ICPS analyses. I would also like to thank John Shepherd of the Department of Urban Archaeology, Museum of London, for providing the archaeological information connected with the glass. Justine Bayley and Dr Julian Henderson provided helpful comments on an earlier draft of this paper. The photograph for Figure 2 was taken by Louise Woodman of the English Heritage Photography Section.

References

Hawthorne, J.G. & Smith, C.S. (1963). *On Divers Arts, the Treatise of Theophilus.* University of Chicago Press, Chicago.

Heyworth, M.P., Hunter, J.R., Warren, S.E. & Walsh, J.N. (1988). The analysis of archaeological materials using inductively coupled plasma spectrometry. In: *Science and Archaeology, Glasgow 1987*, eds. E.A. Slater & J.O. Tate, pp. 27-40. British Archaeological Reports British Series 196, Oxford.

Saleh, S.A., George, A.W. & Helmi, F.M. (1972). Study of glass and glass-making processes at Wadi el-Natrun, Egypt in the Roman period 30 B.C. to 359 A.D. Part I Fritting crucibles, their technical features and temperature employed. In: *Studies in Conservation* 17, 143-72.

Sanderson, D.C.W., Hunter, J.R. & Warren, S.E. (1984). Energy dispersive X-ray fluorescence analysis of 1st millennium glass from Britain. In: *Journal of Archaeological Science* 11, 53-69.

Sayre, E.V. (1963). The intentional use of antimony and manganese in ancient glasses. In: *VI International Congress on Glass: Advances in Glass Technology*, Part 2, eds. F.R. Matson & G.E.Rindone, pp. 263-82. Plenum Press, New York.

Thompson, M. & Walsh, J.N. (1983). *A Handbook of Inductively Coupled Plasma Spectrometry*, 1st edn. Blackie & Son Ltd., Glasgow.

PROVENANCE STUDIES OF SPANISH MEDIEVAL TIN-GLAZED POTTERY BY NEUTRON ACTIVATION ANALYSIS

M.J. Hughes

British Museum, Research Laboratory, Great Russell Street, London WC1B 3DG

Introduction

Medieval Spain had a vigorous ceramic industry producing a wide range of products at many different centres for both the home and overseas markets. For example, it is reported that there were fifty pottery workshops operating in Spain by the middle of the 16th century (Lister and Lister, 1982:45). The tin-glazed pottery of Medieval Spain has attracted the attentions of art historians for some time, especially to the richly-decorated and prized lustrewares of Malaga and Valencia (Frothingham 1951 and Caiger-Smith 1973 and 1985), whose succession of decorative styles have provided much scope for art-historical research. In recent years archaeologists have become interested in the tin-glaze wares for several reasons. The characteristic styles provide useful chronological markers when fragments of such ceramics occur on archaeological sites, and indeed precise context dating on some sites, e.g. London, has assisted in dating some styles more precisely. As well as being found on port sites around southern England and Wales (Hurst 1977), there was an extensive movement of tin-glazed wares to Mediterranean France and Italy. Following the discovery of America there was also large-scale shipment of some types of tin-glaze across the Atlantic. Spanish fleets carried large quantities of these ceramics so that sherds have been found on sites along the Eastern seaboard of the United States including Florida, the Caribbean islands and parts of South America such as Venezuela. These American imports have been the subject of vigorous archaeological and art historical research (Lister and Lister 1982, 1987; Deagan 1988).

This paper is an interim report on a neutron activation analysis project, concentrating on more technical aspects, which includes a number of different provenance studies, some unrelated, but which are all on Spanish Medieval ceramics. The examples selected are meant to illustrate the type of results which are emerging, and future joint publications with the archaeologists/art historians concerned will discuss the detailed linking of the scientific with the art-historical aspects. All the styles of pottery included here have been conveniently described and illustrated by Hurst et al. (1986).

Analytical investigations on Spanish tin-glazed pottery

The present study began as a project to investigate whether chemical analyses could be used to differentiate between the lustreware products of the two main centres at Malaga and Valencia - (Hughes and Vince 1986). Petrology had demonstrated the presence of characteristic schist inclusions in Malaga lustreware, but experience with sherds recovered from excavations by the Department of Urban Archaeology in London (Vince, 1982) had shown that there were difficulties in positive identification. Due to erratic distributions, the schist fragments might not be present in either a broken edge of a sherd or a thin section. Neutron activation analysis proved to be a positive means of identifying the products of these two centres (Hughes & Vince, 1986), and the present study provides an update to the earlier paper in demonstrating how these results have been applied to answering questions of identification. Additionally many more samples from these two centres have now been analysed which reveal a range of chemical composition sub-groups within the ceramic output of the two centres, whose interpretation involves study of their art-historical aspect. Other centres in Spain also produced tin-glazed pottery of which

Seville was the most important regarding overseas exports : a range of ceramics from Seville has been studied in this project.

Some other parallel investigations using scientific techniques are in progress in other laboratories. Earlier work by D'Archimbaud and Lemoine (1980) using x-ray fluorescence analysis showed that distinctions could be made between Malaga and Valencia ceramics using elemental analysis, but the lack of overlapping elements analysed by their technique and ours (apart from potassium, calcium and iron) has prevented our use of their data. In contrast, neutron activation has been applied to Sevillian ceramics found on Spanish colonial sites in America and the Caribbean by Olin *et al.* (1978) and Jornet *et al.* (1985), which are directly comparable to the data produced in this Laboratory (see below) and also that using DC plasma spectrometry (Maggetti *et al.*, 1984). The latest study by the Smithsonian group has been lead isotope analyses of the glazes of Spanish and Mexican maiolica (Joel *et al.*, 1989). Such studies have not however exhausted the rich repetoire of Spanish medieval pottery products but have provided an example of the kind of information that is forthcoming.

Tin-glaze pottery from Seville

Seville was a major producer of pottery; the potters' quarter was in the Triana district on the south bank of the Guadalquivir. In the 16th century its output grew especially in response to the demand for tablewares and coarse wares such as oil jars from the Spanish colonies in the West.

From an analytical viewpoint, Sevillian ceramics have interested us from both questions of identification of specific pieces thought to have been made there, and also for examining the chemical relationship between the various identified types of tin-glaze pottery presumed to have been made there.

As an example of the first point, a question has been recently raised as to whether the tiles in the chapel of Leo X in the Castel Sant'

Angelo in Rome (Figure 1a) which bear the arms of the Medici family were made in Liguria (Italy) or in Spain, as has long been assumed (Wilson 1987:31). Two such tiles are in the collections of the British Museum and it was hoped that analysis of these would settle the question of origin. For comparison, four other tiles from the collections of definite Sevillian origin (on art-historical grounds) were selected (one is shown in Figure 1b to the same scale as 1a, showing a clear size difference), as well as one tile from the Godman collection decorated with a lustre star on a blue background and a fragment of a modern tile made in Seville of local clay. Analyses had also been carried out on contemporary tablewares in several defined styles (e.g. Isabela Polychrome etc, Hurst *et al.*, 1986:54-7). Upon analysis all the tiles, including those from Sant Angelo, fell into such a narrow composition profile that it was clear that all were made of the same clay. By contrast, in another analytical programme undertaken by the Research Laboratory, (on Italian maiolica: Hughes, forthcoming) there were analyses of ceramics from Liguria and they were clearly different from the Sevillian tiles in composition. The analysis therefore demonstrates that the Sant' Angelo tiles are Sevillian, as Ray (forthcoming) has strongly argued on art-historical grounds, attributing them to the workshop of Nicoloso in Triana.

What are the chemical relationships between the many different styles of Sevillian ceramics, for example the series made at Seville during the 16th century? These styles are now commonly called after the names applied in the fundamental study by Goggin (1968) of Spanish colonial pottery found in the Americas: Isabela Polychrome, Columbia Plain (white), Yayal Blue on White and Santo Domingo Blue on White (Hurst *et al.* 1986: 38-67). Examples of all these styles have been made available for our analysis by a number of archaeologists, and although numbers of samples were small it was interesting to investigate how these styles compared with each other in composition. There were more examples of Isabela Polychrome available for analysis than the others: mainly bowls and plates were decorated in this style which often includes a

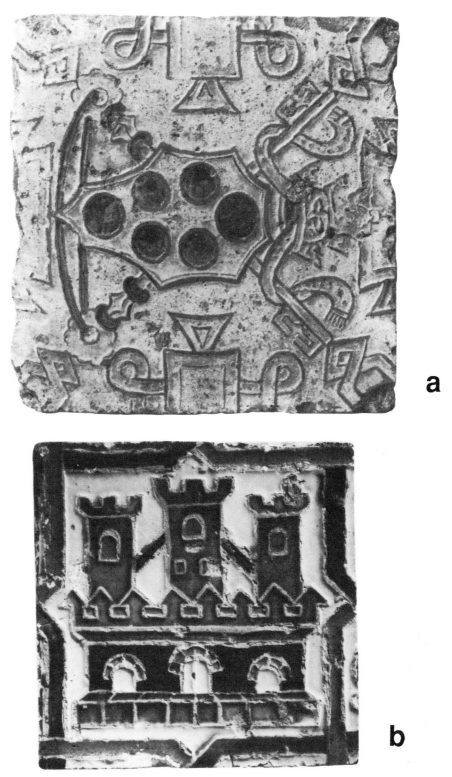

Figure 1 : a) Tile from the chapel of Leo X in the Castel Sant'Angelo, Rome (British Museum Reg.no. MLA1883,11-6,9; Research Laboratory no.24399S) and b) tile attributed to Seville (BM Reg.no. MLA1900,7-18,1; Res.Lab.no.27541T) in blue, green, brown and white glazes for comparison with the Castel Sant'Angelo tiles. Both tiles decorated in the cuenca technique: impressed design with ridges to separate the glaze colours during firing, and both to the same scale (0.75x full size). (Photographs courtesy of the Trustees of the British Museum).

air of blue and purple concentric lines around the rim of the plate, with a broad-brush foliage design in the centre in blue and purple (Hurst *et al*. 1986: 54-7). Some of the samples analysed came from excavations in London although it has been found at a number of coastal sites in Britain including Southampton, Bristol and from a shipwreck in Poole harbour. Analysis revealed a reasonably consistent composition for the Isabela Polychrome which was slightly but clearly different from that of the tiles analysed above. A principal components analysis of the results from the Seville samples (Figure 1)

shows a separation between the tiles and the Isabela Polychrome. Using Mahalanobis distances, four sherds seem to be outliers : three blue on white and one blue on blue sherd. Blue on blue was originally made in Liguria (Genoa) and is known as 'berettino' ware, and was copied in Spain, probably in Seville (Hurst *et al*. 1986, Lister and Lister 1982 p.62). These latter authors (op.cit., p.74) note that 'berettino' occurs as stray surface finds on sites near Seville, and analysis of one blue on blue fragment (Figure 2, no.1) shows it to be quite unlike the other Sevillian ceramics and is therefore probably an Italian berettino.

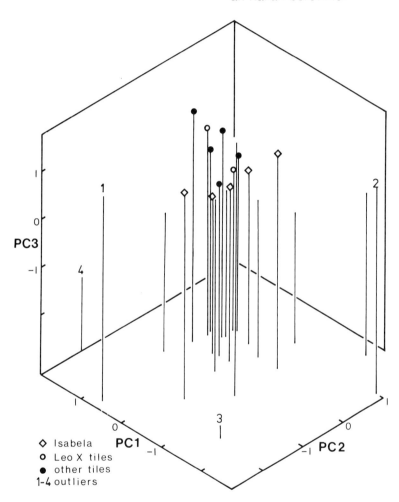

Figure 2 : Isometric principal components plot of the first three components for NAA results on Seville tin-glazed pottery including tiles and Isabela Polychrome. PC1 contains 38.4% of the variance, PC2 27.3% and PC3 14.9% (PC1+2+3 = 80.6%). PC1 is correlated positively with Cr, Na, Sc, Ce, Eu and negatively with K; PC2 is +ve with Cs, Th, Ce, Eu, Hf and K and -ve with Na, and PC3 +ve with Na, Hf, Ce, Lu, Eu and -ve with Cr in descending order of contribution.

Comparison with the analysis of Italian maiolica from Liguria shows a similarity with Ligurian ceramics and this is probably its place of manufacture. Similarly another outlier (Figure 2, no.2) was a fragment much thinner than the other ceramics (ca. 3mm compared with a minimum of ca. 5mm for the others), with a "runny" blue decoration on a white background : it has been suggested that it is probably 17th century Portuguese. The other two outliers are one unidentified blue-painted sherd (3) and one ?Yayal Blue (4).

Interlaboratory comparison of analyses

Quantities of ceramic that appear to be Sevillian have been found in the Americas and groups of these have been analysed by NAA (Olin et al.,1978; Maggetti et al., 1984). Assuming it is of Spanish origin it should be possible to make a direct comparison between Sevillian ceramics analysed in the present work and their published analyses from the Spanish colonies, inter-laboratory effects permitting. Table 1 shows one direct

Table 1. Comparison of the average composition of two groups of Seville ceramics obtained by the British Museum Research Laboratory and Brookhaven National Laboratory/Smithsonian Institution.

Neutron Activation Analyses of 15-16th Century Ceramics from Seville

Element	BMRL (tiles and tablewares (n=15)		Brookhaven Nat. Lab (tablewares) (n=40)	
Ce	64.1	6.6	62.6	14.6
Th	10.9	1.3	8.4	0.64
Cr	85.1	13.6	82.8	6.8
Hf	5.89	1.05	4.20	0.37
Cs	6.34	1.25	5.09	1.04
Rb	115	29	80	22
Sc	11.8	1.5	13.6	0.85
Fe	3.38	0.43	3.56	0.25
Eu	1.28	0.12	1.35	0.13
La	29.5	3.2	33.6	3.4
Na	0.54	0.13	0.79	0.15
Co	12.8	3.8	12.9	3.4

all results are given as the element in parts per million except Fe and Na in %.

Note: The Brookhaven results for thorium (Th), hafnium (Hf), scandium (Sc) and cobalt (Co) have been multiplied by factors of 0.81, 0.84, 1.05 and 0.91 to compensate for inter-laboratory differences established from tests on other samples. Some standard deviations are larger for the BMRL group because it included both tiles and other pottery types whereas the Brookhaven results included only Columbia Plain and Yayal Blue on White from one site (Convento de San Francisco, Dominican Republic: Olin et al., 1978, Table II, Column 1).

comparison, based upon analyses of a group of Columbia Plain and Yayal Blue sherds analysed by neutron activation at Brookhaven National Laboratory (Olin et al., 1978). The method of comparing the analyses between these two laboratories has been described in the earlier paper (Hughes and Vince, 1986) where it was found necessary to adjust the concentrations of scandium (1.05), barium (1.06), hafnium (0.84), thorium (0.81) and cobalt (0.91) by the Brookhaven/British Museum factors shown. The results of the two laboratories, adjusted by these factors, are shown in Table 1 for all the elements analysed in common by both groups. The comparison shows a good agreement between the results, which illustrates the possibility of data exchange between laboratories and also indicates the Seville origin of the Americas ceramics, given the undoubted Sevillian origin of the tiles. Such agreement has in fact been achieved even though different groups of samples of the same type were analysed in each laboratory.

The question of inter-laboratory exchange of data is related to the question of standardisation. Since each laboratory operating neutron activation uses one or perhaps a small number of standard materials to calibrate their procedure in each irradiation, eg. pottery standards, powdered rock standards, then analysis by a laboratory of another, internationally recognised standard material, provides one way forward in the quest for linking together the results of several laboratories. The US National Institute of Standards and Technology (formerly National Bureau of Standards) reference materials have long been recognised as important contributors to standardisation of analytical results in many fields - industrial, chemical, environmental. A widely-used standard in several such fields is Coal Fly Ash (NBS Fly Ash 1633A) which has convenient concentrations of elements appropriate for pottery analysis (its predecessor 1633 is used as a standard by the Smithsonian Institution: Jornet et al. 1985). We have included several portions of 1633A in recent irradiations and the mean figures obtained are given in Table 2, alongside results obtained by Natural History Museum, London analytical group, and

the 'quoted' values for this material. The degree of agreement is good, and the actual composition of the fly ash can be seen to be appropriate for pottery analysis. The question of standardisation is important and its implications have been discussed elsewhere (Harbottle 1982; Hughes et al., forthcoming).

Malaga and Valencia lustrewares and other tin-glazed pottery

The first lustre pottery produced in Europe was introduced into Spain by Arab potters. The Muslim potters of Malaga in the Kingdom of Granada became justly famous for their pottery and it was subsequently made by both Muslim and Christian potters in the Christian town of Valencia. This 'golden pottery' was highly prized both in Spain and elsewhere as the numbers of examples found across Europe testifies. Both lustre and blue-on-white wares (Caiger- Smith, 1973 and 1985) were made from the mid-13th century to the late 15th century at Malaga, and while Frothingham (1951) and others have termed it 'Andalusian', in the light of the analytical programme this is too broad a term for the ceramics which we now know were made around the town of Malaga. Analysis has demonstrated that none of these pieces were made at Granada, as some maintain - results to be published elsewhere. The designs on these ceramics, deriving from and made in an Islamic environment , fill the exposed surface of the ceramics almost completely with intricate designs, often including calligraphic elements. From the late 14th century onwards, similar pottery was produced at Valencia, already a long-established pottery-making centre, by Malaga potters who migrated, it is normally proposed, to avoid the effects of the sea blockade of Southern Spain by Christian ships. At this period, the design spectrum of each centre was very alike and there are considerable difficulties in distinguishing late Malaga/early Valencia ceramics of this type (Hurst, 1977). As the 15th century progressed, Valencia became the dominant lustreware centre of Spain (particularly for exported ceramics), and as the design repertoire developed away from the earlier forms it can be distinguished as Valencian.

Table 2. Intercomparison of Laboratories' results on the US National Institute of Standards and Technology (NBS) Coal Fly Ash no. NBS 1633A.

Element	BMRL (n=6)		NHM (n=20)		Quoted (n=5-10)	
	m	sd	m	sd	m	sd
Ce	163	8	169	5	175	8
Tb	2.53	0.12	2.43	0.13	2.4	0.4
Th	26.4	1.6	22.8	0.3	24.6	1.0
Cr	210	12	189	6	193	5
Hf	7.41	0.4	7.06	0.13	7.2	0.5
Cs	10.8	0.43	10.6	0.2	10.0	0.4
Sc	36.7	2.4	38.7	0.4	38	3
Co	42.7	1.4	42.9	0.6	43	4
Eu	3.80	0.23	3.55	0.12	3.5	0.3
Lu	1.10	0.06	1.07	0.05	1.1	0.3
Ca	1.06	0.05			1.12	0.04
Rb	158	22	132	2	140	12
Fe	9.08	0.63			9.45	0.17
U	8.43	0.69	9.9	0.4	10.4	0.2
Ta	1.80	0.13	1.95	0.09	1.89	0.14
La	65	8	83.5	2.1	83	3
Na	0.166	0.019			0.175	0.012
K	1.80	0.17			1.89	0.06
Sm	17.2	2.6	15.9	0.5	17	2
Ba	1213	61			1400	200
Yb	6.73	0.32	7.74	0.32	8.2	1.3
As	146	15			144	2
Sb	5.81	0.64	7.3	0.2	7.1	0.6

All results are in parts per million as the element except Ca, Fe, Na and K in %.

m = mean and sd = standard deviation.

BMRL = British Museum Research Laboratory.

NHM = The Natural History Museum, London (data coutesy of Dr. C.T. Williams - see also Williams and Wall (forthcoming)).

Quoted = latest quoted figures on this standard - see Gladney et al. (1984).

In an earlier paper (Hughes & Vince, 1986), sherds of known origin from the two centres were analysed, together with a number of ambiguous origin (Hurst 1977) including sherds from excavations in London. The results were evaluated using principal components as a preliminary test followed by cluster analysis using the program CLUSTAN (Wishart 1978, 1982) and finally discriminant analysis to test the hypothesis of chemical difference between the products of the two sites. The cluster analyses were performed using Euclidean distance, with Ward's method followed by the Relocate option and using a significant jump in the error sum of squares as the stop criterion for the optimum number of final clusters. The main discriminating elements were found to be chromium, sodium, iron, scandium, thorium, cerium, hafnium, lanthanum and rubidium and the cluster analysis produced one major cluster for Malaga ceramics and three sub-clusters for Valencian (of 18, 1 and 5 samples). The presence of several clusters of different composition from a single pottery- making centre was intriguing though not unexpected given the known existance of a number of pottery workshops around Valencia including the two most famous of Paterna and Manises. How then are the composition clusters related to the quality of the pieces (i.e. ordinary tablewares vs specially commissioned major pieces) , or can they be identified with specific workshops? Such questions are discussed later in this paper.

Since that date, many more samples of such ceramics, including 'unknowns' have been analysed, principally of excavated sherds found in London (mostly tablewares), and of fine complete vessels (specially comissioned 'quality' pieces) including the superb Godman collection of Hispano- Morseque pottery in the British Museum (Wilson, 1987:28- 31). A subsidiary aim has been to provide identification for ambiguous pieces.

Multivariate statistical analysis

As in the earlier study, principal component and cluster analysis was applied to the data, outliers (samples of very different composition to the rest) were identified and the cluster analysis re-run with these removed and discriminant analysis or mutidimensional scaling then applied to display the relationships between these clusters. This is becoming a standard procedure in our Laboratory for processing NAA ceramic data (Leese et al, 1986). An optimum grouping was established which split the samples into nine clusters, six of which contained only Valencian, and three only Malaga, samples: each cluster contained both samples known to have come from only one production centre together with some 'unknowns'. Since the cluster analysis used the Relocate feature (Wishart,1982), it is not possible to display the results as a dendrogram, which is heirarchically constructed, whereas the Relocate method is fundamentally not heirarchical. In any case dendrograms do not indicate inter-cluster distances, instead, we have used discriminant analysis using the nine cluster groups identified above as input, and the resulting plot of the first two discriminant scores is shown in Figure 3. By definition this plot shows the best separation which can be achieved between these clusters (under certain assumptions) and is therefore a good graphical representation of the relationships between the clusters and indeed the spread of compositions within each cluster. In addition, the output from the discriminant analysis identifies the best discriminating elements between the cluster groups. One cautionary note is that one must not place any reliance upon the discrimination success 'rates', which, when one defines the groups for input in this way, usually come out close to 100%, since one has used the chemical data already to define the groups so inevitably they are the best chemically - separated groups. In effect, the same chemical data is being 'recycled', i.e. used twice over in a circular argument. Abuses of discriminant analysis of this type are occurring in the archaeometric literature because of failure to realise this point. Provided the "success rate" information is ignored, the use of the display plots (e.g. scatter plots of discriminant scores DS1 vs DS2) and identification of discriminating elements is quite legitimate.

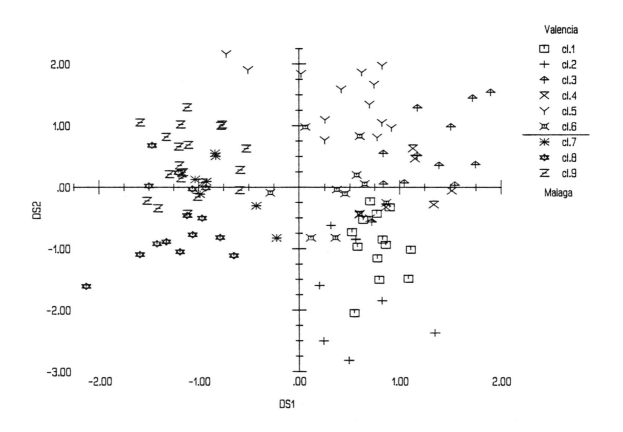

Figure 3 : Discriminant Analysis plot of the first two scores showing the difference between Malaga and Valencia ceramics. DS1 contains 69.1% of the discrimination, and DS2 21.6% (DS1+2 = 96.8%). DS1 is positively correlated with La, negative with Fe and Sc; DS2 is +ve with La, Ce, Fe, Sc and Cs in descending order of contribution.

The six Valencian clusters are on the right (1-6) of Figure 3 while the three Malaga clusters are on the left (7-9). The mean concentrations of 13 elements in these clusters are shown in Figure 4 as a series of 'mountain' plots (Leese *et al*, 1989). The means were first calculated, converted to natural logs, the average for all the clusters was subtracted for each element to reduce the graph's scale and the results plotted for those elements as a connected line. There will be detailed discussion of the individuals in these clusters in forthcoming publications but here some general features may be noted:

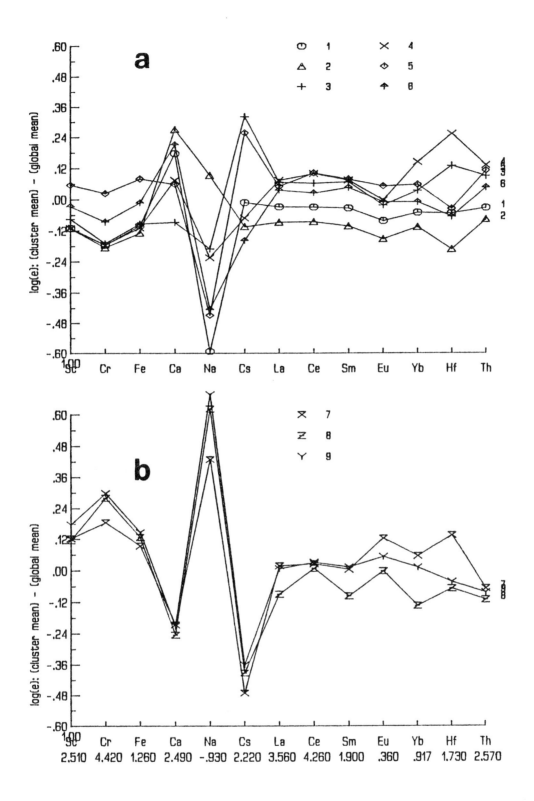

Figure 4 : 'Mountain' plots comparing the mean log(base e) concentrations of the elements in the Valencia (a) and Malaga (b) groups derived from cluster analysis. To avoid scaling problems, the global mean of all the cluster means has been subtracted from each (and is quoted under each element).

1)   In the earlier study, only one Malaga cluster was obtained, whereas there are now three. In fact inspection of the Mahalonobis distances obtained for the earlier clustering showed that a very few pieces were near the edge of the one major cluster, including the fragment of the Alhambra jar, and the addition of further analysed samples has now shown that far from being a 'fringe' member, these "edge" samples constitute part of another analytical group.   These results are in agreement with those of D'Archimbaud et al. (1986) using x-ray fluorescence on blue and lustre pottery found at Granada, who found that analytically, two fragments of 'Alhambra jars' were slightly marginal to their Malaga composition group. Both D'Archimbaud and ourselves find a very different composition for pottery locally made at Granada compared to the Malaga composition.

One of the tiles from the Alhambra (C667-1917 = no.30) also appeared to be on the edge of the Malaga group in the earlier paper and is now joined both other samples in cluster 7. All the Malaga NAA clusters are in fact quite close to each other compositionally, as the mountain plots of Figure 4b show.

2)   In contrast, the six Valencian clusters cover a rather wider composition range (cf. Figure 4a), which is not surprising considering the multiple pottery-making centres around the medieval town of Valencia, including the two main centres of Paterna and Manises. At first sight there appear to be only slight trends visible in the present data to link up clusters with Paterna/Manises production. Some of the ceramics identified as Manises products occur in fact in clusters 1-4 and 6 and Paterna - identified pieces occur in 2,3,5 and 6: i.e. there is considerable overlap. The problem has been tackled before: D'Archimbaud and Lemoine (1980) carried out x-ray fluorescence analyses on such ceramics found on French sites bordering the Mediterranean. They found just two major groups in the Valencian ceramics, distinguished mainly by potassium content, although the larger of the two groups appeared to contain both Paterna and other wares and the smaller group contained (among others) Pula-type wares. Although the

authors mention sub-groups in the larger group, no chemical analyses for these are given so it is not possible to link their data with ours, as well as the fact that only calcium, iron and potassium were measured in common with the present study. Jornet et al. (1985) have made a study of some products of the two centres and find an apparent difference based upon lower caesium and rubidium and higher calcium in the Manises samples but similar concentrations of all other elements. At present our relatively few results tend to support some differences between the products of the two centres, but we do not find as clear-cut a distinction as Jornet et al. (1985). The explanation lies in their selection of samples for analysis; many of their Manises samples are of 16th-18th century date, with only one of ? 15th C date, whereas their Paterna samples are of 13-17th C date and include cockspurs which might be made of less well prepared clay than the pottery. By contrast, many of the Manises samples from the present study are of similar date to their Paterna samples and there is less distinction between the two centres' products at that period. Their microscopic studies and glaze analyses confirm the late dating of their Manises pieces : the glazes are clear, not tin-based (cf. Caiger-Smith, 1973). Such an explanation, involving a shift in chemical composition with time, would be in line with one of the general hypotheses put forward in the present paper.

Turning to the Valencian clusters of Figure 4 it is noticeable that the mountain plots are parallel to each other for a number of elements, i.e. the concentrations are systematically lower or higher from cluster to cluster. This well-known phenomenon (the dilution effect - Mommsen et al. 1988 have proposed a mathematical technique for assessing it) arises either from the addition of a pure diluting temper (e.g. quartz sand) which contains very low amounts of the analysed elements, or more careful clay preparation (levigation tends to raise the concentration levels of most elements), or alteration in the mixing ratios of blended clays. Caiger-Smith (1973: 200) discusses the Valencian potters' use of blended clays, at Paterna a dark red plastic clay was blended with a short white-firing calcite-rich earth, while at Manises a redder blend was made with a smaller quantity of white clay. Jornet *et al.* (1985) have also concluded from their technical study that blended clays were used by the potters, but they find more rather than less calcium in their Manises samples, which are of a later period than those mentioned by Caiger-Smith.

Fine quality vs ordinary tablewares

Examination of the Valencian clusters shows that the very fine quality pieces of mature Valencian style from the Godman collection fall mainly into cluster 1, with two in clusters 2 and 6 and one each in clusters 3 and 4. In contrast, ordinary tablewares occur in all the clusters, both exported pieces found in London and those found on local excavations in Spain. The association of these quality pieces together suggests that they were made by a small number of workshops which also however made more common wares.

Identification of an albarello

One example of the application of the data was to provide information about an albarello whose crowded design (Figure 5) is reminiscent of Malaga production - the use of Arabic designs, for example, whereas in other stylistic aspects the albarello seems related to Valencia. The albarello fails to fit neatly into one category or another and it was hoped that analysis would help. In fact the analysis showed it to belong clearly to the Valencian group and it is extremely close in composition to the average concentrations of cluster 3, in which it was placed by cluster analysis. Of particular interest is the fact that this cluster also contains a transition period piece (item 37) found at Fostat, Egypt: i.e. from the period when Malaga and Valencia were both producing lustreware and before Malaga faded out as a production centre of these. Another albarello in the Godman collection (G581 - no 188) also falls into cluster 3.

An ancient repair

A wing-handled jar in the Godman collection (G619) has a repair to its foot consisting of a specially made piece which reproduces the lost fragment even down to being lustred. While the wing-handle jar has a typical Valencian composition (cluster 1), the repair has a noticeably different composition which however matches with just one other sample which has been analysed, namely a sherd of 18th century pottery from Manises (no.206). The match in composition seems sufficiently clear to conclude that the repair to this outstanding piece was made at Valencia, although precisely when one cannot be certain on the basis of a single matching composition. In the later 18th century the jar belonged to Horace Walpole at Strawberry Hill.

Conclusions

The tin-glazed wares of Seville, Malaga and Valencia can be distinguished from each other by neutron activation analysis. Within the production from each, sub-groups of composition can be identified which appear to correspond to particular types or cultural styles of ceramic which in turn reflect the output of particular workshops, some of which may be contemporary. There is evidence of changes in composition with time, a theme which recurs in elemental analysis studies of pottery-making from many different times and places: e.g. Athens (Fillieres *et al.* 1983) and Lakonia (Attas *et al.* 1982) in Greece, Ephesus in Turkey (Hughes *et al.* 1988), medieval decorated floor tiles from Bordesley Abbey, England (Leese *et al.*, 1989). Such changes in composition needs to be reckoned with when choosing samples for provenance studies: ceramics of the same type and period (i.e. fine with fine) should be considered if close matches between source and unknown are to be achieved. The finer quality wares of Valencia have a restricted range of compositions out of a wider spectrum shared with ordinary tablewares which implies that both types were made in some workshops.

Figure 5 : Albarello (drug jar) in Hispano-Moresque decorative style which on analysis proved to be Valencian (Catalogue number C123-1931; photograph courtesy of the Trustees of the Victoria & Albert Museum).

Acknowledgements

The analytical programme on lustreware grew out of an idea of Alan Vince (City of Lincoln Arch.Unit) and it has expanded to include other type of tin-glazed wares. Many colleagues have been involved in aspects of the project including Hugo Blake (Queen Mary College, London) and John Hurst (formerly of English Heritage) who have taken an active role in moving the project on and Chris Gerrard (Cotswold Arch. Trust) who has collected sherds and clays assiduously to be included in the project. Anthony Ray (who is preparing a catalogue of the Hispano-Moresque pottery at the Victoria & Albert Museum, London) has kept me in touch with his research, and Alfonso Hernandez (Seville) kindly supplied me with sherds of Seville ceramics for analysis. Within the British Museum I wish to thank Tim Wilson (Dept.of Medieval & Later Antiquities - now at the Ashmolean Museum, Oxford) for his role in providing advice and pointing out art-historical questions to be answered and within my own Department of Scientific Research to the Keepers Michael Tite and Sheridan Bowman for their encouragement in the work, and other colleagues for assistance in preparing samples for analysis and advice on the statistical aspects.

References

Attas, M., Fossey, J.M. & Yaffe, L. (1977). Variation of ceramic composition with time: a test case using Lakonian pottery. *Archaeometry* 24, 181-90.

Caiger-Smith, A. (1973). *Tin-Glaze Pottery in Europe and the Islamic World*, Faber & Faber, London.

Caiger-Smith, A. (1985). *Lustre Pottery*, Faber & Faber, London.

D'Archimbaud, G.D. & Lemoine, C. (1980) Les importations Valenciennes et Andalouses en France Méditerranée: Essai de classification en laboratoire. In: *La Céramique médiévale en Méditerranée Occidentale X-XV Sieceles*: Valbonne 11-14 Sept. 1978, Colloques Internationaux du CNRS, 584, 359-72.

D'Archimbaud, G.D., Lemoine, C., Picon, M. & Vallouri, L. (1986). Recherches de laboratoire sur les ateliers medievaux espagnols, In: *Segundo Coloquio Internacional de Ceramica Medieval en el Mediterraneo Occidental* (Toledo, October 1981), 43-45.

Deagan, K. (1987). *Artifacts of the Spanish Colonies of Florida and the Caribbean 1500-1800: Volume 1 Ceramics, Glassware and Beads*, Smithsonian Institution Press, Washington DC.

Fillieres, D., Harbottle, G. & Sayre, E.V. (1983). Neutron activation study of figurines and pottery from the Athenian Agora. *Journal of Field Archaeology*, 10, 55-69.

Frothingham, A. (1951). *Lustreware of Spain*, Hispanic Society of America, New York.

Gladney, E.S., Burns, C.E., Perrin, D.R., Roelandts, I. & Gills, T.E. (1984). *1982 compilation of elemental concentration data for the NBS biological, geological and environmental standard reference materials*, National Bureau of Standards Special Publication no. 260-88, Washington DC.

Goggin, J.M. (1968). *Spanish Majolica in the New World : Types of the Sixteenth to Eighteenth Centuries*, Yale University Publications in Anthropology, LXXII.

Harbottle, G. (1982) Provenience studies using neutron activation analysis: the role of standardisation. In: *Archaeological Ceramics*, eds. J.S.Olin and A.D.Franklin, pp.67-77, Smithsonian Institution Press, Washington DC.

Hughes, M.J. (1989). Provenance studies on Italian maiolica by neutron activation analysis. In: *Papers on Italian Renaissance Pottery*, ed. T. Wilson, British Museum Publications, London.

Hughes, M.J., Cowell, M.R. & Hook, D.R. (forthcoming) Neutron activation analysis procedure at the British Museum Research Laboratory. In: *Neutron Activation and Plasma Emission Spectrometric Analysis in Archaeology*, eds. M.J.Hughes, M.R.Cowell and D.R.Hook, British Museum Occasional Papers, London.

Hughes, M.J. & Vince, A. (1986). Neutron activation analysis and petrology of Hispano-Moresque pottery, In: *Proceedings of the Archaeometry Symposium, Washington 1984*, eds. J.S.Olin and M.J.Blackman, pp.353-367, Smithsonian Institution Press, Washington DC.

Hughes, M.J., Leese, M.N. & Smith, R.J. (1988). The analysis of pottery lamps mainly from Western Asia, including Ephesus, by neutron activation analysis. In: *A Catalogue of Lamps in the British Museum, III Roman Provincial Lamps*, by D.M.Bailey, pp.461-485, British Museum Publications, London.

Hurst, J. (1977). Spanish pottery imported into Medieval Britain. *Medieval Archaeology*, 21, 68 - 105.

Hurst, J.G., Neal, D.S. & van Beuningen, H.J.E. (1986). *Pottery Produced and Traded in north-west Europe 1350- 1650*, Rotterdam Papers VI, Museum Boymans-van Beuningen, Rotterdam.

Joel, E.C., Olin, J.S., Blackman, M.J. & Barnes, J.L. (1989). Lead isotope studies of Spanish, spanish-Colonial and mexican majolica. In: *Proceedings of the Toronto Archaeometry Conference*, ed. R.Hancock, pp.188-95.

Jornet, A., Blackman, M.J. & Olin, J.S. (1985). XIIIth to XVIIIth Century ceramics from the Paterna-Manises area (Spain). In: Ceramics and Civilisation: 1 , ed. D.Kingery, pp.235-55, American Ceramic Society.

Leese, M.N., Hughes, M.J. & Cherry, J. (1986). A scientific study of N.Midlands Medieval tile production. *Oxford Journal of Archeology* 5 (3), 355-70.

Leese, M.N., Hughes, M.J. & Stopford, J. (1989). The chemical composition of tiles from Bordesley Abbey : a case study in data treatment. In: *Computer Applications and Quantitative Methods in Archaeology 1989*, eds. S. Rahtz & J. Richards, British Archaeological Reports S548, pp.241-9, Oxford.

Lister, F.C. & Lister, R.H. (1982). *Sixteenth Century Majolica Pottery in the Valley of Mexico*, University of Arizona Anthropological Papers no.39, Tucson.

Lister, F.C. & Lister, R.H. (1987). *Andalusian Ceramics in Spain and New Spain - a Cultural Register from the Third Century BC to 1700*, University of Arizona Press, Tucson.

Maggetti, H., Westley, H. & Olin, J.S. (1984). Provenance and technical studies of Mexican majolica using elemental and phase analysis. In: *Archaeological Chemistry III*, ed. J.B.Lambert, pp. 151-91, American Chemical Society Advances in Chemistry Series 205.

Mommsen, H. , Kreuser, A. & Weber, J. (1988). A method for grouping pottery by chemical composition. *Archaeometry*, 30 (1), 47-57.

Olin, J.S., Harbottle, G. & Sayre, E.D. (1978). Elemental composition of Spanish and Spanish-Colonial majolica ceramics in the identification of provenience. In: *Archaeological Chemistry II*, ed. G.F.Carter, pp.200-29, American Chemical Society Advances in Chemistry Series, 171, Washington DC.

Ray, A. (forthcoming). Francisco Niculoso called Pisano. In: *Papers on Italian Renaissance Pottery*, ed. T.Wilson, British Museum Publications, London.

Vince, A. (1982). Medieval and Post-medieval Spanish pottery in the City of London. In: *Current Research in Ceramics: Thin Section Studies*, eds. I.Freestone, C.Johns & T.Potter, pp.135-44, British Museum Occasional Paper 32, London.

Willams, C.T. & Wall, F. (forthcoming). An INAA scheme for the routine determination of 27 elements in geological and archaeological samples. In: *Neutron Activation and Plasma Emission Spectrometric Analysis in Archaeology*, eds. M.J.Hughes, M.R.Cowell and D.R.Hook, British Museum Occasional Papers, London.

Wilson, T. (1987). *Ceramic Art of the Italian Renaissance*, British Museum Publications, London.

Wishart, D. (1978,1982) *CLUSTAN User Manual Version 1C release 2 and Supplement*, Program Library Unit, University of Edinburgh

NEUTRON ACTIVATION ANALYSIS OF SCOTTISH MEDIEVAL POTTERY: A GRAIN SIZE STUDY

A. Inscker[1] & J. Tate[2]

[1]Department of Archaeological Sciences, University of Bradford, Bradford, BD7 1DP.

[2]Department of Conservation and Analytical Research, National Museums of Scotland, Chambers Street, Edinburgh, EH1 1JF.

Introduction

Previous provenance studies undertaken at our laboratories have included fairly coarse grained pottery where large inclusions and apparent inhomogeneities in the texture are clearly visible (Macsween, 1984). Sampling such pottery is problematic since it is hard to be convinced that any crude sampling strategy of 'avoiding' or 'including' inclusions will really give reproducible results. This problem is particularly acute on museum material where only small samples can be removed and where there is insufficient material to undertake further mixing and subsampling.

The current project was undertaken with two aims: firstly to analytically compare pottery from four medieval sites in central Scotland, and secondly to try and determine any significant changes in composition (as determined by NAA) from samples of different grain size taken from the same sherd. It was hoped that this study might help in establishing sample sizes and selection criteria necessary to complement published work on minimum sample sizes (eg. Hancock, 1983).

Samples

Pottery sherds from four medieval sites were provided by Dr David Caldwell of the NMS. These sites were Kelso Abbey (Tabraham, 1984), Bothwell and Coulston (Brooks, 1978) and Throsk (Caldwell & Dean, 1981 and 1986). Each of the sites was thought to be producing its own ceramics. The material from Kelso and Throsk came from individual pits on the excavations, and the Throsk samples were entirely of wasters from the process of pottery production, and were therefore felt with some confidence to represent one manufacturing site. The sherds were generally of a fairly fine grain fabric, many with green or brown lead glazes. Sherds from Coulston were of a coarser texture that the other three sites.

All sherds were sampled by first removing the glaze and surface material from an area at one end to a depth of about 1 mm using a diamond impregnated file. The remaining inner part of the sherd was then broken off and crushed with an agate mortar and pestle. For some samples the resulting material was sieved into a number of different grain size fractions.

Sample Preparation and Irradiation

The pottery samples were dried in an oven at $350^{\circ}C$ for half an hour, prior to sealing them in 2 mm internal diameter 'spectrasil' quartz tubing. Each sample tube was 30 mm long and contained around 80 mg of sample, the weight of sample tube and sample being determined as accurately as possible before and after sealing, which was done using an oxygen/butane flame. It was found that some loss of sample may occur on sealing if the sample tube was allowed to get too hot: for this reason an air gap of about 10 mm was left at the top of each powdered sample.

Two batches of samples were prepared, each was irradiated at the Scottish Universities Research Reactor Centre at East Kilbride at a quoted neutron flux of 3.8×10^{12} n $cm^{-2}s^{-1}$, for the first batch for six hours and the second batch for five hours. For these irradiations the samples were packed into aluminium 'A' cans, 81 in the first irradiation and 61 in the second.

Flux monitors and standards

Edinburgh 'E4' pottery standards were included in each of the two batches, prepared exactly as the pottery samples. Because of concern about unevenness of the neutron flux across the 'A' can, eight stainless steel flux monitors were also included in the second batch, their precise position within the container being recorded.

Counting

All samples were counted using a high-purity germanium detector coupled to an Ortec amplifier and 4 K MCA emulation card in an IBM PCAT (an EG&G ACE system). This system has a quoted resolution of 1.88 KeV at 1.33 MeV and a relative efficiency of 22%.

On return from irradiation the individual quartz sample tubes were sealed within polythene or glass containers and their gamma spectra measured approximately one week and four weeks after irradiation.

Samples were located accurately in a fixed geometry co-axially above the detector. Because all samples were in standard sized phials filled to approximately equal amounts, no counting corrections were made for sample geometry. In the first count the samples were located higher above the detector than in the second, longer, count where the activity was considerably lower.

Peak areas from the isotopes listed in Table 1 were determined from these data.

Table 1. Isotopes used in each counting period.

Isotope	Half Life		Energy (KeV)	Isotope	Half Life		Energy (KeV)
SHORT				LONG			
140 La	40.22	H	1596.7	51 Cr	27.7	D	320.1
153 Sm	46.44	H	103.2	181 Hf	42.5	D	482.1
239 Np	2.35	D	277.6	134 Cs	2.06	Y	795.8
177 Lu	6.71	D	208.4	46 Sc	83.8	D	889.3
122 Sb	2.72	D	564.1	182 Ta	115	D	1221.4
175 Yb	4.21	D	396.3	60 Co	5.26	Y	1173.5
24 Na	15.03	H	1368.3	152 Eu	13.4	Y	1407.9
198 Au	2.69	D	411.8	59 Fe	44.6	D	1291.6
76 As	26.32	H	559.5	233 Pa	27.0	D	312.6
46 Sc	83.8	D	1120.5				
59 Fe	44.6	D	1291.6				

Calculation of Elemental Compositions

Peak areas for each of the isotopes from the above table were obtained by setting regions of interest on the MCA and using the MAESTRO software (which estimates a linear background from 3 data points on either side of the peak) to calculate the peak parameters. Multiple peaks, or those on complex backgrounds, were not used. As a basic

'quality control', each spectrum was examined by eye on the display to check that the regions of interest calculated and the linear background extrapolation were not adversely effected by adjacent peaks. The raw data were corrected for decay for each isotope and the integrals were then normalised by sample weight. The figures from the E4 standards were averaged and the averages used to convert the integrals of the unknown samples to ppm or % composition of each element using the known composition of the standards. At this stage no corrections were made for sample interferences. Counting errors (based on signal to background statistics) were also recorded at this stage for each peak. (These procedures were carried out using basic programs GETROI, PPM and READPPM.)

The results from the E4 standards showed considerable scatter which on investigation was considered most likely to have been caused by the neutron flux not being homogeneous across the aluminium A can.

This was confirmed by the results from the flux monitors in the second experiment, where the fall-off in neutron flux could be roughly mapped from the known location of the flux monitors in the irradiated sample can. As can be seen from Figure 1 the estimated flux variation was surprisingly large. Since the position of each individual sample phial in the aluminium can was not recorded it was not possible to correct each sample for flux variations. Consequently all the calculated compositions were normalised to a standard composition for one element to allow intercomparison of the data.

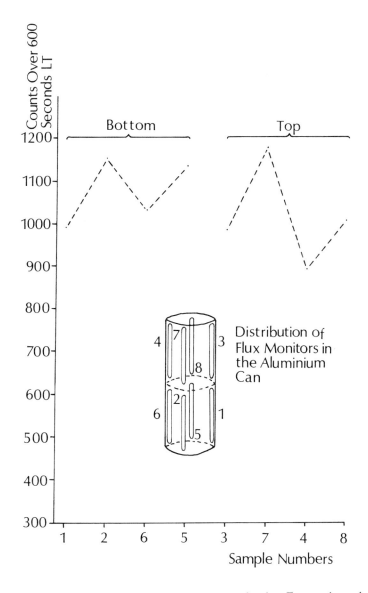

Figure 1. Flux variation across sample container as measured using Fe monitors (weight normalised).

Results

Grain Size Effects

As indicated above, for a number of sherd samples had been prepared of different grain size fractions. Because of the desire to obtain compositional information about samples from each site, and the physical restriction on the number of samples in each 'A' can for irradiation, in the end it was possible to include only a small number of samples. It is clear that the range of grain sizes used was not great enough for other than very preliminary results. Figure 2 shows the calculated composition for Co, Sm and Cr for a number of different sherds each of which had been separated into grain size fractions in the ranges (microns) d < 106; 106 < d < 250; 250 < d < 355; 355 < d.

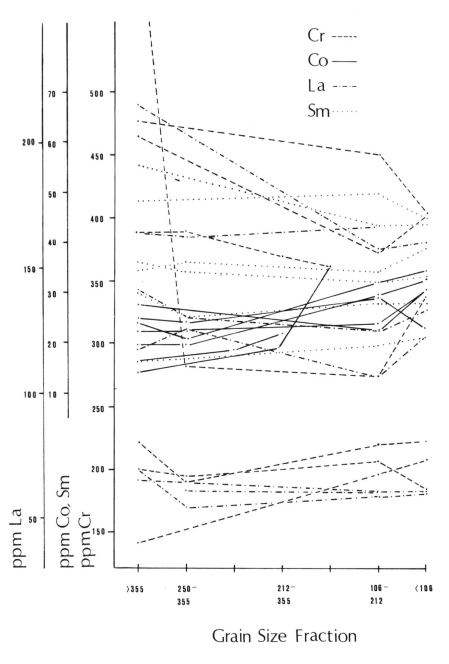

Figure 2. Composition changes with grain size.

72

Whilst these results, and those for other elements, have too large uncertainties to draw any firm conclusions, it does seem that in some cases there are trends indicating a correlation between calculated composition and grain size. The most notable trend is for Co where there is an apparent increase in composition with decreasing grain size.

We have not yet been able to examine the mineralogy of the various grain fractions to see whether the trace element content relates to different mineral types, or whether it may be correlated with elements leached into or out of the pottery.

Cluster Analysis

The compositional data were compared using a number of statistical techniques available using the CSS-2 software package running on a HP Vectra PC. Techniques used were cluster analysis (Wards Method and single and complete linkage, K-means clustering)and factor analysis. The results of these techniques were broadly similar, and are illustrated in Figure 3 where factor analysis results are plotted on a scatterplot. In Figure 4 the results of K-means clustering of the same data (with input of 4 clusters, using 16 elements) are presented.

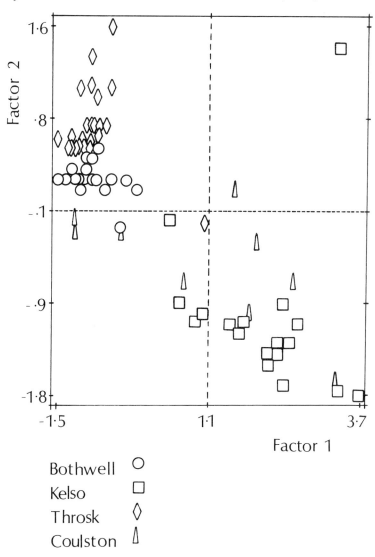

Figure 3. Scatterplot of the analytical results.

The results of these analysis indicate that:

-   Material from Throsk and Bothwell are tightly grouped and well distinguished from Kelso. They are also distinguished from one another, although compositional averages are similar.

-   Kelso has a considerably different average composition.

-   Samples from Coulston have a wider compositional variation and cannot clearly be distinguished from Kelso. Cluster analysis does however separate them from the Throsk group and largely from Bothwell.

-   Coulston material has the largest analytical spread. This is probably because the pottery was markedly more varied in texture than material from the other three sites.

Conclusion

This project has shown that the visually similar medieval pottery from three sites in central Scotland can indeed by distinguished by their trace element content, as indicated in Figures 3 and 4. Pottery from the site at Coulston is more problematic:however, it is exactly this pottery which has the coarsest texture, demonstrating clearly the need for more investigation of the grain size effect.

GROUP 1:

16 17 18 19 21 22 23 24 25 26 27 28 29 31 32 33 79   64 65 67 68 70 72

Kelso   Coulston

GROUP 2:

34 35 36 37 38 39 40 41 42 43 44 45 46 47 48 49 50 51 52 53 54 55 56 57 58 59 60 61 62 63 76 77 78

All from Throsk

GROUP 3:

1 2 3 4 5 6 7 8 9 10 11 12 13 14 15 73 74 75   20   66 69 71

Bothwell   Kelso Coulston

GROUP 4:

30   Kelso (NB. This sample is an outlier, caused by high Sb content, believed to be due to some glaze being included in the sample).

Figure 4: Results of K-means cluster analysis (numbers refer to sample identity).

References

Brooks, C.M. (1978). Medieval pottery from the kiln at Coulston, East Lothian, In: *Proceedings of the Society of Antiquaries of Scotland* 110, pp 364-403.

Caldwell, D.H. & Dean, V.E. (1981). The Post-Medieval pottery industry at Throsk, Stirlingshire. In: *Scottish Pottery Historical Review*. (no vol/page no).

Caldwell, D.H. & Dean, V.E. (1986). Post-Medieval Pots and Potters at Throsk, Stirlingshire. In: *Review of Scottish Culture*. (no vol/page no.)

Hancock, R.G.V. (1983). The effect of sherd sampling size on analytical reliability. In: *Proceedings of the 22nd Symposium on Archaeometry*, Bradford, 1982. (no page no.)

Macsween, A. (1984). Neutron Activation Analysis of Barvas Ware: handthrown Pottery from the Hebrides. Postgraduate Dissertation, University of Bradford.

Tabraham, J. (1984). Excavations at Kelso Abbey. In: *Proceedings of the Society of Antiquaries of Scotland* 114, pp.365-404.

THE ANALYSIS OF GLASS FROM COPPERGATE, YORK, BY INDUCTIVELY COUPLED PLASMA SPECTROMETRY.

C.M. Jackson, J.R. Hunter and S.E. Warren

Department of Archaeological Sciences, University of Bradford, Bradford, BD7 1DP.

Following recent work at Bradford using ICPS to study the elemental compositions of archaeological glass from the late Roman to medieval periods at sites such as Saxon Southampton and Winchester, a study of chemical analysis was initiated on the excavated glass from the site of Coppergate, York, a multi-period urban site covering a similar period.

Excavations at Coppergate started in 1976; the earliest levels from outside the legionary fortress suggest that the area was used by the Romans for trade until 400 AD. Later levels show sparse evidence for Saxon occupation, but the archaeology supports the historical evidence that the Vikings were resident in York during the period 850-900 AD. The glass recovered consists of window and vessel fragments, and waste deposits, lying in the vicinity of, but not necessarily contemporary with, a hearth sealed within Viking contexts. This, allied with the discovery of an area of intense burning, paved with Roman tiles led to the interpretation that this was a site of glass production.

An archaeomagnetic date for the hearth, by Mrs Hammo Yassi of Newcastle University (P Addyman, pers. comm.), gave a date of around 860 $\pm$ 20 years AD and showed the putative glass furnace to be of the Viking/Anglian period, although the tiles were said to be Roman in origin, presumably reused. The phasing of the site, according to pottery evidence studied by The York Archaeological Trust, also supports this date, the relevant phase (phase 3) being from 850-900 AD (Table 1). The tiles lining the 'furnace' exhibited layers and droplets of vitrified material.

Table 1. Phases of the site at Coppergate, York

Phase 1	before 400 AD
Phase 2	400 - 850 AD
Phase 3	850 - 900 AD
Phases 4-6	later than 900 AD

(D. Tweddle, pers. comm)

Unlike the furnace at Moorgate, London (Heyworth, this volume), the furnace at Coppergate is of a more open form and is likely to be one where the glass was heated and melted in pots, as suggested by the numerous finds of pottery from around the kiln area (Bayley, 1983). These pottery fragments exhibit glass droplets and dribbles on the inner and outer surfaces, and around the lip. These glassy coatings do not seem to be merely glazes; most are quite thick, up to 1 cm, implying that the original pottery vessels were glass-melting crucibles. This is supported by other, rather scant evidence for glass melting in the form of independent irregularly-shaped glassy lumps.

The best evidence however, that glass may have been made from raw materials is a partly vitrified fragment, glassy on the top surface but grading through white partly-reacted quartz to almost unaltered sand at the bottom. Both the white frothy material and the glassy state were taken for analysis by ICPS to see how their chemical constituents differed and how the glassy component compared to the rest of the glass fragments found on the site.

The bulk of the excavated glass has been dated typologically to the Roman period by Hilary Cool of the Romano-British Glass project, the assemblage having a character typical of domestic assemblages found on many Romano-British sites. The vessel and window material consisted of the more common blue-green colour, some fine colourless glass, and various yellow-green and yellow-brown fragments. These finds, although mainly of the second to third century, could be dated typologically to within the late 1st to 4th centuries - the known period of occupation of the Romans in York. This assemblage together with the glaze-like fragments, the glassy waste and the furnace itself would suggest that glass manufacture, either from the raw materials or from cullet, was carried out at some period at York.

Methods

The ICP system was a Philips PV8050 emission spectrometer incorporating a PV8490 ICP source unit sited in the Department of Geology at Royal Holloway and Bedford New College, Egham, UK. This is a polychromator system using a spectrometer equipped to measure a large number of fixed spectral lines simultaneously.

A sliver was cut from each glass sample to be analysed with a diamond-tipped saw and cleaned with a diamond-tipped drill to remove surface contamination. After grinding to a fine powder in a ball mill, 100 mg was weighed into a PTFE beaker. Hydrofluoric acid was added to the sample to remove the silica as a volatile tetrafluorosilicate, together with perchloric acid to dissolve the metals. The sample was then evaporated to dryness and made up to 10 mls with hydrochloric acid (Thompson and Walsh, 1983). As indicated, this preparation does not allow the direct determination of silica, which is later derived by subtraction.

The solution was introduced into the ICP torch, as an aqueous aerosol, where the light emitted by the constituent elements is converted to an electrical signal by a photomultiplier. The intensity of this signal is then compared to previously measured intensities of known elemental concentration and the new concentrations computed. The number of elements determined in a single sample was 32, using an initial 100 mg of powdered glass. These are Al_2O_3, Fe_2O_3, MgO, CaO, Na_2O, K_2O, TiO_2, P_2O_5, MnO, Pb, Sb, Ba, Co, Cr, Cu, Li, Nb, Ni, Sc, Sr, V, Y, Zn, Zr, La, Ce, Nd, Sm, Eu, Dy, Yb and SiO_2.

124 glass samples were analysed from Phase 3 of the Coppergate site. Although dated to the Anglo-Scandinavian period, that is 850-900 AD, the glass forms were dated typologically from the late 1st to 4th centuries. This suite of glass comprised window and vessel sherds of various colours, and associated waste material.

Results and discussion

Chemical analysis showed the glass to be of the durable soda-lime-silica type (Sanderson & Hunter, 1980), indicated by typical concentrations of 70.0% silica, 7.0% CaO and 18.5% Na_2O, which is characteristic of glass found in Britain and western Europe in both Roman and Anglo-Saxon periods. This can be seen in Table 2 where the Coppergate mean values are compared to those of a typical soda lime silica glass incorporating an evaporite alkali source.

When analysed, the glass vessel and window sherds formed a group which could be distinguished chemically from the glass adhering to the pottery. The glassy deposits on the pottery were higher in iron and potassium and lower in sodium (Table 3).

Table 2. Mean glass compositions for Roman and immediate Post-Roman glass.

	Type III (From Sanderson and Hunter, Figure 7) (%)	Phase 3 Coppergate Glass (%)
SiO_2	67.7 ± 2.8	70.0 ± 1.6
Na_2O	17.3 ± 2.4	18.6 ± 1.6
K_2O	1.1 ± 0.8	0.9 ± 0.4
MgO	0.9 ± 0.5	0.6 ± 0.1
CaO	7.2 ± 1.2	6.5 ± 1.0
No. of Samples	159	99

Table 3. Glass from Phase 3, Coppergate showing compositional differences between glass adhering to pottery and window and vessel glass.

	Average Composition (main discriminants)	
	Glassy Waste from Pottery (%)	Vessel and Window Glass (%)
Fe_2O_3	1.2 ± 0.6	0.5 ± 0.2
K_2O	1.4 ± 0.6	0.7 ± 0.3
Na_2O	16.7 ± 2.2	18.5 ± 1.6
Number of samples	24	99

Three possible explanations for these differences are:

1.   The glass was contaminated by the pottery during the physical separation of the two with a diamond tipped drill.

2.   There may be some diffusion of certain elements between the glass and pottery when the glass was in a molten state or that leaching depleted the glass of certain elements (Saleh, 1972; Velde, 1990)

3.   If (2) did not occur, then the fragments analysed could have been part of the material used to produce the glassy deposits in the pottery, but the converse is not true, vessel fragments compositionally similar to the glassy deposits were unevidenced.

The single sample of fused silica and glass analysed (Table 4) contained, as expected, much higher levels of silica in the partially vitrified layer than in the glassy layer together with much lower calcium and soda levels. The fused glassy part of the sample was well within the typical compositional range of chemical values of those of the fragment group, indicating that glass may well have been produced from the raw materials. The evidence from a single sherd is by no means conclusive, but may provide a basis for further work such as thermoluminescence analysis for dating the sample.

Table 4. Compositions of fused silica and window and vessel glass from Coppergate.

	Partially vitrified glassy residue (%)	Window and vessel glass mean composition (%)
SiO$_2$	79.6	70.0
Na$_2$O	8.6	18.5
CaO	3.5	6.8
	(1 sample)	(99 samples)

Colourless glass

The vessel and window fragment group was generally chemically homogeneous, although the fragments consisted of various typological forms within the Roman period. However, within this group the colourless glass could be differentiated, firstly on the basis of a higher level of antimony, which has a mean value of 0.35% compared to the other glass sherds which have a mean antimony value of 0.15%. Antimony is a strong decolouriser known to have been widely used in Roman times (Sayre, 1963). It can secondly be differentiated on the basis of lower levels of

phosphorus, aluminium and calcium. Table 5 shows the main discriminating elements and indicates that colourless glass was of a different, probably higher, quality to the rest of the glasses made and as such only specially selected cullet may have been added to the glass mix, or very pure, refined sand. Possibly this was a specialised product, which may have originated from a source different from that of the other glass at Coppergate. These differences have also been seen in the analysis of several forms of colourless glass from the Roman site of Colchester (Heyworth, M., pers. comm.).

Table 5. Composition of colourless and blue-green glass from Coppergate showing main discriminants.

	Colourless glass (%)	Mean of blue-green glass (%)
Sb	0.4 ± 0.1	0.2 ± 0.1
P$_2$O$_5$	0.04 ± 0.01	0.11 ± 0.03
Al$_2$O$_3$	2.0 ± 0.2	2.5 ± 0.3
CaO	5.8 ± 0.5	7.1 ± 0.9
MnO	0.1 ± 0.1	0.6 ± 0.4
Number of samples	38	61

Light blue and light green glasses

There are no striking chemical differences between the light blue and light green glasses, but there are elemental trends which have also been seen from previous glass analyses (Hunter & Heyworth, in press). Iron, an impurity found in the raw materials of glass production, is usually responsible for the green colour in glass, while manganese is a known decolourant (Henderson, 1985). When both are present, an equal or excess amount of iron relative to manganese is often associated with a light blue colour, whilst a lower ratio is found in light green glasses.

Table 6. Iron:manganese ratios.

	Iron : Manganese ratio	
	Light blue glass Fe_2O_3 : MnO	Light green glass Fe_2O_3 : MnO
Coppergate, phase 3	1.2 : 1	0.7 : 1
Saxon Southampton*	1.6 : 1	0.9 : 1

* Data taken from unpublished work by D. Whithorn, University of Bradford.

Colour can equally be influenced by firing conditions, with oxidation producing a greenish rather than a bluish hue but it is significant to note that the ratios seen at Coppergate, although not as marked as those of Southampton, follow similar trends. This might be explained by the use of a different alkali component for the blue and green glasses respectively, either intentionally or unintentionally. The low levels of manganese found in the glass suggest that it was not being deliberately added as a decolourant.

Further work on the analysis of phases 1 (before 400 AD) and 4 (after 900 AD) at Coppergate is beginning to show similar patterns to the work done on glass from Phase 3, namely the separation of the glassy waste from the glass fragments, and of the colourless glass from the blue-green glass.

These are preliminary findings which require confirmation by further work on material from Coppergate and other sites of the same period.

A cross comparison of this data with that from previous work undertaken at Bradford from the sites of Winchester, Saxon Southampton and Repton (Hunter and Heyworth, in press; Jackson, 1987), demonstrates the overall compositional similarity between the individual groups. This may indicate that there is nothing compositionally unique in the glass found at York as can be seen in Table 7, which also shows how the Coppergate results compare to the general composition of first millennium AD glass and from other programmes (eg. Type III glass) (Sanderson and Hunter 1981). There are, of course, pointers to differences but the overall pattern seems to be similar.

Table 7. Means of soda-lime-silica glass compositions from first millennium British sites

	Mean Glass compositions (%)				
	Southampton*	Winchester*	Repton*	Coppergate	Type III (Sanderson and Hunter, 1981, Table 7)
Al_2O_3	2.82	2.52	2.63	2.56	n/a
Fe_2O_3	1.16	1.04	0.94	0.68	n/a
MgO	0.83	0.84	0.75	0.60	0.89
CaO	7.18	6.62	6.82	6.50	7.24
Na_2O	15.6	16.5	15.7	18.6	17.3
K_2O	1.14	0.64	1.44	0.87	1.12
TiO_2	0.15	0.21	0.13	0.10	n/a
P_2O_5	0.52	0.09	0.79	0.09	n/a
MnO	0.49	1.08	0.52	0.43	n/a
Pb	0.64	0.08	n/a	0.07	n/a
Sb	0.23	0.06	n/a	0.25	n/a
SiO_2	69.3	69.1	70.6	70.0	67.6
Number of samples	271	102	44	124	159

n/a = not available.
*M. Heyworth, pers.comm.

In summary, the typological glass evidence points to the manufacture of glass in the Roman period for the Coppergate site. The date of the furnace may well be due to refiring at around 800-900 AD. The few fragments of partly vitrified glassy waste imply the manufacture of glass from raw materials and it is postulated that this may have taken place during the Roman period, although the evidence is extremely scant. Despite the glass being found in Anglo Saxon layers it is suggested that as the site was littered with glass dated from 1st to the 4th centuries, relatively evenly dispersed between phases, and that most of the glass found in the later phases was residual.

References

Bayley, J. (1983). 'Interim Report ..... Through the Looking glass'. *Interim; Bulletin of the York Archaeological Trust*, Vol. 9(1), 27-29.

Henderson, J. (1985). The raw materials of early glass production. *Oxford J. of Archaeology* 4(3), 267-292.

Heyworth, M. Analysis of Roman glassworking evidence from London. This volume.

Hunter, J.R. and Heyworth, M.P. The Glass from Saxon Southampton, forthcoming.

Jackson, C.M. (1987) The importance of rare earth elements in the analysis of Saxon glass by inductively coupled plasma optical emission spectroscopy. Unpublished MA dissertation, University of Bradford.

Sanderson, D.C.W. and Hunter, J.R. (1980). 'Major element glass type specification for Roman and Medieval glass'. *Revue d'Archaeometrie* 3, 255-264.

Saleh, S.A., George, A.W. & Helmi, F.M. (1972). Study of glass and glass-making processes at Wadi El-Natrun, Egypt in the Roman period 30 BC-359 AD. Part I. Fritting crucibles, their technical features and temperature employed. *Studies in Conservation* 17, pp143-172.

Sayre, E.V. (1963). The intentional use of antimony and manganese in ancient glasses. In: VI International Congress on Glass: Advances in Glass Technology, Part 2, eds. F.R. Matson and G.E. Rindone, pp.263-82. Plenum Press, New York.

Thompson, M. and Walsh, J.N. (1983). *A Handbook of Inductively Coupled Plasma Spectrometry*, 1st edition. Blackie and Son Ltd., Glasgow.

Velde, B. (1990). Alumina and calcium oxide content of glass found in western and northern Europe, first to ninth centuries. *Oxford J. of Archaeology* 9(1), 105-117.

CLUNIA: PRODUCER AND RECEIVER CENTRE OF HISPANIC TERRA SIGILLATA

P. de Palol[1], J.M. Gurt[1], F. Tuset[1], C. Planas[1], J. Buxeda[1], M.A. Cau[1], X. Alcobé[1].

[1]Dept. Prehistòria, Història Antiga i Arqueologia, Facultat de Geografia i Història, Universitat de Barcelona, C. Baldiri i Reixac, s/n, 08028, Barcelona.

[2]Serveis Científico-Tècnics de la Universitat de Barcelona, C. Martí i Franquès, s/n, 08028, Barcelona.

Introduction

The production of Hispanic Terra Sigillata is divided into two periods. First, the Hispanic Terra Sigillata (T.S.H.) that was produced during the Early Empire whose workshops are distributed throughout the Península Ibérica. Second, the Late Hispanic Terra Sigillata (T.S.H.T.) that was produced during the Late Empire and the workshops, with its products, are mainly distributed on the North Meseta, Lusitania and the Upper Valley of the Ebro river. The T.S.H.T. wares were essentially systematized by the studies of Palol, in base of the ceramic products of the Late Roman Villa of Pedrosa de la Vega (Palencia, North Meseta) (Palol & Cortés, 1974). In this context, the Roman city of Clunia, capital of a 'Conventus Iuridicus', a probable Tiberian foundation which obtained the status of Colony possibly with Galba under the name of Colonia Clunia Sulpicia (Figure 1), seems to be a Late centre of production (Palol, 1984).

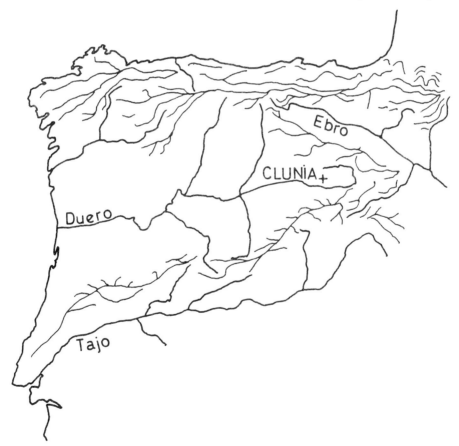

Figure 1. Location of the city of Clunia within the Duero Valley.

The example of Clunia

A continuous production of local tradition coarse ware has been well-recorded at the city of Clunia, dating between the first and fifth century A.D. The workshops, localized and excavated by Palol, were only ones to supply painted fine wares to the city during their active period. In the case of T.S.H., the city was mainly supplied by productions of La Rioja but we should not exclude the possibility, supported by several archaeological data, of a hypothetical autochthonous production of Hispanic Terra Sigillata, mainly in the third century A.D. Such supposition, also referred to the previous period, is verified on the Late Empire; small workshops would be placed inside re-used Early Empire buildings in which we find waste pieces and mould fragments. Nevertheless, we have to take into consideration the fact that neither kilns nor refuse dumps of the quoted workshops have been found yet.

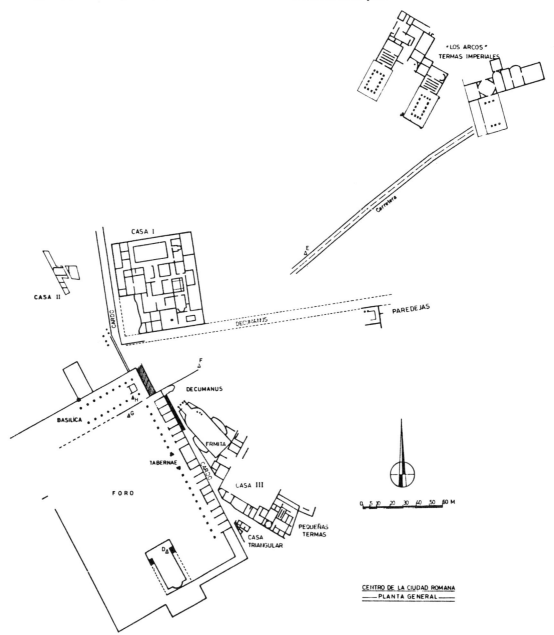

Figure 2. Map of Clunia with Los Arcos.

84

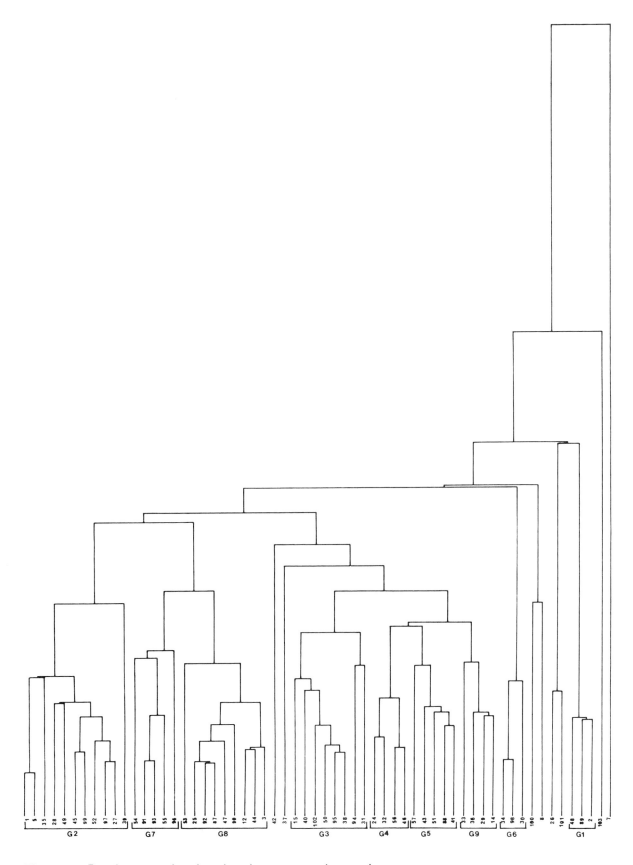

Figure 3. Dendrogram showing the nine groups observed.

Table 1. Concentration values ($\bar{m}$ is the mean value of each group and s is the standard deviation). All numbers are in % except Ba, Rb, Zr and Sr (in ppm).

elements / groups		Fe_2O_3	Al_2O_3	MnO	TiO_2	MgO	CaO	K_2O	SiO_2	Ba	Rb	Zr	Sr
G1	$\bar{m}$	4,69	16,16	0,043	0,56	1,25	24,15	2,62	47,12	480	156	135	189
	s	0,22	0,58	0,006	0,03	0,13	1,34	0,13	1,70	52	18	5	25
G2	$\bar{m}$	5,29	16,09	0,046	0,87	0,50	7,66	2,19	64,09	366	141	229	135
	s	0,29	0,91	0,005	0,05	0,11	2,21	0,20	2,61	106	34	13	11
G3	$\bar{m}$	6,31	20,54	0,052	0,84	1,00	6,22	3,62	58,89	519	202	182	134
	s	0,26	0,51	0,013	0,07	0,23	2,65	0,12	2,10	101	22	15	12
G4	$\bar{m}$	5,16	17,57	0,042	0,70	1,29	16,66	3,23	52,08	468	170	147	162
	s	0,06	0,20	0,005	0,04	0,10	2,52	0,16	2,61	73	5	4	24
G5	$\bar{m}$	4,76	16,82	0,052	0,69	1,01	10,57	3,36	59,40	480	174	176	115
	s	0,15	0,26	0,004	0,03	0,29	0,85	0,08	2,41	76	6	9	8
G6	$\bar{m}$	4,79	16,36	0,063	0,82	0,96	7,17	3,54	63,70	455	177	202	122
	s	0,14	0,54	0,006	0,03	0,04	0,39	0,05	0,52	45	12	16	9
G7	$\bar{m}$	6,03	18,34	0,022	0,98	0,64	1,74	3,00	67,02	380	160	244	85
	s	0,13	0,75	0,004	0,07	0,07	0,29	0,13	1,46	48	11	6	4
G8	$\bar{m}$	6,47	19,87	0,036	1,00	0,50	4,60	2,43	61,64	397	155	226	130
	s	0,30	0,64	0,007	0,03	0,08	1,10	0,13	1,70	70	18	14	11
G9	$\bar{m}$	5,41	18,04	0,045	0,84	0,70	6,72	3,10	61,96	401	171	209	119
	s	0,27	0,81	0,006	0 02	0,09	1,44	0,18	1,42	27	13	12	11

Immersed in this archaeological problem, we first decided to initiate the characterization of T.S.H.T. wares found on the hypothetical areas of production and, second, corroborate the existence, or not, of a Late Fine Pottery production within the city of Clunia. A previous formal analysis had showed several associations. We began, in the first phase, with the hypothetical area of production called Los Arcos (Figure 2), the Early Empire Public Baths re-used as a habitat area in some sectors, and as possible workshop in the remainder area (Palol, 1982). For this purpose sixty samples were selected, all of them representatives of the typological and decorative diversity of the whole T.S.H.T. ensemble present at Los Arcos. Samples were systematically analysed by the XRF technique, quantifying the major and minor elements by means of glassy pills and the trace elements by means of powder pills. The X-ray powder diffraction method using the CuKα radiation, was applied on several sherds.

Analytical Results

Chemical compositions were processed with the BMDP statistical package, especially the routines 1D, 2D, 6D, 2M, 4M and 7M (Dixon, 1983). Because of the wide ranges of some element concentrations, multivariate statistical analyses were done over the standardized values of the elements from all the samples as a rapport of Al_2O_3.

The Dendrogram from Fe_2O_3, Al_2O_3, MnO, TiO_2, MgO, CaO, K_2O and SiO_2 concentrations, by centroid linkage method is shown in Figure 3. In this measurement we excluded P_2O_5 and NaO because of its fluctuations. The results show nine groups (Table 1), and other samples not well-situated.

On the archaeological point of view, this great dissimilarity is difficult to explain, but does not state less problems under the analytical point of view. Other facts must be emphasised; the clearest one would be the wide range of CaO concentrations, with variations between 0.45% (sample 103) and 25.49% (Sample 2). We also emphasize SiO_2

concentrations, with highest values in most of the samples, that arise to 72.8% in sample 103.

These results may suggest a possible variation of compositions, influenced by the fluctuations on the abundance values of CaO and SiO_2.

However, the TiO_2,K_2O scattered plot (Figure 4) shows two regression lines constituted by groups G2 and G8 on one hand, and groups G3, G5 and G6 on the other hand, whilst groups G7 and G9 rest on a medium position. Groups G1 and G4, both with higher CaO concentrations than the other groups, seem to follow the G3, G5 and G6 regression line.

On that score, more interesting and clear is the result of the Principal Component Analysis that, excluding marginal samples from the Cluster (samples 101, 103, 26, 37 and 7), has been obtained from Fe_2O_3, Al_2O_3, TiO_2, MgO, CaO, K_2O, SiO_2, Ba, Rb, Zr and Sr. The analysis has been developed over the preassigned values of BMDP4M routine, appearing three factors (Table 2) that can be interpreted as follows: the first factor, with other CaO and Sr positive contributions and minor negative contributions of Zr, would be a content factor of Calcite, where Sr appears associated with CaO; the second factor, of more difficult interpretation, is specially conditions by the positive contribution of K_2O, Rb, and Ba; finally, the third factor shows an inverse ratio between Al_2O_3 and SiO_2, interpreted as the quartz temper factor, present in the pottery.

The localization of the samples on the first and second factor axes (Figure 5) shows a similar situation as in Figure 4. First, groups G2 and G8 on one hand and groups G3, G5 and G6 on the other hand, had a tendency towards association on intermediate values of the first factor but as opposed poles of the second factor. Second, groups G7, G9, G1 and G4, the last one in less measure, show an intermediate position on the second factor but a great dispersion on the first, due to the excessive CaO ranges.

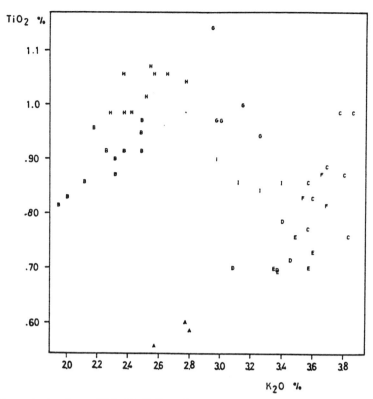

Figure 4. Scattered plot from TiO_2/K_2O (in %).

	FACTOR 1	FACTOR 2	FACTOR 3
Fe_2O_3	-0.544	-0.468	-0.090
Al_2O_3	-0.244	-0.004	0.947
TiO_2	-0.634	-0.531	-0.419
MgO	0.464	0.651	-0.199
CaO	0.930	0.103	-0.256
K_2O	0.076	0.871	-0.235
SiO_2	-0.419	-0.146	-0.859
Ba	0.081	0.744	0.229
Rb	-0.087	0.750	0.389
Zr	-0.753	-0.536	-0.193
Sr	0.837	-0.178	0.083
VP	3.301	3.151	2.227

Table 2. Correlations between the variables used in the PCA analysis and the resulted factors. VP is the variance explained by each factor.

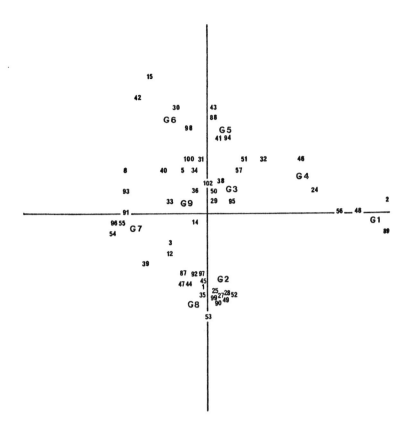

Figure 5. Score plot with Factor 1 in X axis and Factor 2 in Y axis.

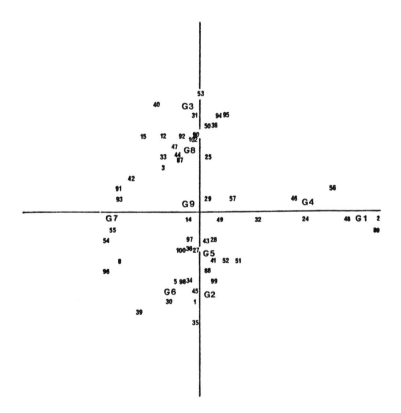

Figure 6. Score plot with Factor 1 in X axis and Factor 3 in Y axis.

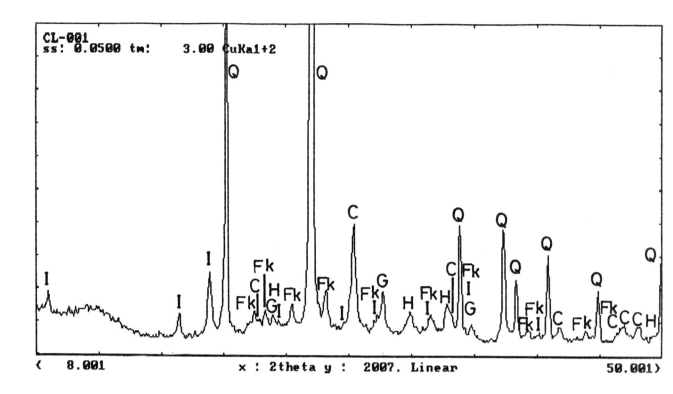

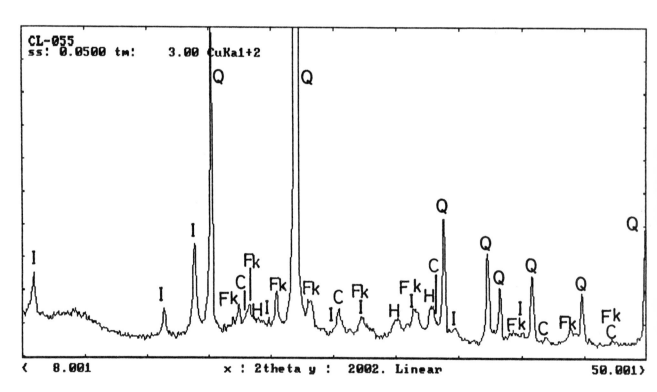

Figure 7. XR diffractograms from samples 1 and 55 (Q=quartz; C=calcite; G=gehlenite; I=illite; H=haematite; Fk=K-feldspars).

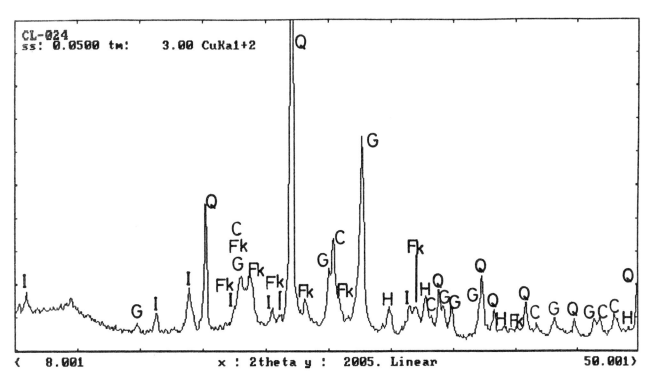

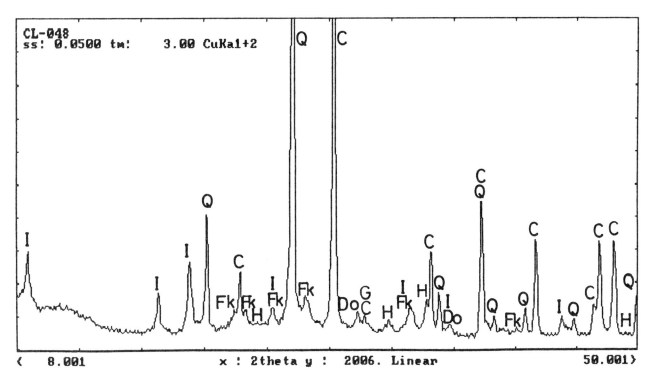

Figure 8. XR diffractograms from samples 24 and 48 (Q=quartz; C=calcite; G=gehlenite; I=illite; Do=dolomite; H=haematite; Fk=K-feldspar).

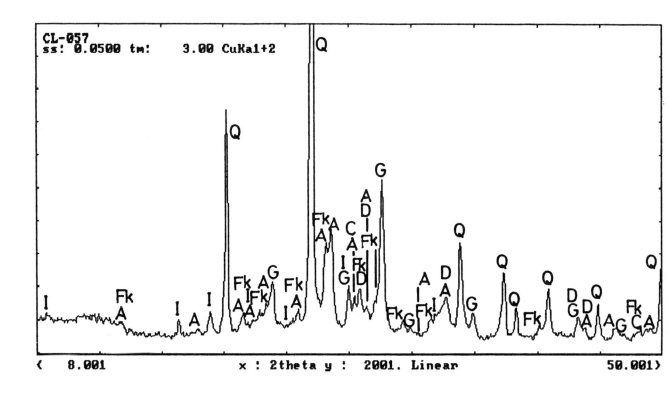

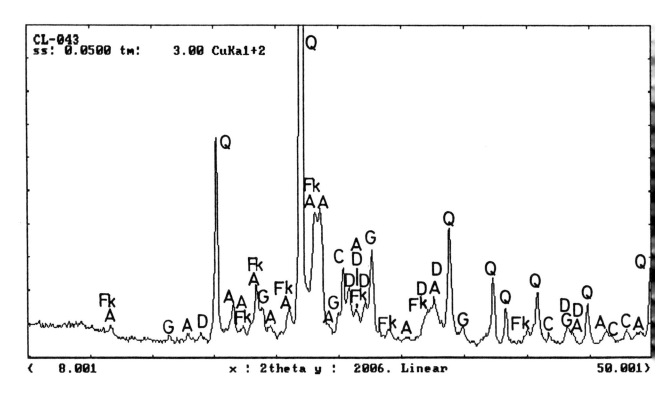

Figure 9. XR diffractograms from samples 57 and 43 (Q=quartz; C=calcite; G=gehlenite; I=illite, Fk - K-feldspar; A=anorthite; D=diopside).

However, the plot of the first and third factor axes (Figure 6) allows us to understand the diversity showed on the Dendrogram (Figure 3). The third factor, expressed as a quartz temper factor, separates the association of the quoted groups, placing on opposite poles G8 in relation to G2 on one hand, and G3 in relation to G5 and G6 on the other hand. In spite of that, the addition of the third factor has not modified the position of G7, G9, G1 and G4.

In both diagrams we see several samples that change its position with regard to the group they belonged (samples 15, 94, 40, 5, 39, 53, 57 and 51). On the analytical point of view, this fact allows us to understand the variations on the Dendrogram when modifying some of the technical conditions of its elaboration as the characterization point of view allows us to identify samples with marginal compositions with regard to the group they belonged.

The scarcity of analysed samples, together with the different variables that seem may be participating on the clays, makes difficult an optimum interpretation of the results discussed until this point.

The results of XRD obtained from samples 1, 3, 24, 30, 34, 38, 43, 46, 48, 50, 51, 55, 57, 90, 96, 97 and 99, let us identify quartz, in large amounts in almost the whole samples, that we may relate with the SiO_2 observed on XRF. Such majority of quartz (Figure 7), because of its abundances, masks the other mineral phases, making its observation difficult. We have only observed abundant calcarious phases in samples 24, 46 and 48 (Figure 8), the first belonging to group G4 and the latter to group G1, fact that is in accordance with XRF results.

On the other hand, the presence of low temperature phases (calcite, illite, haematite ....), together with the weak presence of gehlenite (clearly observed in samples 24 and 47; in smaller degree in 43, 30, 50, 90 and 99 and, finally, as a minor phase in 1, 3, 34, 38, 46, 48 and 97) and the absence of anorthite and other High Temperature Phases (diopside, sanidine ...), they all seem to suggest a not

very high firing temperatures. In fact, only in samples 57 and 43 we have clearly identified anorthite and diopside, together with gehlenite as a significant phase (Figure 9).

Finally, point out that the considerable amounts of quartz excessively homogenize the samples to a complete comparison with the results from XRF.

Final Evaluations

The analytical results show the existence of nine groups, though we shall not discard, due to low number of samples, that G2 and G8 on one hand, and G3, G5 and G6 on the other hand, can correspond with two great associations. These are internally distinguished by its amount of SiO_2. The absence of definitive validation of archaeological criteria, like waste samples, is another problem when interpreting the results.

Despite that, comparison with the previous archaeological survey seems to confirm the existence of the group we call 'individual motives with message', in the manner of those found on Late Roman C wares). Moulds with this kind of motives are also found at Clunia, and constitute a group together with the vessels cited above. The cited archaeological group is present inside the ensemble of groups G3, G5 and G6.

Finally, we think that a greater sampling would allow us to conclude the comparison with archaeological data, appraising the producer and receiver role of the city of Clunia during the Late Empire.

Acknowledgements

Financial support of the Excma. Diputación Provincial de Burgos, for the excavation project, and of the Comisión Interministerial de Ciencia y Tecnología, for the investigation project (PB85-0086), are gratefully acknowledged. We are indebted to Servei d'Espectroscòpia dela Serveis Científico-Tècnics de la Universitat de Barcelona, specially to the Direc. Dra. Baucells and Dra.

Roure, where the samples were analysed. We thank Dr M. Picon, Laboratoire de Ceramologie, C.N.R.S., Lyon, for helpful comments.

References

Dixon, W. (ed.) (1983) *BMDP Statistical Software, 1983, printing with additions*, University of California Los Angeles Press, Los Angeles.

Palol, P. de (1982). *Guía de Clunia*, 5ª edición, Burgos.

Palol, P. de (1984). Clunia, cabeza de un convento jurídico de la Hispania Citerior o Tarraconense. In: *Historia de Burgos*, t. I, pp.395-428, Burgos.

Palol, P. de & Cortés, J. (1974). La villa romana de la Olmeda. Pedrosa de la Vega (Palencia). Excavaciones de 1969-1970, *Acta Arqueológica Hispánica*, 7, Madrid.

NEUTRON ACTIVATION ANALYSIS AND THE BELL BEAKER FOLK

F. Rehman[1], V.J. Robinson[1], G.W.A. Newton[1], S.J. Shennan[2].

[1]Department of Chemistry, Manchester University, Manchester, M13 9PL.

[2]Department of Archaeology, University of Southampton, SO9 5NH.

Abstract

The Beaker culture appeared in various places throughout Central and Western Europe during the 2nd and 3rd millennia BC. The Bell Beakers after which the culture is named exhibit a uniform pattern of form and decoration throughout this region. The study of ceramic exchange patterns can provide an independent basis for defining inter-regional connections and this is particularly important in the period immediately before the beginning of the Early Bronze Age when only small quantities of metal were deposited in graves and hoards. The relevant period is the Bell Beaker phase in Central Europe (ca. 2000 BC) and it is for this reason that Bell Beaker ceramics are of considerable importance. Recent studies of the Bell Beaker assemblages have led to the suggestion that these vessels were part of a group of artefacts introduced from outside into local assemblages as indicators of higher status, rather than the products of a wandering Bell Beaker folk.

To test this hypothesis and study ceramic exchange patterns for this period, a detailed study by neutron activation analysis (26 elements per sample) of Bell Beaker pottery has been carried out. The chemical composition of Bell Beaker sherds has been compared with the pottery thought to be of local origin from a number of sites over a significant part of the geographical range of the beaker culture. The regions studied are Austria, Bohemia, Bavaria, Moravia and Hungary. A detailed statistical analysis of the results of 567 samples has been carried out using several techniques in order to identify reference sources of local material to which Bell Beakers can be related.

Introduction

Neutron activation analysis of trace elements is a well established method for provenancing studies and the present paper deals with its application to the problem of finding a relationship of Bell Beaker pottery with different find sites.

This pottery is found in a number of regions; over an area stretching from Ireland to Hungary and North Africa to Scotland (see map, Figure 1). The distribution of Bell Beakers in Central Europe is characterised by markedly discontinuous groups clustering in natural geographical regions and divided from each other by mountains or highlands, but connected by the larger rivers such as Elbe, Morava and Danube.

In Czechoslovakia the chief concentrations are in Bohemia and Moravia, where many hundreds of burials are known. In Bohemia, with the largest group of all, the Beaker cluster is in the northern half of the country.

The Moravian region extends southwards across the Danube to include lower Austria. There are related concentrations of Beakers in the Saale basin in Bavaria, extending as far westwards as the upper Rhine and Switzerland, and they seem to be generally related to the large group in Bohemia. The outlying group in the region around Budapest in Western Hungary marks the eastern limit (see map, Figure 2).

In every one of these regions there is a uniform pattern to the Bell Beakers and their associations. Difference between one region and another can be expected to be brought out more sharply in the future, but for the time being it is the similarities which have

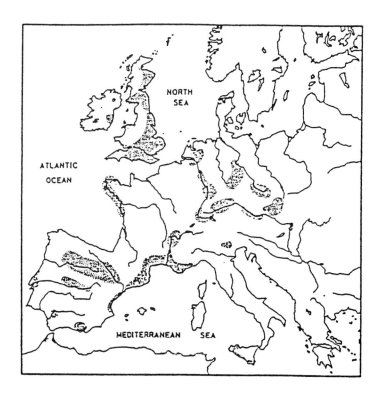

Figure 1. Distribution of Bell Beakers in Europe.

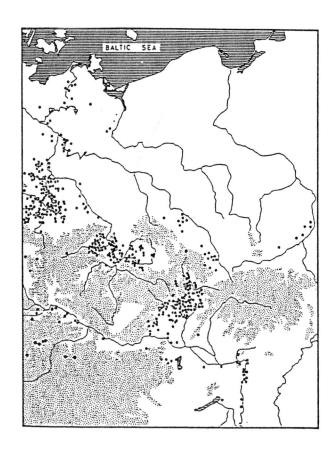

Figure 2. Distribution map showing the eastern limit of the Bell Beaker culture.

attracted general attention and interest. Figure 3 reveals the similarity of beaker patterns from different sites.

In order to investigate the development of political centralization, a specific attempt was made by Shennan (1986) to see if evidence of a regional hierarchy was visible in differences between cemeteries on the basis of the quantity and variety of goods deposited in the graves, but no indication of this emerged. In fact it is the Decorated Bell Beaker which suggests that it was a pottery which had an unusual status and a special function. One reason for the rapid acceptance of Beakers may have been due to their attractiveness as fine, well-made pots produced by part-time specialists, although the really important element in the movement of the Bell Beaker assemblages far across Europe was probably the possession of the technological skills to exploit the more complex copper ores that are commonly found in the west (Harrison, 1980). In a number of areas including Bohemia, Moravia and Slovakia, there are fortified sites of the later Early Bronze Age (*ca.* 2000 BC) (see map, Figure 4). With the exception of the work of Kruk and his colleagues in Southern Poland (Shennan, 1986), almost no systematic work has been carried out and published for the area of Europe.

The samples

A total of 546 samples from Austria, Bavaria, Bohemia, Hungary and Moravia, and a set of 21 clay samples from Austria were analysed. The samples were obtained by one of us (SJS), who is reasonably sure that the samples which are not Bell Beakers are of local origin. The archaeological details about each sample will be published separately for each of the specific study areas, and are outside the scope of this paper.

Experimental

Samples were prepared by homogenising each fragment in an agate mortar after cleaning the surfaces by a diamond drill. The clay samples were fired before the sampling procedure. 200 mg of each sample was neutron irradiated at the Universities Research Reactor, Risley, at a flux of 10^{12} ncm^{-2}s^{-1} for 3 min, 10 min and 7 hours for short-, intermediate- and long-lived isotopes respectively. The induced activities were measured using lithium drifted germanium Ge(Li) detector. Table 1 lists the details of the isotopes measured. Concentrations were obtained by comparison with a standard clay (Podmore Red).

Results

The data on element concentration in individual samples is not possible to list here, however it is available on dBaseIII and can be obtained from the Department of Chemistry, Manchester University. All the concentrations are corrected (Hoffman, 1989) to the basis of the Perlman standard, because of its wide use in this field.

Dealing with errors

There are errors associated with each measurement which are due to weighing, counting geometry, flux irregularities, statistical errors, sampling errors (because of inhomogeneity of sherd) or dilution effects when the pot is either deliberately or naturally 'tempered' with minerals such as sand or limestone which are low in trace elements. Measurement errors can be best estimated by multiple analysis of sherds to include sampling errors (a microphotograph of an Austrian sherd is shown in Figure 4, illustrating heterogeneity).

The Mommsen dilution procedure (Mommsen *et al.*, 1988) has been modified to simultaneously deal with dilution errors and systematic errors such as different exposure to neutron flux, or different counting geometry. The details of the procedure will be published separately. The program is on the Manchester University main frame computer and is called 'Factor'. All the sample concentrations used in this study are corrected for such errors.

97

Statistical treatment of the results and discussion

The data has been analysed by various multivariate procedures using the CLUSTAN package (Wishart, 1988) and the Mommsen dilution programme. It was decided at the outset that the results should be investigated in five sections each corresponding to one of the major geographical regions studied. The reason was that at present our version of Ward's method of hierarchic fusion cannot deal with more than 500 samples at a time, and secondly cluster analysis does not give satisfactory results when used on populations with large numbers of groups (more than 6).

Table 1. List of Isotopes Measured.

ISOTOPE	ENERGY (KeV)	HALF-LIFE	IRRADIATION
Al-28	1779	2.246 m	SHORT
Ca-49	3084	8.72 m	SHORT
Ti-51	320	5.8 m	SHORT
V-52	1434	3.75 m	SHORT
Mn-56	1810	2.58 h	SHORT,INTERMEDIATE
Dy-165	95	2.35 h	SHORT
U-239	75	23.5 m	SHORT,LONG
Np-239	277	2.36 d	LONG
Ba-139	166	82.7 m	SHORT
Na-24	1368	15.03 h	INTERMEDIATE
K-42	1524	12.36 h	INTERMEDIATE
Eu-152	121,344.3	9.3 h	INTERMEDIATE,LONG
		13.33 y	
Sc-46	889.2	83.8 d	LONG
Cr-51	320.1	27.7 d	LONG
Fe-59	1099.3	44.5 d	LONG
Rb-86	1076.6	18.66 d	LONG
Cs-134	795.8	2.06 y	LONG
La-140	487.0	40.28 h	LONG
Ce-141	145.4	32.5 d	LONG
Sm-153	103.2	46.7 h	LONG
Yb-175	282.5	4.19 d	LONG
Hf-181	482	42.4 d	LONG
Pa-233	311.9	27.0 d	LONG
Ta-182	1221	115.O d	LONG
Lu-177	208.4	160.9 d	LONG

Initially the RELOC method was used on all 546 sherds (21 clay samples not included). The results obtained are not very meaningful due to the complex nature of 5 populations, but with hindsight it was perhaps unrealistic to expect separations with five heterogeneous populations present each having a wide range of concentrations. The comparison of mean and standard deviations of each element in 5 major geographical groups of samples is given in Table 2.

98

Figure 3. Illustrating the similarity of Bell Beakers from widely different sources.

a) Bell Beaker from a grave in Wiltshire.

b) Bell Beaker from Hungary.

Figure 4. Microphotograph (x250), showing the heterogeneity of an Austrian Sherd.

c) Bell Beakers from Central Germany.

Table 2. The comparison of mean and spread of individual elements in 5 sets of samples.

Element	Austrian Group 245 samples			Bavaian Group 53 samples			Bohemian Group 124 samples			Hungarian Group 17 samples			Moravian Group 107 samples		
Sc	16.0	± 2.59	(25.9%)	13.9	± 1.82	(13.1%)	14.3	± 2.64	(18.5%)	13.8	± 1.27	(9.20%)	13.1	± 2.20	(16.8%)
Cr	156	± 63.4	(40.8%)	113	± 13.4	(11.8%)	111	± 19.7	(17.7%)	111	± 24.6	(22.2%)	124	± 57.2	(46.0%)
Fe	6.69	± 3.81	(57.0%)	3.69	± 0.82	(22.1%)	4.24	± 0.92	(21.8%)	3.96	± 0.39	(9.90%)	3.97	± 0.67	(16.9%)
Co	52.9	± 72.1	(136%)	12.7	± 2.69	(21.2%)	14.7	± 5.61	(38.2%)	14.1	± 2.76	(19.6%)	13.1	± 3.42	(26.1%)
Rb	110	± 31.4	(28.7%)	110	± 28.2	(26.0%)	121	± 29.0	(23.9%)	99.4	± 28.8	(29.0%)	98.5	± 18.7	(19.0%)
Sb	10.4	± 25.1	(241%)	1.00	± 0.36	(35.9%)	2.40	± 2.74	(114%)	0.78	± 0.17	(21.7%)	0.92	± 0.41	(45.0%)
Cs	15.8	± 6.12	(38.7%)	12.2	± 3.03	(24.8%)	16.1	± 7.57	(47.0%)	18.4	± 14.5	(78.8%)	10.2	± 2.92	(28.6%)
La	41.3	± 8.36	(20.2%)	38.2	± 6.71	(17.6%)	42.6	± 9.47	(22.2%)	30.6	± 4.20	(13.7%)	35.8	± 4.35	(12.2%)
Ce	84.5	± 14.8	(17.5%)	76.0	± 13.7	(18.0%)	85.7	± 16.6	(19.4%)	68.4	± 8.36	(12.2%)	72.0	± 9.22	(12.8%)
Sm	15.3	± 10.4	(68.0%)	12.3	± 3.88	(31.5%)	9.88	± 1.46	(14.8%)	7.45	± 0.83	(11.1%)	8.75	± 1.42	(16.2%)
Eu	1.63	± 0.60	(36.7)	1.44	± 0.32	(22.3%)	1.70	± 0.25	(14.7%)	1.30	± 0.21	(15.8%)	1.87	± 0.78	(41.8%)
Yb	0.94	± 0.23	(24.8%)	0.82	± 0.15	(18.0%)	0.92	± 0.14	(15.5%)	0.80	± 0.12	(13.6%)	0.87	± 0.15	(17.0%)
Lu	0.55	± 0.13	(23.5%)	0.48	± 0.08	(17.4%)	0.55	± 0.08	(15.3%)	0.40	± 0.06	(14.3%)	0.49	± 0.08	(15.8%)
Hf	7.39	± 2.10	(28.4%)	6.88	± 2.34	(34.0%)	11.5	± 2.70	(23.5%)	7.21	± 2.00	(27.7%)	9.15	± 2.07	(22.6%)
Ta	1.35	± 0.35	(26.0%)	1.26	± 0.35	(28.1%)	1.89	± 0.85	(45.0%)	1.14	± 0.12	(10.1%)	1.28	± 0.26	(20.0%)
Pa	13.2	± 2.43	(18.4%)	12.9	± 2.78	(21.6%)	13.1	± 2.14	(16.3%)	10.0	± 0.77	(7.70%)	11.4	± 1.53	(13.4%)
U	3.56	± 2.04	(57.3%)	3.22	± 0.78	(24.2%)	3.44	± 0.98	(28.6%)	2.37	± 0.34	(14.4%)	2.84	± 0.96	(33.8%)
Al	9.63	± 1.43	(14.8%)	8.93	± 1.22	(13.7%)	8.56	± 1.47	(17.2%)	8.33	± 1.03	(12.4%)	8.93	± 1.65	(18.5%)
Ti	0.34	± 0.04	(12.0%)	0.27	± 0.03	(12.5%)	0.37	± 0.08	(21.9%)	0.29	± 0.03	(8.70%)	0.31	± 0.04	(13.1%)
V	112	± 17.3	(15.5%)	91.2	± 11.2	(12.3%)	101	± 17.8	(17.7%)	91.6	± 6.96	(7.60%)	94.9	± 19.1	(20.1%)
Dy	4.63	± 0.98	(21.2%)	4.34	± 0.84	(19.4%)	4.58	± 0.81	(17.7%)	3.36	± 0.48	(14.3%)	4.12	± 1.01	(24.5%)
Na	0.51	± 0.34	(66.6%)	0.69	± 0.63	(90.8%)	0.66	± 0.34	(51.7%)	0.73	± 0.31	(42.1%)	0.91	± 0.65	(71.1%)
K	1.83	± 0.75	(40.8%)	1.66	± 0.57	(34.6%)	1.85	± 0.66	(35.6%)	1.44	± 0.42	(29.2%)	1.60	± 0.55	(34.4%)
Mn	682	± 431	(63.1%)	332	± 221	(79.4%)	307	± 198	(64.4%)	393	± 158	(40.2%)	307	± 162	(52.6%)
Ca	2.51	± 4.56	(181%)	1.24	± 0.98	(79.4%)	1.23	± 0.73	(59.7%)	2.02	± 0.92	(45.3%)	1.12	± 0.94	(83.6%)
Ba	894	± 697	(78.0%)	965	± 538	(55.8%)	1227	± 732	(59.6%)	630	± 131	(20.7%)	1329	± 993	(74.7%)

101

In the context of pottery provenancing, it has been observed that the measurements on clays in some regions can produce erroneous assignments because they are characterised by high correlations between elements (Robinson, 1987). When data is highly correlated, it is necessary either to have large numbers of sherds from individual sites to define correlations precisely, or to use approximate methods to deal with correlation effects. Fortunately, in the present case, correlations among elements did not usually appear to be very large except for the Moravian data set, which were corrected before subsequent statistical analysis.

It is clear from Table 2 that by and large the 5 major groups are significantly different from each other. Detailed investigations showed that there is little or no evidence of inter-regional exchange of ceramics. However strong evidence has been found for local trading of Beaker ceramics in most of the regions studied (as already mentioned these results will be published separately for each region).

Comments on the individual groups

The main features of archaeological interest can be summarized as follows:

1. The division of Austrian samples into several distinct clusters leaves no doubt that there are significant compositional variations along the river (Salzach) valley, from Obereching to Falkenstein Kriml. But the grouping of individual samples from each find sites was only partially successful. This is mainly due to the heterogeneity of samples so that variations in a single site are large and comparable with those between sites. However there is a reasonable separation among the several sites near Salzburg.

2. The Hungarian group is clearly divided into two clusters both corresponding to different find sites even though the geographical distance between these sites is not large. One Bell Beaker (from a different near-by site) is present in one of these defined clusters, suggesting the possibility of local exchange.

3. The Bavarian finds are clearly divided into 4 sub-groups, three of which belong to three different sites. One cluster is a mixture of samples from two sites. Since this group is chemically well defined the obvious explanation seems to be a common but yet unidentified origin.

4. 71 samples out of 107 Moravian data sets were selected from 4 well-represented sites since the other sites were represented by only a small number of sherds (2 or 3). It was necessary to pre-define these groups to correct for significant correlations among the lanthanide elements. These are divided into 4 sub-groups, two of them corresponding to find sites. Two compositional groups contain a mixture of two sites. One of the compositional groups includes all Bell Beakers, which again supports the idea of local exchange.

5. Among Bohemian samples there is reasonable separation between two find sites but the structure of sub-groupings indicates the presence of more than one production site within the region. Many of the find sites were represented by only one or two sherds, and these were either outliers or associated in ill-defined sub-groups. Nothing significant could be deduced about these groups since 1 or 2 samples are insufficient to define a cluster.

6. Comparison of locally collected clay samples with pottery from the same area generally produced rather disappointing results, in that clay and pottery compositions were significantly different. This offers no support for the hypothesis of local

production but also of course, it does not rule it out.

Conclusion

In summary, the present work has produced clear evidence for local manufacturing sites for Bell Beaker ware in which most of the exchange appears to be local, and there is no longer any need to postulate that these artefacts are products of a wandering Bell Beaker folk. However, there is no direct evidence for provenance since no kiln sites are known and the local clay compositions are generally distinct from the pottery. The conclusions here have been reached by comparison of Bell Beakers with pottery of presumed local composition.

Acknowledgements

We would like to thank the staff of the University Research Reactor, for their help with irradiations. One of us (FR) would like to thank S&T, Pakistan for financial support. We thank Dr. S. Hoffman for writing FACTOR program and her continuous help in the statistical analyses.

References

Harrison, R.J. (1980). *The Beaker Folk.* Thames and Hudson Ltd.

Hoffman, S. (1989). Personal communication (to be published). Manchester, UK.

Mommsen, H., Kreuser, A. & Weber, J. (1988). A method for grouping pottery by chemical composition. *Archaeometry* 30(1), 47-57.

Robinson, V.J. (1987). Dealing with highly correlated data - the case of Spanish amphorae. (to be published).

Shennan, S.J. (1986). Central Europe in 3rd millennium BC; an evolutionary trajectory for the beginning of the European Bronze Age. *J. Anthropological Archaeology* 5, 115-146.

Wishart, D. (1988). *CLUSTAN User Manual,* University Press, Edinburgh.

THE COMPOSITION OF THE MATERIALS OF FIRST MILLENNIUM B.C. CYLINDER SEALS FROM WESTERN ASIA

M. Sax

Research Laboratory, British Museum, London, WC1B 3DG, U.K.

Introduction

Cylinder seals developed alongside the cuneiform system of writing sometime after 3500 B.C. They remained in use until about 400 B.C. when alphabetical scripts were adopted and with them stamp seals. Cylinder seals were usually perforated longitudinally and a design in intaglio was carved around the cylindrical face. The seal was rolled in damp clay as an indication of the owner's authority. Each historical period in Mesopotamia created its distinctive styles of seal engraving.

After a period of several hundred years at the end of the second millennium about which very little is known, the seal was used from the tenth century to the seventh century by neo-Assyrians in northern Mesopotamia and by neo-Babylonians in south west Mesopotamia. Further west, within the cuneiform sphere of influence, they were also used by the Syrians. In 539 B.C. Mesopotamia became part of the Achaemenid empire and the use of the cylinder seal, which was decreasing with the adoption of alphabetical scripts, was revived for a further 150 years or so; the seal was used by neo-Elamites who occupied an independent kingdom in south west Iran from ca. 750 B.C. This paper is concerned with the identification of the material of a large collection of first millennium cylinder seals comprising 361 neo-Assyrian, neo-Babylonian and contemporary Syrian seals, 11 neo-Elamite seals and 74 Achaemenid Persian seals. The work described forms part of a major project involving the analysis of some 2000 cylinder seals from the collections of the British Museum. Material from other periods in the collections has been (or is being) examined including that from Uruk to early Dynastic periods (Collon in preparation), from Akkadian, post Akkadian and Ur III periods (Bimson and Sax, 1982), from Isin-Larsa and old Babylonian periods (Sax, 1986), and peripheral second millennium seals (Porada in preparation).

The seals are chiefly composed of simple mineral assemblages and can be precisely described in mineralogical terms; other natural materials such as ivory as well as metals and other synthetic materials were also used. Overall the results from this programme of analysis will provide a relatively comprehensive and representative description of the materials used for the seals with information on geographical and temporal variations. The results show a chronological evolution in the use of materials in the 3000 year period: higher proportions of harder stones were used with advancing time. By the first millennium B.C., the period under consideration here, quartz, one of the hardest materials worked, was the most popular material for seals. Selection of suitable materials is likely to have primarily arisen from availability but also from properties such as hardness and colour; the increase in hardness with time suggests a dependence on technical development.

Although the alluvial flood plains of Mesopotamia are not entirely devoid of natural resources, most of the decorative stones used for the seals were imported from abroad. Precise provenancing of the materials of the seals is, in most cases, not possible because a number of sources are feasible for each material type. However, much useful information about the changing patterns of trade to Mesopotamia during the 3000 year period may be gained by study of the temporal and geographical variations in the materials used.

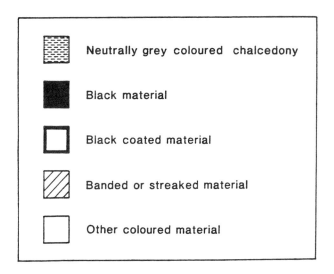

Figure 1. Key to figures 2, 4, 5, 6 & 7.

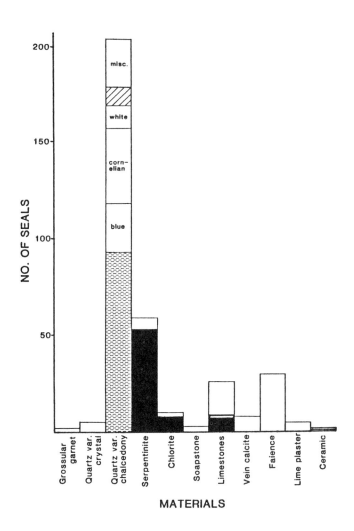

Figure 2. The materials of 361 neo-Assyrian, neo-Babylonian and contemporary Syrian seals excluding single examples of diorite, lapis lazuli, tenorite, mudstone, gypsum, bone or ivory and glass.

Scientific procedure and nomenclature

Qualitative identification of the materials was by x-ray powder diffraction (XRD) analysis, using a Debye-Scherrer camera, of a minute sample usually removed from a blemish on the seal. This was supplemented in some cases by non-destructive elemental analysis using x-ray fluorescence (XRF) and occasionally, when a more detailed analysis was necessary, by examination in the scanning electron microscope (SEM) equipped with an energy-dispersive x-ray analyser. Much of the quartz material was identified non-destructively by density measurement. Further useful information was gained from examination under a low powered binocular microscope.

Because of the large number and variety of quartz seals present in the collection it was necessary to formalize a system of nomenclature for quartz (Sax & Middleton, in preparation) drawing on published literature including Frondel (1962) and Berry & Mason (1959). The varieties of the three types of quartz, crystal (macrocrystalline) chalcedony and jasper (both microcrystalline), were identified by simple observations of colour, translucency, presence of banding, spotting or inclusions and, finally lustre. For instance, of the microcrystalline varieties, cornelian is red brown and translucent, agate is distinctly banded while jasper is opaque.

Neo-Assyrian, neo-Babylonian and contemporary Syrian seals

The range of materials used for 361 neo-Assyrian, neo-Babylonian and Syrian seals is listed in Table 1 and summarised graphically in Figure 2. The key to figures 2, 4, 5, 6 and 7 is shown in Figure 1. Chalcedony, a microcrystalline form of quartz and one of the hardest materials worked at this time, predominates as the most common material for the seals (see Figure 3a, a typical chalcedony seal). Of the 204 chalcedony seals, 93 are neutrally-coloured in generally pale shades of grey and brown, very typical of naturally occurring chalcedony (most of this material is strongly translucent and very little

is clouded white); 39 seals are of the variety, cornelian (including some large and fine examples); 25 are of chalcedony coloured in shades of blue (including some fine examples); 12 are of chalcedony in shades of white; 10 are of agate and 25 are miscellaneous examples of chalcedony including brown, green, pink, violet and turquoise chalcedonies, two deliberately etched chalcedony seals and also some jasper seals.

Other hard materials used include two fine examples of translucent pale green microcrystalline grossular garnet, a little macrocrystalline quartz and single examples of diorite and lapis lazuli.

Of the softer materials, serpentinite was most commonly used; 59 seals are composed of it (see Figure 3b). This material is largely dark in colour, appearing black apart from four pale green and two red examples. Twenty-five seals are composed of limestones; they are mainly coloured in black, in browns or neutral shades of grey and in cream (some of the latter have been coated with a form of carbon to appear black). Other soft materials include chlorite, vein calcite, soapstone and single examples of tenorite (natural copper oxide), gypsum, mudstone and ivory or bone.

Synthetic materials were also used; they include faience, glass, ceramic and lime plaster. While 30 faience seals survive in generally poor states of preservation, only a small fragment of extensively weathered opaque blue glass survives. A small sample from one of the lime plaster bodies was examined in the SEM and appears to be composed of lime plaster worked over a low grade limestone core.

Achaemenid and neo-Elamite seals

The ranges of materials used for 74 Achaemenid and 11 neo-Elamite seals are also listed in Table 1 and summarised graphically in Figure 7. Chalcedony again predominates as the most common material for the Achaemenid seals. Of the 44 chalcedony seals, 14 are banded (see Figure

106

Table 1. The materials of the first millennium seals

Material (Mohs' Hardness)	Neo-Ass. + Neo-Bab. + Syrian	Neo-Assyrian	Neo-Babylonian	Syrian	Neo-Elamite	Achaemenid
Quartz var. chalcedony (6.5-7)	204	58	65	9	4	44
Serpentinite (4-6)	59	44	2	3	-	-
Limestone (3-4)	26	14	3	6	4	20
Chlorite (2-4)	10	3	1	6	-	-
Vein calcite (3)	8	-	4	1	1	2
Quartz var. crystal (7)	5	1	1	-	-	-
Soapstone (1)	3	2	-	-	-	-
Grossular garnet (6-7.5)	2	1	1	-	-	-
Jaspilite (5-6.5)	-	-	-	-	-	1
Diorite (ca.6)	1	-	1	-	-	-
Haematite (5-6)	-	-	-	-	1	-
Lapis lazuli (ca. 5.5)	1	-	-	-	-	1
Tenorite (3.5)	1	-	-	-	-	-
Gypsum (2)	1	1	-	-	-	-
Mudstone (ca.2.5)	1	-	-	1	-	1
Bone/Ivory	1	-	-	-	-	-
Gold	-	-	-	-	-	1
Faience	30	5	9	10	1	-
Glass	1	-	-	-	-	2
Lime plaster	5	2	-	2	-	1
Ceramic	2	-	-	1	-	1
Totals	361	131	87	39	11	74

Figure 3. Lifesize photographs of: (a) Chalcedony (cornelian) neo-Assyrian seal, cat. no. 89145 with impression of intaglio in modelled style; (b) Black serpentinite neo-Assyrian seal, 89586, with impression of intaglio in linear style; (c) Artificially dyed brown and white agate Achaemenid seal, 89448, with impression of intaglio in modelled style; (d) Achaemenid bead-shaped seals, left to right, 129596 (etched blue agate), 113878 (brown and white agate), 89696 (jasper onyx) and 135161 (crudely dyed brown and cream agate).

3c and d): they include nine brown and white agates (at least two and perhaps as many as four have been artificially dyed, probably by the sugar technique (Nassau, 1984), a cornelian onyx, a neutrally coloured agate, a single blue agate etched white with blue banding and two jasper seals. The banded and the plain chalcedonies are shown separately in Figure 7. The plainer chalcedony includes ten seals predominantly neutral in colour (although four of these are streaked with darker veining/inclusions/banding), ten blue seals, four cornelian seals, single examples of jasper, light green and very dark green chalcedonies and possibly as many as three deliberately etched seals. A further seal is of jaspilite which is composed of banded jasper and haematite.

Limestones were the preferred softer material: 20 seals are composed of them. Some of this material is black but most is coloured in brown shades including nine seals which are strikingly streaked (discontinuously banded). Sparry vein calcite, an alternative soft material, was used for two seals. There are also single examples of lapis lazuli, mudstone and gold. Synthetic materials include single examples of brown and amber glasses, carbon coated lime plaster and a ceramic body.

The eleven neo-Elamite seals include a similar range of materials but in different proportions (see Figure 7). There are four seals each in chalcedony and limestone; while one chalcedony seal is banded (a brown and white agate), three of the limestones are streaked in brown shades. There are single examples of sparry calcite, haematite and faience.

Geographical and temporal variations

A high proportion of the 361 first millennium seals can be separately attributed on stylistic grounds (Collon, in preparation) to the neo-Assyrians (131 seals), to the neo-Babylonians (87 seals) and to the Syrians (39 seals) (see Table 1). The materials used in the three geographical areas are summarised in Figures 4-6. The materials used by the neo-Elamites (11 seals) in a fourth geographical area are summarised in Figure 7 as also are those used during the later Achaemenid period (74 seals).

Prior to the Achaemenid expansion the results show there to be a correlation of material with geographical area. Thus in Assyria both hard and soft materials (mainly serpentinite and some limestone) were used with only a low proportion of synthetic materials while in Babylon hard materials were used almost exclusively with only low proportions of soft and synthetic materials. In Syria, in contrast to Mesopotamia, there appears to have been a paucity of good quality stone available, both hard and soft, so that good quality quartz and sepentinite was supplemented by the use of chlorite and synthetic materials; chlorite has a distinct plane of cleavage making it a less satisfactory material for the seals. Finally, rather later in south west Iran, the neo-Elamites used hard and soft materials (mainly limestones) but the small number of samples analysed are unlikely to be fully representative.

The chalcedony used in Mesopotamia was typically neutral grey, blue or red brown (cornelian) in colour while much of the chalcedony used in Syria is, atypically, coloured white: 56% of the Syrian chalcedony seals are white as compared to 5% of the neo-Assyrian and 6% of the neo-Babylonian seals. In Mesopotamia there is a correlation between good quality chalcedony (unflawed) and good workmanship; also, individual workshops, defined on stylistic ground (Collon, in preparation), appear to have used a range of variously coloured chalcedonies except for a few who specialised in cornelian. The natural materials, both hard and soft, used in Mesopotamia during the neo-Assyrian and neo-Babylonian periods are variously coloured while contemporary Syrian materials are generally darker; those used in south west Iran include a high proportion of brown, banded and streaked examples.

The Syrians appear to have used higher proportions of soft and synthetic materials and a correspondingly lower proportion of hard materials than was apparently used in Mesopotamia. The soft materials include a high proportion of chlorite. The higher

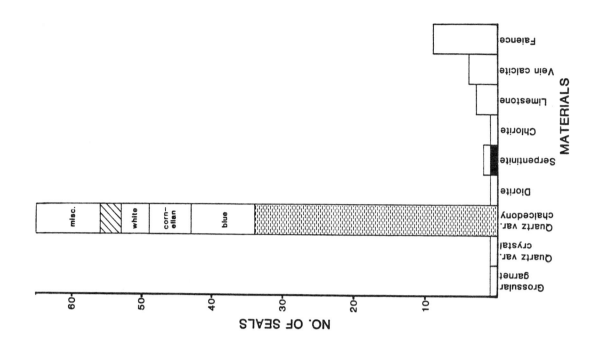

Figure 5. The materials of 87 neo-Babylonian seals.

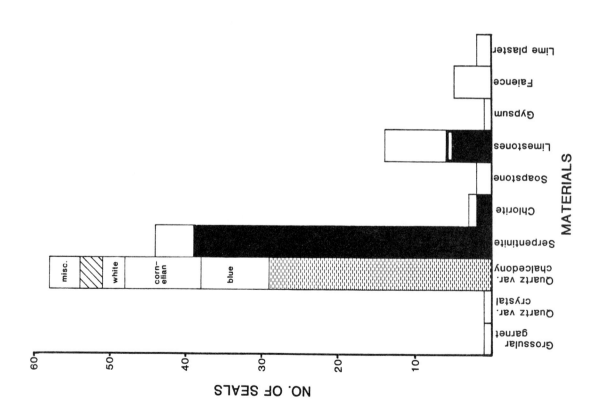

Figure 4. The materials of 131 neo-Assyrian seals.

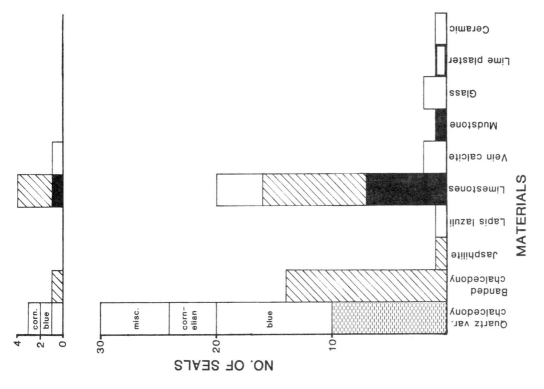

Figure 7. The materials of: (above) 11 neo-Elamite seals excluding single examples of haematite and faience and (below) 74 Achaemenid seals excluding one gold seal.

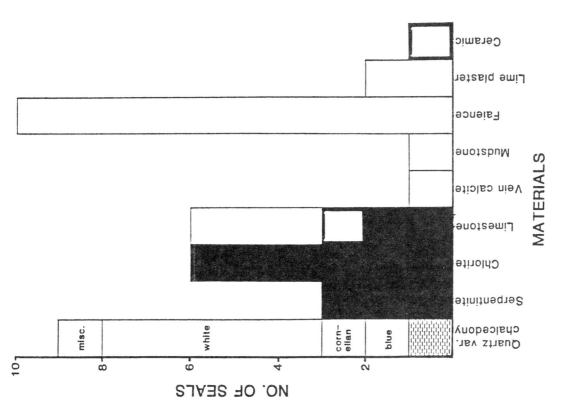

Figure 6. The materials of 39 Syrian seals.

111

proportion of generally poorly preserved and unattractive faience bodies in the Syrian group may in part be related to the fact that all this material (with the exception of one seal) was excavated as opposed to having been acquired from collectors, who would perhaps have favoured the more attractive seals.

After the expansion of the Achaemenid empire, chalcedony continued to prevail as the most popular material for the seals although softer materials (principally limestones) were also used. The choice of materials during this period appears to have been influenced by colour and the presence of banding: shades of brown were particularly favoured. Although the Achaemenids appear to have used a higher proportion of harder material than the neo-Elamites, there is almost complete concordance of the materials used by the two peoples (see Figure 7) suggesting that the Achaemenids largely continued with the traditional seal materials of the neo-Elamites in south west Iran.

Trading patterns

The various materials used for the seals during the first millennium are likely to have originated from a variety of geographical sources including igneous, metamorphic and sedimentary deposits. The identification of sources is unfortunately beyond the scope of this paper and in most cases would be difficult to establish. Nevertheless some inferences may be made from the materials of the seals and from their geographical and temporal variations as to their directions of trade. The wide variety of materials used for the first millennium seals presumably could have originated from Eastern Turkey via Syria, from India by land or sea and also from the Zagros mountains, the Iranian plateau and around the Persian gulf.

As noted already there appears to have been a paucity of good quality natural material in Syria suggesting limited access to mineral resources: approximately a quarter of the Syrian material is of low grade natural materials (six chlorite seals, two coarse grained limestone seals and a mudstone seal), most of which are likely to have been locally available, and a third of it is of synthetic materials (including, in addition to ten faience seals, two lime plaster seals and a carbon coated ceramic seal, apparently in imitation of stone). White chalcedony was used only infrequently in Mesopotamia but over half the Syrian seals are white suggesting that different sources of raw materials were being exploited. The white chalcedony appears to have been more readily available to the Syrians and may have been traded from the Taurus Mountains in eastern Turkey. The use of different materials by the Syrians and the Mesopotamians is in marked contrast to the situation between the neo-Elamites and the Achaemenids where there is almost complete concordance of materials.

Turning to a second area of interest, that of India and Pakistan: contact with the Indus was renewed when the Achaemenids extended their empire eastwards around 500 B.C. The lava flows of the Deccan and Rajmahal Traps in India have long been renowned as sources of many varieties of quartz, particularly of cornelian and agate, and India has a long tradition of ornamental bead manufacture dating back at least to the mid-third millennium (Reade, 1979). Seven of the Achaemenid quartz seals, (four brown and white agates, two banded jaspers and an etched blue agate) are distinctly barrel-shaped and appear to have been reworked from beads (see Figure 3d) suggesting that they may have been imported from India. Hence there is reasonable evidence to suppose that the banded and the blue chalcedony, much used by the Achaemenids, and to a limited extent by the neo-Elamites, was imported from India.

There is some evidence to suggest that some of the chalcedony of the preceding neo-Assyrian and neo-Babylonian seals was likewise traded from India in that brown and white agate similar to the Achaemenid material was used for four seals, one of which is bead-shaped.

Further evidence supporting trade links from India/Pakistan in the neo-Assyrian and neo-Babylonian periods is provided by the early use of grossular garnet for two seals. XRD analysis showed these bodies to be composed of grossular and idocrase while the colour of small vivid green inclusions was shown by XRF analysis to be due to the presence of chromium. References (Brown & Dey, 1955; Dana, 1892; Deer et al., 1982 and Thiagarajan, 1961) to such material and the mineral collections of the Natural History Museum in London were searched for possible sources of grossular occurring with idocrase. Such searches may provide only limited information due to the possibility that the ancient source was exhausted in antiquity. Although there are reported occurrences of grossular/idocraase in central Europe and in Russia mainly in the Urals around Perm, the occurrences reported in Pakistan and India seem more likely sources for the material of these seals from Mesopotamia. Samples of massive green grossular with idocrase and chromium rich inclusions, from the area of Malakand, north Pakistan, in the Natural History Museum mineral collections (ref. BM 1985, MI 34520) are of similar general appearance to the seals; lenses of a similar hard and compact greenish white material with green speckles which are composed of grossular with idocraase and which have been sold locally as jade, occur in the Malakand area (Qaiser et al., 1970). There are also references to the occurrence of idocrase near Quetta in Baluchistan. Many varieties of of garnet occur widely and abundantly in India: for instance grossular has been reported from Danta in Gujerat and from Mysore in southern India and a jade-like rock has been reported from Madras. Similarly idocrase (of unknown colour) has been reported from Tonk in Rajasthan and in the collections of the Natural History Museum there is a block of massive green idocrase (ref. BM 1985, MI 8314) labelled as coming from India. However, no clear evidence for the precise location in India of suitable material for the seals was found.

The brown limestones, both plain and streaked, which were used by both the neo-Elamites and the Achaemenids were perhaps available to the neo-Elamites from a source in or near South west Iran; the colours of these limestones are comparable to those of eight of the earlier limestone seals (including three neo-Assyrian, a neo-Babylonian and two Syrian seals) suggesting that some or all of the earlier brown limestones may have been traded from the same source.

Conclusions

XRD analysis has been used to identify the materials of almost 450 first millennium seals and shown them to be composed of a variety of materials. Chalcedony, one of the hardest materials worked during this time predominates as the most popular material both prior to and during the Achaemenid period in all areas except Syria, and to a lesser extent, in south west Iran (where it was worked by the neo-Elamites). Softer minerals (particularly serpentinite and limestones) and synthetic materials (including faience, glass, ceramic and lime plaster) were also used. There is a correlation between good quality material and well executed workmanship. The results show a correlation of material with geographical area: in Assyria both hard and soft minerals were used; in Babylonia, chalcedony was used more or less exclusively; in Syria good quality natural materials, either hard or soft, do not appear to have been readily available and synthetic materials were more commonly used. During the later Achaemenid period, hard and soft minerals were used and there is almost complete concordance of these materials and neo-Elamite materials. The choice of material in the Achaemenid period appears to have been influenced by colour and banding: brown agates were particularly popular and these were imitated with artificially dyed chalcedony, with brown, sometimes streaked limestones and with brown glasses.

The relatively limited quantities of materials used for the seals are likely to have been obtained from a variety of available sources. Trade with India during the Achaemenid period is documented and is supported by the materials of the seals; there is also evidence to show that the materials may have been traded from India/Pakistan earlier in the first millennium during the neo-Assyrian and neo-Babylonian periods, while trade into Syria appears to have been limited.

Acknowledgements

I am very grateful to Dr Dominique Collon, Department of Western Asiatic Antiquity, for her continued advice and encouragement on this project and also to Prof. Edith Porada. I would like to thank several of my colleagues for their support and assistance, in particular Dr A P Middleton, Dr I C Freestone and Dr R M Cowell. I would also like to thank John D Rouse for information regarding the occurrence of grossular garnet in India/Pakistan and the Natural History Museum for allowing me to examine their mineral collections.

References

Berry, L.G. & Mason, B. (1959). *Mineralogy*. Freeman, San Francisco.

Bimson, M. & Sax, M. (1982). The materials of the seals. In: Collon, D. *Catalogue of the Western Asiatic seals in the British Museum, Cylinder seals 2, Akkadian, post Akkadian, Ur 3 periods*, pp. 12-14. British Museum Publications, London.

Brown, J.C. & Dey, A.K. (1955). *India's Mineral Wealth*, 3rd edn. Oxford University Press.

Dana, E.S. (1892). *The System of Mineralogy*, 6th edn. Kegan Paul, Trench, Trubner & Co., London.

Deer, W.A., Howie, R.A. & Zussman, J. (1982). *Rock-forming Minerals, Vol. 1A, Orthosilicates*, 2nd edn. Longman, London.

Frondel, C.F. (1962). *Dana's System of Mineralogy, Vol 3, Silica Minerals*, 7th edn. Wiley, New York.

Nassau, K. (1984). *Gemstone Enhancement*. Butterworths, London.

Qaiser, M.A., Mansoor Akhter, S. & Khan, A.H. (1970). Rodingite from Naranji Sar, Dargai ultramific complex, Malakand, West Pakistan. *Mineralogical Magazine* 37, 735-38.

Reade, J. (1979). *Early Etched Beads and the Indus-Mesopotamia Trade*, British Museum Occasional Paper, no. 2. British Museum Publications, London.

Sax, M. (1986). The materials of seals. In: Collon, D. *Catalogue of the Western Asiatic Seals in the British Museum, Cylinder Seals 3, Isin-Larsa and Old Babylonian Periods*, pp. 4-9. British Museum Publications, London.

Thiagarajan, R. (1961). Indian precious stones. *Bulletins of the Geological Survey of India, Series A, Economic Geology* 18.

INVESTIGATIONS ON ANATOLIAN OPAQUE GLASSES

B. Tugrul

Istanbul Technical University, Institute for Nuclear Energy, Istanbul, Turkey.

Introduction

The opaque glasses which are the subject of this research originate from a number of sites and belong to different periods. Three methods of non-destructive testing have been used; X-radiography, neutron activation analysis (NAA) and light microscopy.

The principal method used is X-radiography which has widespread applications in art and archaeology. The application of X-radiography to glasses is novel. However due to the addition of metallic oxides and metallic compounds it is applicable in this case (Tugrul, 1988). The production of good quality radiographs are discussed by Tugrul and Atik (1987).

In addition to the programme of radiography artefacts from the Enez site were analysed by optical microscopy and neutron activation analysis (Tugrul, 1985).

Studies

Different attributes of opaque glass artefacts can be measured by X-radiography. Firstly, physical dimensions can be easily determined which would otherwise be difficult due to the shape of the vessels. For example, the white opaque glass bottle from Biga-Cataltepe (Inv. Nr. 737c) has an interior diameter of 20 mm, a body diameter of 40 mm and a height of 73 mm (conventional measurement was difficult because the bottle exterior had a wavy surface). The bottle is dated to the first or second century AD. Figure 1 shows a conventional photograph and a radiograph of the bottle.

Determination of the vessel wall thickness can be easily made from the radiograph of the bottle. The wall thickness at the neck and bottom are approximately 2 mm, but is only 1 mm in the main body. In the radiograph some stress cracks can be seen in the sides of the neck of the bottle.

In addition to the determination of physical dimensions, other characteristics of glass vessel construction are illustrated by the X-radiograph. The second example of this vessel type is a vase from the joint construction of Istanbul Archaeological Museums (Inv. Nr. 73.5c). The vase has a height of 73 mm, and a body width of 35 mm to 45 mm. It is made of both dark blue and white opaque glasses with a glazed layer overlain by guilded decorative figures.

The thickness of the antique vase can be measured at approximately 2 mm from the radiograph (Figure 2). Furthermore, as determined by visual examination, the handle rings are not the same height.

When the glass base was investigated from the radiograph, some light and dark points were clearly seen. The dark points represent gas porosity and the light points represent the inclusions. Both of these faults can be seen on the radiograph. Therefore, it can be said that the base of the glass vase was not homogeneous. Moreover, there appeared to be stress cracks in the handle ring and a crack in the plinth foot.

A further study was made of a knob, made of dark blue and coral red glass belonging to the first or second century BC (Inv. Nr. 76.70c). The knob has a diameter of 23.5-24 mm and a height of 24 mm. Both the photograph and radiograph of the artefact can be seen in Figure 3. The radiograph is taken from the upper side of the knob, and shows a regular arrangement of glass bars.

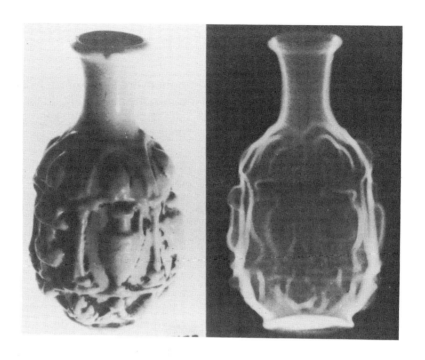

Figure 1. Photograph and radiograph of the bottle.

Figure 2. Photograph and radiograph of the vase.

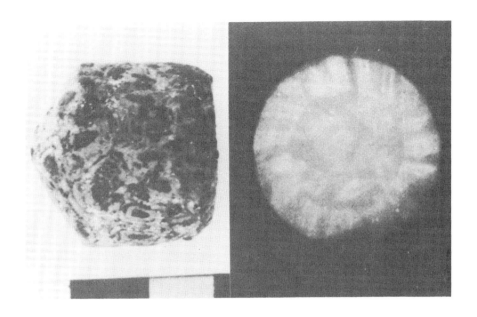

Figure 3. Photograph and radiograph of the knob.

Figure 4. Photograph and radiograph of the alabastrum.

117

X-radiography can also be applied to restored glasses. An example of this can be seen in the study of an alabastrum of white and lilac coloured glasses (Inv. Nr. 268c). This has a neck diameter of 35 mm, a body diameter of 38 mm and a height of 119 mm. The alabastrum dates from the sixth to fourth centuries BC, and was found at Myrina-Kalaba. The vessel and its radiograph are shown in Figure 4 (Tugrul & Atik, 1987).

On examination of the radiograph, some differences in the thickness of the glass can be seen; varying from 2 to 3.5 mm in the body area. It can also be seen that parts of the alabastrum have been fixed together during restoration and that the fit between the pieces is very good. The quality of the conservation work is therefore seen to be of a high standard.

X-radiography was then applied to some glass fragments from the Enez site, belonging to the third century AD. Those which showed porosity or cracks were also investigated by light microscopy. Furthermore, some glass fragments were analysed by neutron activation analysis (NAA), at the TRIGA Mark-II Training and Research Reactor at Istanbul Technical University.

Several elements were determined including manganese, sodium, potassium, zinc, copper, mercury, tin and barium.

Conclusion

The study has shown opaque glasses can be studied by X-radiography in conjunction with other types of analysis, and here three different methods have been used for the evaluation of the artefacts. The techniques used are both non-destructive and rapid, and therefore it is possible to obtain information similar information from opaque glasses to that obtained from transparent glasses.

Acknowledgement

I wish to thank the Archaeological Museums of Istanbul and Assoc. Prof. Dr. Sait Basaran for allowing me to examine the artefacts, and Istanbul Technical University - Institute for Nuclear Energy for supporting me. I am grateful to the Central Laboratories for Restoration and Conservation of Istanbul for helping me in my study and the use of their workshop for some radiographic exposures.

References

Tugrul, B. (1985). Useage of Nuclear Techniques in Archaeometry. *Turkish Scientific and Technical Research Council, Archaeometry Unit Colloquium, Istanbul (in Turkish), Proceedings (1986)*, pp. 12-29.

Tugrul, B. & Atik, S. (1987). Evaluation of ancient opaque glasses by X-ray radiography. In: *Conference on Science and Archaeology, Glasgow, Proceedings (1988)*, BAR British Series 196, pp. 145-152.

Tugrul, B. (1988). Evaluation of some Anatolian antique opaque glass artefacts by the X-ray radiography technique. In *I. International Anatolian Glass Symposium, Istanbul, (in Turkish), Proceedings (1990)*, pp. 37-39, 109-110.

METALLURGICAL ANALYSIS AND THE INTERPRETATION OF MODES OF PRODUCTION: GOLD BRACTEATES IN ANGLO-SAXON KENT

C.J. Arnold

Department of Extramural Studies, University College of Wales, 9 Marine Terrace, Aberystwyth, Dyfed, SY23 2AZ.

In 1981 the subject of gold bracteates from sixth-century Kent was reviewed in two major papers (Hawkes & Pollard, 1981; Bakka, 1981). Both papers principally sought resolution of chronological problems and there was a large measure of agreement between them. The differences arise from a consideration of the degree of wear on the objects in one paper but not the other. Hence there is a measure of agreement regarding the date of deposition of some of the significant grave-groups but variance about the time of manufacture (Hawkes & Pollard, 1981, 370). The concern here is not with the dates ascribed to these objects (more recently discussed by Hines, 1984, 20-22, 216), but with the methodology employed, and in particular the use of the results of X-ray fluorescence and metrological analysis.

Bracteates are thin circular gold pendants decorated with anthropomorphic and zoomorphic designs, produced by stamping with a negative die, and suspended by attached loops. The majority view is that the Kentish types under consideration date to the sixth century. The recent papers place much emphasis on the decorative style used on the objects although it may be argued that ornamental metalwork cannot be placed into an order until the factors governing manufacture, use and burial are known or postulated. Inherent in such an argument is the notion that the economic and social context and role of such metalwork is as important as its chronology and sequence. The data-set involved is complicated and confusing. The writer believes that when there is a detailed data-set to consider such ornamented artefacts may tell us a great deal about the society involved and the roles that such artefacts played. Indeed such questions are perhaps fundamental to understanding their chronology. It is essential therefore that all the available data is considered, and that all realistic models for manufacture, use and deposition are examined. This has not been the case. There are a series of equally probable additional models to those which have been put forward which may greatly affect the dates ascribed to, and the interpretation of, the Kentish bracteates. Such alternatives must be demonstrated to be incorrect or unlikely before a single interpretation can be accepted.

All Kentish D-bracteates are believed to be related to three Scandinavian prototypes (Bakka, 1981, 13) yet there is considerable divergence of opinion about the nature of that relationship. Hawkes & Pollard believe that the first Kentish bracteates were brought from Jutland (ibid. 320, 327, 352) by aristocratic settlers and imply that the rest were imported (ibid. 320). Hence examples found in France which have exact and close links with Kentish examples are viewed as re-exports (ibid. 320, 340). Webster (1977) has stated the belief that the mechanism for their movement was trade. Bakka believes they demonstrate 'continued contacts with the old homeland' (1981, 11). One problem with such explanations is the absence of any die-links between Kentish and Scandinavian examples. It is perhaps for this reason that Bakka implies Kentish manufacture when pointing out that the non-Scandinavian imitations form a large proportion of the total and, therefore, that the 'supply of good standard Scandinavian originals was sufficient to meet the demands of fashion' (ibid. 28). Hines has more strongly argued in favour of Kentish manufacture (1985, 216). Some writers cannot accept this because gold is thought to have been in short supply 'before the middle of the sixth century' (Hawkes & Pollard, 1981, 342) whereas others emphasise that some gold clearly was available (Hines 1984, 216), although in reality the actual quantity cannot be calculated.

The question of the origin of the Kentish bracteates and the range of opinion expressed on the subject emphasises how the available data, or at least the use that has been made of that data, is not suitable to answer the questions posed. Indeed the data-set is very difficult. There are no known die-links between Kent and Jutland and, indeed, 'die-links are very uncommon amongst bracteates, except where they are from the same hoard or grave' (Hawkes & Pollard, 1981, 320). This problem has led to research into the nature of the dies themselves (Axboe, 1975). Because none actually survive and the number of surviving bracteates manufactured with each die is consistently low, it is assumed that the dies were not made of a durable material, such as metal, but perhaps of hardwood or ivory. However, given that the range of variation in the dating of particular examples is as little as one or two generations the relative production life of dies of different materials becomes irrelevant.

The classifications of the Kentish D-bracteates and their chronologies have been constructed using a number of types of data governed by a series of, usually, unstated rules.

Die-linkage

It is assumed that once cut the die has a short life and is then disposed of, or at least not used again. To be sure that this was the case, however, requires an understanding of why and at what rates decorative styles changed, an understanding which we do not possess. On the basis of the available evidence, that individuals travelled with bracteates seems as likely as their movement with the dies to make them. They may also have carried an adequate memory of the pattern(s) to make the dies when and where necessary. The acceptance of both possibilities has implications for the chronologies. For instance, within the methodology of Hawkes & Pollard there are two dates of interest, manufacture and deposition. The date of manufacture of the

die is rarely considered and clearly it need not be as close to the date of a bracteate's manufacture as has been assumed.

Similarity

A 'near relative' stylistically is assumed to be a little further separated chronologically than die-linked examples. Many of the problems pertaining to die-linkage apply here also, as are questions regarding the rate of change of decorative style.

Metal analysis

When the metal composition is very similar the bracteates are taken to be from the same workshop and/or made at the same time (eg. Hawkes & Pollard, 1981, 341). How the distinction is made between the same workshop and different workshops using the same metal source is not clear. It is an important distinction as it may have implications for the contemporaneity of die-linked examples.

Wear

The degree of wear is taken to reflect the length of time an artefact was in use. The question of the frequency of use is ignored (a point also made by Hines, 1984, 21). Hence it is implied that objects displaying little wear must be nearly new when buried and *vice versa*. There is, in fact, confusion and inconsistency in the arguments used by the writers (eg. Hawkes & Pollard, 1981, 343, 347). The frequency and length of time with which an object was used, how it was actually used and the nature of the particular alloy, are important elements of the equation when using wear as an indicator of age when buried. Hawkes and Pollard seem to suggest that a worn object in a grave was acquired when the person was young, or was inherited, a fresh object acquired when the person was nearer death (1981, 338).

These measures are applied selectively to generate a series of 'rules' for the establishment of relative and absolute chronology: chronologically close and die-linked pieces with similar metal compositions were made in the same workshop and are contemporary; those with different amounts of wear but which are die-linked reflect 'hoarding'; and those with the same metal composition but occurring in different cemeteries are all made in Jutland from the same batch of metal.

At the heart of one paper (Hawkes & Pollard, 1981) are the bracteates from Finglesham graves D3 and 203. The bracteates 8 and 9 from D3, and two pendants accompanied by the bracteates 11 and 12 from grave 203 are accepted as having the same metal composition. The main difference is that 8 and 9 are very worn, whereas 11 and 12 are not. On the basis of the accompanying grave-goods they are considered to have been buried at different times, D3 being an earlier grave than 203 by a generation. As it is assumed that they were made at the same time, it is suggested that the pendants were made by reusing part of a stock of bracteates. Of the remaining bracteates, two were used by the woman buried in grave D3 while the others were stored and later passed on to the person buried in grave 203. It is implied that the inheritance took place late in the latter's life and, presumably all were made in Jutland and the bracteates exported together, by or to the person in grave D3. A third bracteate, 10, in grave D3 was made with a different die but was presumably exported with the rest as it has the same metal composition. Hawkes found it 'impossible to imagine the Finglesham bracteates being made in Jutland, in pairs separated by as much as fifty years, and then finding their way into the hands of the same family in Kent' (ibid. 340) and argues that they were all made in Jutland, those from grave 203 being stored. Of course, it has nowhere been demonstrated that the incumbents of graves D3 and 203 were related, nor does this assumption have to be the main support of the argument. They may have been made at different times if either the source of the metal or the die existed for a number of years; this may have occurred on either side of the North Sea. Bakka's shorter chronology, ignoring the wear

on 8 and 9 from grave D3, would place them all into a single generation because he also assumes that a die would only be used for that length of time (Bakka, 1981, 22).

There are a series of plausible alternatives which should also be considered: that dies alone may have been imported into Kent; the dies may have been manufactured in Kent; other bracteates or metalwork may have been reworked to produce the pendants, the bracteates and their loops, or both; some of the bracteates may have been robbed from an earlier grave. A variety of implications flow from the possibilities. The worn bracteates 8 and 9 may have been made earlier than 11 and 12 if the die was inherited. This may seem less likely because they have the same metal composition, but, as we shall see, bracteates from other cemeteries made from different dies are grouped together with those from Finglesham because they have the same composition, a point not pursued by Hawkes. A common source of metal may have existed and it need not be assumed that the bracteates with the same composition were made at one operation.

If bracteates 8, 9, 10, 11 and 12 from Finglesham, of which all but 10 are from the same die, are 'uniform in composition' (Hawkes & Pollard, 1981, 364), and made from the same metal standard (ibid. 341), then so also are 2, 3, 4, 5 and 20 (Figure 1), a conclusion reached by Pollard (ibid. 364) but not discussed by Hawkes. This, by the same set of assumptions, would be explained as the result of them all having been made from the same metal source in Jutland, although the same evidence is used elsewhere with Finglesham to justify manufacture in the same workshop. Hoarding of part of the grave-group of bracteates is again invoked for Bifrons grave 29 because the typologically later of the four bracteates, number 4, is more worn than the earliest. Two of the bracteates from this grave, 2 and 3, have the same metal composition, different weight values and similar degrees of wear. Number 4 and its die companion 6 from Bifrons grave 64, have different metal compositions and degrees of wear. This would be taken to suggest that they were not made from the same metal

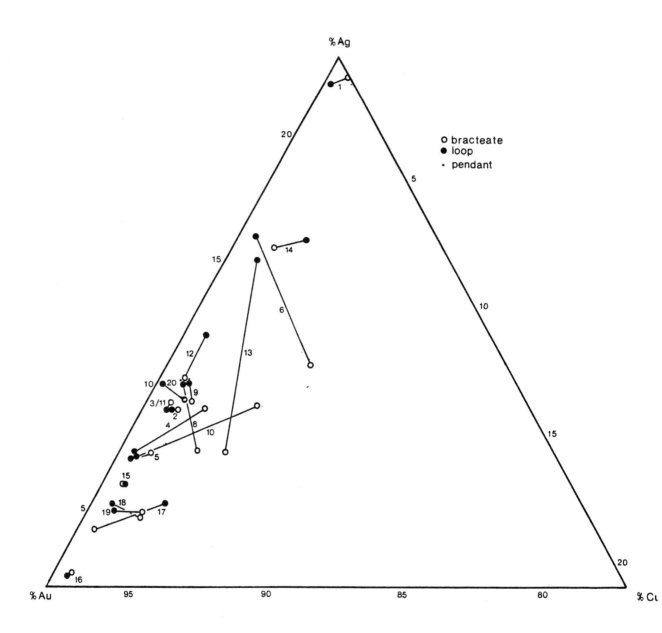

Figure 1. Ternary diagram showing the proportions of different metallic elements in Kentish bracteates.

source or at the same time, whereas 2 and 4, the one typologically earlier than the other, have the same composition.

If the relationship between the bracteates and their metal composition is examined in detail the situation is actually far more complex than we are led to believe. Bracteates 17, 18 and 19, and 4 and 6 are made from the same die and have similar metal compositions. While this may seem a very tidy relationship it is not the case with all die-linked examples, as not all the loops appear to be made with the same alloys as their bracteates.

Thus the loops of 8 and 9 have a similar composition to bracteate 12 from Bifrons grave 29 and the pendants from Finglesham 203, but the composition of the metal of the loop of bracteate 12 is quite different. Similarly the loops of 4 and 10 are similar to bracteate 5 and its loop, but the composition of the bracteates 4 and 10 are quite different. Bracteates 17, 18 and 19 from Sarre grave 4 have, in general terms, the same metal composition and are die-linked, but 14, 15 and 16 from the same grave have different compositions and are not die-linked. Some bracteates have similar compositions in their loops, for instance 2, 3 and 11, whereas others are quite different, for instance 6, 10, 13.

This may suggest that the reworking of bracteates and their loops, both together and separately was common. It also emphasises that contemporaneity need not be demonstrated by two objects having the same metal composition. Alternatively, it could be suggested that the source of alloy for bracteates and loops need not always be the same. The bracteate itself was possibly more likely to be re-worked than the loop as the latter conforms to a standard geometric design whereas the former is highly decorated and variable. Die-linked bracteates in the same grave but with different metal compositions, in all or in part, would be impossible by Hawkes' reasoning.

The weights of the bracteates are given by Hawkes & Pollard but are not brought into the main argument. If the goldsmith was producing a pair or trio of die-linked

bracteates for one owner, together and in the same manner, it might be predicted that they would have similar quantities of metal. Clearly if the re-use of loops was to be entertained, there would be considerable confusion and it would be preferable if the bracteate and its loop could be assessed separately. Bracteates 17, 18 and 19 have very similar values ranging from 2.644 g to 2.684 g and are from the same die. The same pattern might be expected from those from Finglesham graves D3 and 203 if they were, as Hawkes suggests, made at the same time, although due allowance must be taken for the different degrees of wear. 11 and 12 have very similar weight values (1.547 g and 1.512 g) whereas 8 (2.120 g) and the odd 10 (2.094 g) have closer values than the die-linked 8 and 9 (1.976 g), although the total range is only 0.144 g. This might lead to the suggestion that 11 and 12 were made together but 8 and 9 were made at different times and not necessarily in the same place. If the variance between 8 and 9 is put down to differential wear, this may at least indicate that they were not used in the same manner.

The available data begins to suggest that we should imbue the dies with the same degree of resilience and mobility as is given to the bracteates themselves. From a methodological point of view it might be argued that wear is an extremely dubious method of assessing the age of an item, and when taken in conjunction with the metal analyses requires special pleading for any sense to be made if the assumptions are to be adhered to. The problem may lie in the assumptions. If the dies were curated and used only when was appropriate it would go far to explain the wear discrepancies. The storage of the dies seems as likely, at present, as that of the bracteates themselves. The metal analyses do not overturn the suggestion; die-linked bracteates from the same grave show a stronger tendency to have the same metal composition than die-linked bracteates from different graves. The implications of the metal analyses are not as straightforward as has been suggested. If the dies were curated, it would have to be admitted that typological study could only treat the development of the dies. All of the other parameters have to be considered when calculating the dates of manufacture and deposition and may also lead to an understanding of the manner in which society actually used the dies and the products. The dangers of assuming that bracteates would be buried in typological order, and when not this would be indicated by wear, could, of course, be explored much further.

The data-set in this case is relatively poor, but despite this, the full range of possibilities needs to be explored using all of the available data if we are going to do justice to the archaeological record over and above the mere dating of contexts.

References

Axboe, M. (1975). A non-stylistic approach to the gold bracteates. *Norwegian Archaeological Review*, 8, 63-68.

Bakka, A. (1981). Scandinavian-type gold bracteates in Kentish and continental grave-finds. In: *Angles, Saxons and Jutes*, ed. V.I. Evison, pp. 11-38. Oxford University Press, Oxford.

Hawkes, S.C. & Pollard, M. (1981). The gold bracteates from sixth-century Anglo-Saxon graves in Kent, in the light of a new find from Finglesham. *Frühmittelalterliche Studien*, 15, 316-70.

Hines, J. (1984). The Scandinavian Character of Anglian England in the Pre-Viking Period, *British Archaeological Reports* 124, Oxford.

Webster, L. (1977). Brakteaten. 1. Archäelogisches. 2. England. *Johannes Hoops: Reallexikon der Germanischen Altertumskunde* 3, 341-42.

PROCESSES IN PRECIOUS METAL WORKING

J. Bayley

Ancient Monuments Laboratory, English Heritage, 23 Savile Row, London W1X 1AB.

Recent excavations in England have produced a variety of evidence for the working of the precious metals, gold and silver. This ranges in date from Iron Age to post-medieval but the majority of the material discussed here is of Roman or late Saxon date. This is only a preliminary report on the finds and the processes they represent as work on them is still in progress.

Before precious metals could be melted, cast and smithed to shape they had to be refined and then, if necessary, mixed with a controlled amount of base metal to produce an alloy of the desired degree of fineness; although very pure gold was used, most silver contained more than minor amounts of other metals. There are three refining processes that were used to purify precious metals - removing base metals from silver, removing base metals from gold, and separating gold from silver - and archaeological evidence for all three has now been recognised.

Refining silver

Silver was refined by the process known as cupellation. The impure metal was melted with an excess of lead and heated under oxidising conditions. The lead was oxidised to litharge (PbO) which acted in two ways. It reacted with the base metals mixed with the silver, oxidising them, and also reacted with these other metal oxides, forming fusible compounds (Percy, 1870) which then separated from the melt. Silver does not react with litharge and so is left behind when all the base metals have been oxidised and removed.

Cupellation can be carried out on both large and small scales, depending on the quantities of metal involved. Where amounts are large, a special hearth was constructed such as that at Silchester recognised by Gowland (1900). At the other end of the scale, small amounts of silver were tested for purity (assayed) on shallow dish or disc shaped vessels known as cupels. Theophilus, writing in the 12th century AD, describes making cupels of bone ash (Hawthorne and Smith, 1979) but the earliest surviving bone ash examples come from the site of the old Royal Mint in the Tower of London and date to the 16th century; they are about 20 mm across (Fig 1.1). The advantage of using bone ash was that it was a neutral material which did not react with the litharge but just absorbed it. This provided very efficient separation of the litharge from the silver so a very high proportion of the silver was recoverable.

Earlier cupels are known, but they are mainly made of fired clay and are often described in the literature as 'heating trays', a name applied before their function was positively identified (eg. Bayley, 1982). They vary in size but are typically 30-40 mm in diameter and their fabric is reduced fired (Figs 1.2 and 1.3). The litharge which formed during the cupellation process reacted with the fired clay, producing a lead-rich vitreous surface on the cupel which is often coloured red by traces of copper which the litharge had removed from the silver. There is normally a circular depression near the centre of the vitrified surface of the cupel which marks the place where the droplet of refined silver solidified. Replication experiments by Foley (1981) reproduced this feature which had been noted in late Saxon examples. The apparent contradiction in cupels appearing reduced fired when they have been used for cupellation, an oxidising process, is explained by variations in the the degree of oxidation; lead is more easily oxidised than most other metals (Bayley, 1988).

A few larger cupels have been recognised from various late Saxon sites in Winchester (Bayley and Barclay, 1990). They are about 150 mm across and have a distinct central

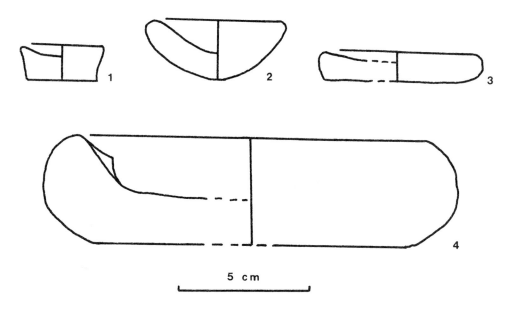

Figure 1. Sketch drawings of cupels. 1: Bone ash (Tower of London), 2: Ceramic (16-22 Coppergate, York), 3: Ceramic (Flaxengate, Lincoln), 4: Ceramic (Wolvesey Palace, Winchester).

depression about 100 mm across that has been eaten out of the clay fabric (Fig 1.4). The exact way these large cupels were used is not yet known but it is unlikely to be a coincidence that the central hollow is of roughly the same size as many of the 'litharge cakes' which are relatively common finds.

Litharge cakes are known in both Roman and late Saxon contexts (eg. Tylecote, 1986). They are usually 100-150 mm in diameter, 20-30 mm thick and have a slightly concave upper surface, sometimes with a slight central depression (Fig 2). They are varied in appearance, with colours ranging from grey to buff, green and orangey-red. Most colours are due to the lead oxides which make up the bulk of the material, though the green is produced by the corrosion of copper trapped in the litharge. Analysis of these objects regularly detects lead with variable amounts of copper also present. Silver is detected too, but only on the concave upper surface, which brings to mind Pliny's comment (Bk XXXIII, 95)

that during cupellation "... the silver floats on top like oil on water ..." (Bailey, 1929, 95). Silver is actually denser than litharge so further work is needed to explain fully the mechanisms by which the litharge cakes formed.

Refining gold

Gold can also be refined or assayed by cupellation as, like silver, it does not react with litharge. Ceramic cupels with traces of gold on them are known in Roman and late Anglo-Scandinavian contexts but they normally have far lower levels of lead present. This is probably because gold is far less reactive than silver and can be refined just by oxidising the melt. This process would require higher temperatures than cupelling silver as pure gold melts at 1063°C while the melting point of silver is 960°C. The more refractory nature of the fabrics used to make cupels for gold refining shows they were purpose-made and an even more refractory material has

126

Figure 2. Fragment of a litharge cake, diameter 120 mm (16-22 Coppergate, York).

Figure 3. Part of cupel of fused quartz chips with trapped gold droplets, diameter 55 mm (16-22 Coppergate, York).

127

been used to make a group of cupels from the Coppergate site in York (Fig 3); they were made of quartz chips about 1 mm across which had been formed into blocks and fluxed so the chips were set in a glassy matrix which stuck them together (Bayley, forthcoming).

Separating silver from gold

Separating gold from silver is known as parting. Nowadays this can be done by bubbling chlorine gas through the molten bullion for several hours (Dennis, 1963) or, on a smaller scale, by putting the metal into nitric acid which dissolves out the silver (providing it is present as the major constituent), leaving behind the gold which is insoluble. Nitric acid is produced by distillation and so was unknown until late medieval times (Taylor & Singer, 1957); before then other, more complex parting processes had to be used. Both Agricola, writing in the 16th century AD (Hoover & Hoover, 1950), and Theophilus describe a solid state parting process whose use is consistent with the nature of the recently identified archaeological evidence (see below). Notton (1974) has performed replication experiments following a description dating from the 2nd century BC of a similar Egytian process.

This parting process involved hammering the metal out into thin sheets, cutting it into pieces, placing them into a pottery vessel interleaved with a 'cement', sealing the vessel by luting on a lid and then heating it for a considerable time at a temperature just below the melting point of the metal. Theophilus' description of the cement is detailed and clear; Hawthorne and Smith's (1979) translation reads:

"... break into tiny pieces a tile or piece of burnt and reddened furnace-clay and when it is powdered, divide it into two equal parts by weight and add to it a third part of salt of the same weight. It should then be lightly sprinkled with urine and mixed so that it does not stick together but is just moistened."

The active constituent in this mixture is the salt (NaCl) which Hoover & Hoover (1950) say is decomposed at high temperatures in the presence of silica and alumina (the burnt clay) to give hydrochloric acid and, probably, free chlorine which react with the silver, forming silver chloride (AgCl). The urine is acidic and would probably aid the decomposition of the salt. Moisture, presumably from the combustion of the fuel, is another unspecified but vital reactant. Silver chloride is volatile and so was removed from the metal and absorbed by the powdered fired clay and, to a lesser extent, by the walls of the vessel used. The silver could be recovered by smelting the cement. If the original metal contained copper as well as silver and gold, the parting vessel would tend to acquire a copper-coloured alkali glaze.

Parting vessels, which must have been used in the way described by Theophilus and Agricola, have been recognised from Roman contexts in Chichester and Exeter and from Anglo-Scandinavian contexts in York and Lincoln. All have rather different appearances but they have a number of common attributes. They are generally oxidised fired and the inner surface of the vessel has a pale pinkish purple colour, in contrast to the orangish red of the clay fabric. Analysis of the surface by X-ray fluorescence (XRF) detects considerable amounts of silver and, on occasion, traces of gold and/or copper. Thin layers of a sandy deposit, the remains of the cement, are sometimes found inside the vessels and traces of a luted-on lid are normal.

The parting vessels from Chichester (Fig 4.1) date from the 1st century AD and were originally published as possible fritting vessels (Bayley, 1978). Their initial identification was based partly on their association with a group of crucibles containing red enamel and partly on the presence of sandy deposits adhering to the rim and inside of some of the 21 fragments; analysis of small areas on two of the pieces had failed to detect any significant levels of metals. Once parting vessels had been recognised from other sites, re-examination and further analysis indicated that these crudely made, flat bottomed dishes had been used as parting vessels. They have diameters of about 130 mm and typically are a pinkish purple colour on the inner surface; XRF analysis almost universally detects silver.

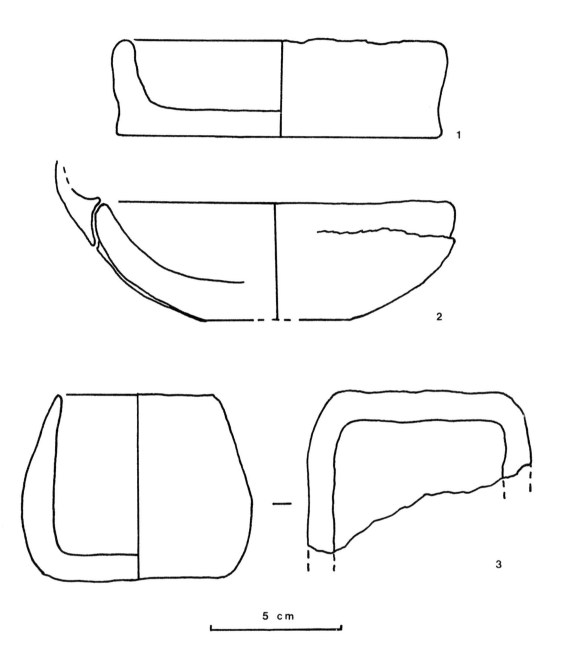

Figure 4. Sketch drawings of parting vessels. 1: Roman (Chapel Street, Chichester), 2: Roman (Frienhay Street, Exeter), 3: Anglo-Scandinavian (22 Piccadilly, York).

The Exeter finds are from Frienhay Street and come from a late 1st century ditch fill. A nearby mid-late 1st century military dump interestingly produced two shallow crucibles apparently used for cupelling silver (Bayley, 1989). The seven fragments of parting vessels are from oxidised-fired, lidded, shallow vessels around 150 mm in diameter which had sandy deposits and the typical pinkish purple colour on the inner surface (Fig 4.2). XRF analysis detected silver, and in one case gold too. The outer surfaces were vitrified with colours ranging from pale to deep (copper) green.

The finds from York are several dozen sherds that come from two nearby sites, 16-22 Coppergate and 22 Piccadilly, from contexts that date to the 10th and 11th centuries. These parting vessels are cuboid with thin sheets of luted on clay acting as lids (Fig 4.3). Their outer surfaces normally show some slight vitrification but there is no trace of copper colouring. Silver is universally detectable on the inner surfaces which are a pale greyish purple colour as the vessels are made of an almost iron-free clay. Some sherds have traces of the sandy 'cement' adhering to the inner surfaces (Bayley, forthcoming).

The finds from Lincoln come from Anglo-Scandinavian deposits on several sites in the lower city. Those from Flaxengate were originally thought to be alkali glazed Islamic pottery because of the bright turquoise (copper coloured) glaze they carried (Adams, 1979). The recognition of further sherds among the Saltergate finds prompted a reconsideration. XRF analysis detected both silver and gold on many of the pieces from both sites so an unspecified metallurgical use was postulated (Bayley in Gilmour, 1988). Continuing work and comparison with the finds described above suggest that these too are parting vessels. A full description and discussion will appear in due course (Bayley et al., forthcoming).

Conclusion

The archaeological finds described above demonstrate that evidence for a wide range of metallurgical processes survives, though it has not always been correctly identified. Some apparent anomalies have been resolved but further work is necessary before a full explanation of all the observed details is possible.

References

Adams, L. (1979) Early Islamic pottery from Flaxengate, Lincoln. Medieval Archaeology 23,218-9.

Bailey, K.C. (ed. and tr.) (1929) The elder Pliny's chapters in chemical subjects, Part I. Edward Arnold, London.

Bayley, J. (1978) The Roman crucibles from Chapel Street. In: Chichester Excavations 3, A. Down, pp. 254-5. Phillimore, Chichester.

Bayley, J. (1982) Non-ferrous metal and glass working in Anglo-Scandinavian England: an interim statement. In: PACT 7 (Proceedings of the second Nordic conference on the application of scientific methods in archaeology), pp. 487-96. Council of Europe, Strasbourg.

Bayley, J. (1988) Non-ferrous metal working: continuity and change. In: Science and archaeology, Glasgow 1987, eds. E.A. Slater & J.O. Tate, pp. 193-202. British Archaeological Reports, Oxford.

Bayley, J. (1989) Evidence for non-ferrous metalworking from Frienhay Street, Exeter, Devon. Ancient Monuments Laboratory Report No 55/89 (unpublished).

Bayley, J. (forthcoming) Non-ferrous metalworking at 16-22 Coppergate. Council for British Archaeology for York Archaeological Trust, London.

Bayley, J. & Barclay, K. (1990). The crucibles, heating trays, parting sherds, and related material. In: *Object and Economy in Medieval Winchester*, ed. M. Biddle, pp. 175-197. Clarendon Press, Oxford.

Bayley, J., Foley K. & White, R. (forthcoming) *Metal and glass working in early medieval Lincoln: the evidence from Flexengate and other sites.* Council for British Archaeology for City of Lincoln Archaeology Unit, London.

Dennis, W.H. (1963) *A hundred years of metallurgy.* Duckworth, London.

Foley, K.M. (1981) Glass- and metalworking industrial material from Anglo-Scandinavian Lincoln. Unpublished. MSc dissertation, University of London Institute of Archaeology.

Gilmour, L.A. (1988) *Early medieval pottery from Flexengate, Lincoln.* Council for British Arcaheology for the Trust for Lincolnshire Archaeology, London.

Gowland, W. (1900) Remains of a Roman silver refinery at Silchester. *Archaeologia 57*, 113-24.

Hawthorne, J.G. & Smith, C.S. (tr.) (1979) *Theophilus: On divers arts.* Dover Publications, New York.

Hoover, H.C. & L.H. (tr.) (1950) *Georgius Agricola's De Re Metallica.* Dover Publications, New York.

Notton, J.H.F. (1974) Ancient Egyptian gold refining: a reproduction of early techniques. *Gold Bulletin* 7(2), 50-6.

Percy, J. (1870) *Metallurgy, Vol. III: Lead including extraction of silver from lead.* John Murray, London.

Taylor, F.S. & Singer, C. (1957) Pre-scientific industrial chemestry. In: *A History of Technology, Vol. II.* eds. C. Singer, E.J. Holmyard, A.R. Hall & T.I. Williams, pp. 347-74. Clarendon Press, Oxford.

Tylecote, R.F. (1986) Litharge from a second century pit (Site DG, Pit 21) in Frenchgate, Doncaster. In: *The archaeology of Doncaster 1. The Roman civil settlement*, P.C. Buckland and J.R. Magilton, p. 196. British Archaeological Reports, Oxford.

THE PROPERTIES OF ARSENICAL COPPER ALLOYS: IMPLICATIONS FOR THE DEVELOPMENT OF ENEOLITHIC METALLURGY.

P. Budd & B.S. Ottaway

Department of Archaeological Sciences, University of Bradford, Bradford BD7 1DP, England.

Introduction

The archaeometric investigation of prehistoric metalwork has, at least until recently, concentrated on compositional analysis with the object of determining provenance. The approach has resulted in the compilation of an extensive database of information which continues to prove problematic to interpret in archaeological terms. A major limitation of the compositional approach is its inability to provide detailed information on the manufacturing technology of artefacts. This information is of paramount importance in characterizing the production methods used in prehistoric societies. The nature of arsenical copper production in the Eneolithic of central and south-eastern Europe may be taken as a case study into the potential of metallographic studies to extend the range of information available from the investigation of prehistoric metalwork.

As a result of a number of large scale analytical programmes it is now widely accepted that copper artefacts containing appreciable amounts of arsenic are a feature of the archaeological record of the Eneolithic and Early Bronze Age periods in Europe (Ottaway, 1989). What is less clear from the compositional information alone is whether such artefacts were the products of chance or design.

Attempts to interpret the arsenical copper phenomenon have concentrated on the advantageous properties of the material over pure copper for the manufacture of tools and weapons with the implication that alloy was deliberately produced and the alloyed metal known to be superior (Charles, 1967; 1974; 1979 & Slater, 1972). However, Charles (1967) has also indicated that randomly mined ores from enargite and tennantite deposits, containing up to 9% arsenic, can result in

smelted metal containing considerable arsenic. This raises the possibility of the unintentional production of arsenical copper, although, so far, we have no conclusive evidence for the prehistoric mining of arsenic rich copper ores.

Much of the difficulty in assessing the technological significance of early arsenical copper results from attempts to distinguish between impure copper and alloyed arsenical copper on compositional grounds alone. Initially analysts tended to select rather arbitary "cut-off" levels, typically of 1% arsenic above which artefacts were deemed to be "arsenical". Later attempts made use of the more sophisticated statistical treatment of analytical data to discrimitate between alloyed and un-alloyed material (Ottaway, 1982, 131). Broadening the scope of analytical investigations to consider metallurgical data allows more complex criteria to be used. Northover (1989) has pointed out that the term alloy "...implies deliberate control of composition for the modification of properties". With sufficiently detailed knowledge, both of the mechanical properties of copper-arsenic alloys, and of the manufacturing histories of "arsenical copper" artefacts, there is the potential to distinguish between those artefacts where the properties of arsenical copper were exploited in manufacturing, which can be regarded as true alloys, and those where they were not.

The availability of detailed information on the mechanical properties of copper-arsenic alloys of archaeological interest cannot be taken for granted. Although there are numerous references to the "advantageous properties" of arsenical copper in the archaeometallurgical literature there has been little fundamental research on the properties of copper-arsenic alloys of the compositions often encountered in archaeological material.

This paper is the first interim report on an ongoing study (Budd, forthcoming) designed to explore the mechanical and heat treatment properties of alloys of up to 12% arsenic. The study specifically concentrates on the properties of arsenical copper which would have been noticeable to, and exploitable by, prehistoric metalworkers. This paper concentrates on the casting and cold working properties of arsenical copper. Publication of results relating to the heat treatment of arsenical copper is anticipated (Budd, in press).

Experimental procedure

Copper-arsenic alloys of compositions greater than approximately 0.5% As have never been used in engineering and few such alloys have ever been commercially produced. In order to investigate the properties of arsenical coppers of archaeological compositions it was necessary to create a range of alloys. The range was extended beyond the compositions normally found in archaeological artefacts to include compositions above the limit of solid solubility of arsenic in copper (7.96 wt.% As) to ensure the presence of the Cu-Cu_3As eutectic in at least some of the samples. Accordingly seven alloys of between 0.461% and 11.92% arsenic, as well as pure copper were prepared.

The alloys were produced from commercial pure metal stock in a clay-graphite crucible heated in a coke-fired shaft furnace. Quantities were calculated to give billets of approximately 1kg. The copper was melted under a charcoal cover and the arsenic, wrapped in sheet copper to prevent volatilization, was then added to the melt. In simulation of Eneolithic practice the alloys were cast into open moulds in this case manufactured from a naturally bonded (water bonded) sand-clay mixture.

In a second experiment aimed at extending the investigation of the casting properties and quantifying arsenic loss in casting the sand-cast billets were halved and approximately 500g samples were re-melted and chill cast in open steel moulds. As well as producing sounder castings for cold working experiments it was hoped that casting conditions would be fairly close to those obtaining in an open stone mould. It has been suggested that the cooling curves for stone and metal moulds are similar (Staniasek & Northover, 1982).

Standard sample preparation techniques were used to prepare both the sand-cast and the chill-cast billets for metallographic examination by both optical and scanning electron microscopy. Complete cross-sections were examined in all cases. Samples from both casting series were prepared for neutron activation analysis. The use of a large sample size (approximately 100mg) extracted with a high speed twist drill penetrating to the centre of the casting ensured a representative analysis of the metal.

Following sampling for compositional analysis a chill-cast billet of each composition was machined and used for cold rolling. Samples were removed for metallographic examination and hardness testing following nine successive reductions in thickness.

Casting properties.

All melting and alloying took place in a crucible with a charcoal cover. Conditions were highly reducing and there was no evidence of arsenic loss. In casting, where the molten metal is in contact with the air, conditions are clearly more oxidizing. Under these circumstances arsenic loss is more likely. Although not noted in lower arsenic compositions the solidification of alloys of 8.01 and 11.7% As was accompanied by the evolution of white vapour. Subsequent X-ray diffraction analysis of the residue which condensed onto the mould surface showed the vapour to be the arsenious oxide, As_2O_3. The material is extremely toxic and the experiment highlights the potential hazard of casting high arsenic material.

Consideration of the analytical data (Table 1) suggests that, despite the loss of arsenious oxide from the higher arsenic material, the loss of arsenic is negligible even when casting into open moulds. The results are more likely to reflect sample heterogeneity and analytical error than genuine compositional differences.

Perhaps the most immediate effect of adding arsenic is the colour change of the cast metal.

Arsenic causes a notable whitening that is certainly detectable at the upper end of the range of compositions reported from Eneolithic contexts (approximately 4%). Alloying additions above about 8% arsenic produce a silver coloured metal which would have been very easily distinguished from pure copper.

Table 1. Arsenic content of reference material when alloyed, after casting in sand and re-casting in chill moulds. Arsenic contents of the sand cast and chill cast billets were determined by neutron activation analysis. The high value for the sand cast 8% arsenic alloy is probably due to sample heterogenity resulting from inverse segregation.

Alloying Addition	Sand Cast	Chill Cast (Re-cast)
0.50	0.461	0.467
1.00	1.00	1.005
2.01	2.03	2.01
4.00	3.92	3.90
5.99	6.02	6.06
8.03	8.44	8.10
12.20	11.92	11.70

All figures are in weight % arsenic.

One of the effects of arsenic in copper which has received particular attention in the archaeological literature is the ability of arsenic to act as a deoxidant in copper. Charles (1967) argued that dissolved oxygen, which would undoubtedly be a feature of copper cast in open moulds, would lead to the formation of a copper-cuprite grain boundary eutectic on solidification of the metal. Such a grain boundary eutectic would embrittle the casting and have a disastrous effect on the cold workability. The addition of arsenic would, he proposed, lead to the preferential formation of arsenious oxide, insoluble in the copper, which would separate as a sublimate.

Although it is quite possible that the oxygen content of some early castings would have been high the assumption that this would in all cases result in eutectic formation is not necessarily valid. There is evidence to suggest that the structures of oxygenated coppers are often heterogeneous and that eutectic formation is highly dependant on cooling rate. Hanson et al. (1923) compared two 0.282% oxygen castings from the same crucible and whereas in a slowly cooled casting the eutectic was well developed in a more rapidly cooled specimen it was practically irresolvable. Furthermore, it is uncertain that the addition of relatively small amounts of arsenic would result exclusively in the formation of arsenious oxide. Hanson & Marryat (1927) noted the presence of 'cuprous oxide' in copper-arsenic alloys up to 1.87% arsenic. Northover (1989) has pointed out that, whereas low oxygen contents in copper can lead to embrittlement, larger oxygen contents present less of a problem with oxides distributed throughout the metal as discrete inclusions deformable in cold work which do not generally lead to crack formation. Northover suggests oxygen uptake during pouring would be considerable in prehistoric casting and has reported hyper-eutectic (>0.39% oxygen) compositions in experimental castings.

Very little data is available on the effects of impurities, such as antimony, silver, nickel, bismuth and cobalt, which are known to occur in Eneolithic copper artefacts, on the mechanical properties of the metal. A number of elements are effective deoxidants in smaller quantities than arsenic. Allen (1930) found iron a more effective deoxidant than arsenic which, even at compositions of 0.6%, did little to improve the casting properties of hydrogen and oxygen rich copper.

Both Charles (1967) and Slater (1972) acknowledged that deoxidized castings, although not embrittled, would present disadvantages in terms of increased porosity due to the rejection of dissolved hydrogen on solidification. The equilibrium between the two gases in solution having been upset by the removal of oxygen. However, Charles felt that such porosity would not effect the workability of the material and would weld up satisfactorily on hammering.

It is true that porosity caused by the rejection of hydrogen would not result in the formation of an oxide film on the walls of blowholes so that such pores would weld up on cold working. However, it seems unlikely that deoxidation by arsenic could be sufficiently thorough to remove oxygen from the melt as rapidly as it would be dissolved. Any free oxygen would combine with hydrogen to form water vapour when the gases were rejected on solidification. Allen's experiments with deoxidants reveal that even with the efficient removal of cuprous oxide by the addition of 0.2% iron oxygen was readily absorbed into the top part of the casting and steam porosity resulted (Allen,1930). Porosity encountered in archaeological copper is far more likely to be due to steam than hydrogen.

Although the copper cast in this investigation contained a small quantity of iron (<0.2%), so that direct comparisons with Eneolithic material cannot be made, the effects of oxygen on copper cast in simulation of early metallurgical practice is currently the subject of further study.

All of the experimental arsenical copper castings were porous, the sand-cast material particularly so. Pores varied greatly in size, but were of the spherical and semi-spherical shape characteristic of steam porosity. Density determinations show no trend in the overall soundness of the castings (see Fig. 1) and there is certainly no suggestion of a relationship between arsenic additions and reduction in density. The soundness of castings probably relates to the specific detail of casting (temperature of melt, pouring rate etc). Initial metallographic examination of the material suggests that there may be a relationship between increasing arsenic concentration and reducing pore size although this may simply be due to the increased solidification time of the higher arsenic alloys.

Metallographic examination of copper-arsenic alloys reveals substantial micro-segregation and a radical departure from the composition - microstructure relationship which might be expected from too literal an interpretation of the phase diagram (Fig. 2). Such non-equilibrium freezing, to which copper-arsenic alloys are particularly subject, due to the steep gradient of the solidus and liquidus lines, results in a cored dendritic microstructure. Copper-rich dendrites are the first to solidify leaving arsenic rich inter-dendritic liquids which remain liquid for longer. Solidification of the arsenic-rich liquid can take place very much faster than diffusion in the deposited copper-rich solid so that the heterogeneous structure will be preserved on solidification of the metal.

The practical effect of micro-segregation is that an alloy of lower arsenic composition than the limit of solid solubility (7.96wt.% As) cooling below the eutectic temperature (685oC) will have remaining arsenic-rich interdendritic liquid which will solidify as a

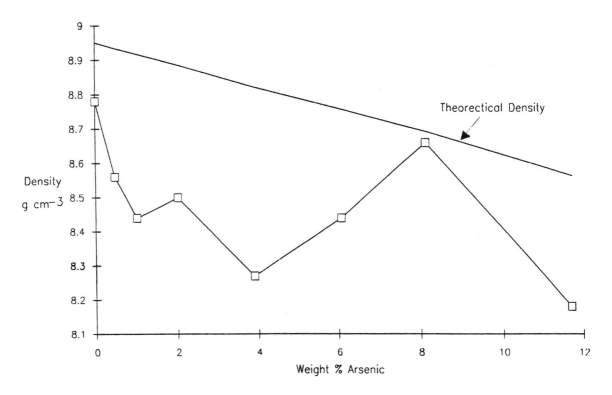

Fig. 1. The density of chill cast reference alloys. The theoretical density for perfect sound copper-arsenic castings shown for comparison.

copper-copper arsenide (Cu$_3$As) eutectic (20.8 wt.% As) rather than as primary solid solution. Shalev & Northover (1987) have put the lowest arsenic composition capable of producing a two-phase alloy in "normal cooling" at 2.5% As. In examining the range of alloys cast in this study it is clear that small islands of Cu$_3$As are present in alloys of 2% As and that where islands are large enough eutectic structure can be seen within them. Furthermore, small islands of Cu$_3$As can be found between heavily cored dendrites in 1% As alloys. Similar islands have been observed in archaeological material of this composition, although they were interpreted as inclusions (Faber-Morse & Gordon, 1986).

The presence of copper arsenide in alloys as low as 1% arsenic clearly demonstrates that the effect of a second phase on the properties of arsenical copper cannot be disregarded in material of archaeological interest. The formation of the second phase depletes the solid solution in arsenic and, given that the mechanical properties of such alloys are determined principally by the properties of the solid solution, archaeologists should be aware of the dangers of relating bulk compositional analysis to mechanical properties.

In addition to micro-segregation arsenical copper alloys are also subject to considerable macro-segregation, that is compositional heterogeneity over areas larger than single dendrites. The mechanism for macro-segregation in arsenical copper is usually argued to be inverse segregation the theory of which has been usefully summarized by Chalmers (1964) and discussed in an archaeological context by Charles (1973). As solidification takes place by the growth of columnar crystals from the mould wall micro-

136

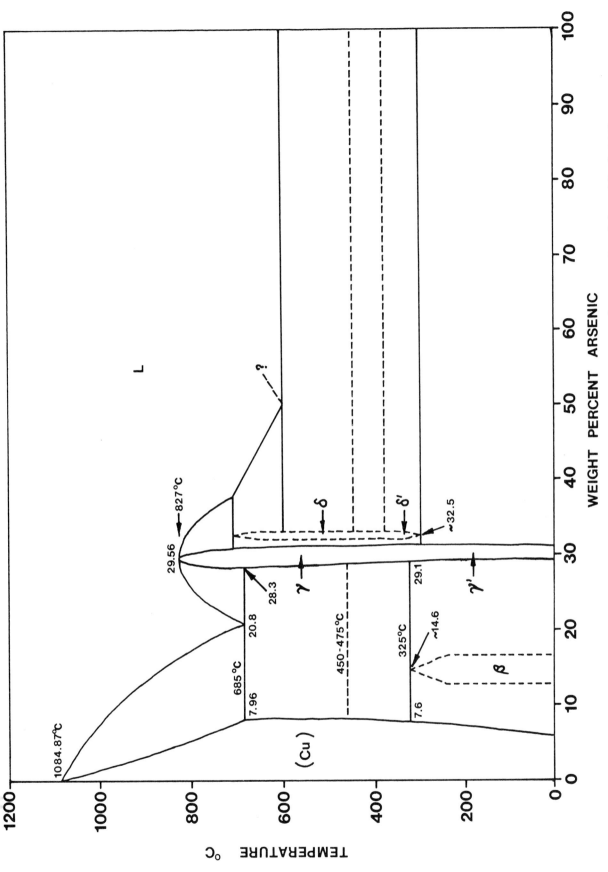

Fig.2. The copper-arsenic equilibrium phase diagram, after Subramanian & Laughlin (1988).

137

segregation results in solute enriched liquids being present at the solid/liquid interface as well as between the growing dendrite arms. Inverse segregation is thought to result from a "sucking back" of these liquids due to pressures caused by the contraction of dendrites as a result of the change in specific volume from liquid to solid. Solute-rich liquids can be exuded at the surface of the casting due to contraction between the solidifying casting and the mould wall.

Charles (1967) has emphasized the dependance of inverse segregation on cooling rate. The more rapid the solidification the greater the dendritic coring and the more directional the intercrystalline shrinkage.

Examination of the higher arsenic material from the current study revealed considerable evidence for inverse segregation. The final structure takes the form of cored solid solution dendrites interspersed with eutectic. Although the eutectic is dispersed throughout the structure there is clear enrichment at the surfaces of the casting typically taking the form of a more-or-less continuous layer. Inverse segregation is far more pronounced in the chill cast material where segregate layers a few hundred microns thick form on both the open surface and surfaces adjacent to the mould walls. In the slow cooled, sand cast billets segregate layers do not form adjacent to the mould walls, which are both porous and highly thermally insulating. However, a eutectic layer can clearly be seen on the relatively rapidly cooled top surface.

Although surface segregate layers were not observed in alloys of compositions similar to most European Eneolithic material the presence of such layers in material of higher arsenic content may be important in some cases. The surface eutectic is nearly 21% As and has a bright silver appearance, it is harder than the solid solution, although more brittle and difficult to work, and it is not easily removed by heat treatment (Budd, in press). It is of interest to note that inverse segregation could have been encouraged, even in alloys of modest arsenic content (less than 8%), by the adoption of casting techniques involving rapid solidification. The effect would have been most noticeable in relatively thin castings such as dagger blades. This rather extreme form of surface enrichment also sounds a note of caution to researchers engaged in surface analytical techniques and the analysis of corrosion products.

Cold working properties.

Perhaps the greatest advantage claimed for arsenical copper over pure copper in the manufacture of ancient artefacts is the superior cold workability of the arsenical material, that is the ability for copper alloyed with arsenic to achieve a greater hardness for a given degree of cold working. Indeed it has been claimed that arsenical copper needed to be work hardened as adding arsenic did not significantly alter the hardness in the as-cast state (Faber-Morse & Gordon, 1986). It was argued that this apparent lack of solution hardening was one of the major reasons for the replacement of arsenical copper by tin-bronze (McKerell & Tylecote, 1972).

In fact examination of the hardness of copper-arsenic alloys in the as-cast condition shows that there is some solution hardening (Fig. 3). Although the effect would be minimal in most arsenical copper of archaeological interest it should not be disregarded in all cases; when, for instance, considering the higher arsenic material reported from the Caucasian Early Bronze Age (Selimkhanov, 1962). An 11.7% As alloy in the as-cast state can be harder than pure copper cold worked to a 25% reduction in width.

Experimental data derived from cold rolling experiments on the range of alloys produced in this study graphically demonstrates the effect of arsenic additions on work hardenability (Figs. 4 & 5). It is notable that pure copper (containing no arsenic and under 0.2% iron) is sufficiently ductile to be cold rolled to reductions of over 75% and that additions of around 1% arsenic or less do little to improve the hardenability of the copper. Alloys of 8.1 and 11.7% arsenic work hardened very rapidly, but were extremely brittle. Surface cracking became apparent even with the most modest degree of cold

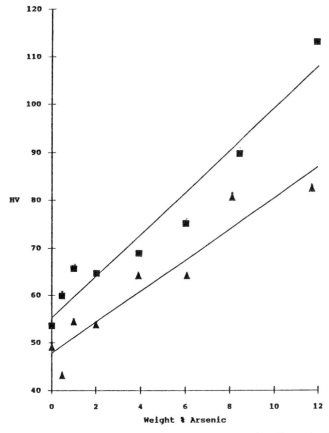

Fig. 3. The hardness of sand cast and chill cast copper-arsenic alloys in the as-cast condition.

work (5% reductions in width). Both of these alloys fractured at reductions of just over 31% having attained a hardness of 180HV.

Alloys in the range 2 - 6% arsenic represent by far the most successful range of compositions in terms of cold working material in as-cast condition, being sufficiently ductile to give reductions in width of 60-80% without cracking and giving hardness values in excess of 200HV at the upper end of the compositional range.

Conclusions.

The investigation of the properties of arsenical copper is ongoing, however some preliminary conclusions based on the study of the casting and cold working properties of copper-arsenic alloys can be offered.

There is little evidence that copper containing about 1 or 1.5% arsenic, which is frequently referred to in the archaeological literature as arsenical copper, would have offered any real advantages to prehistoric metalworkers engaged in the production of artefacts. Copper-arsenic alloys of less than approximately 2% arsenic offer little improvement over copper in terms of work hardenability. Smaller quantities of arsenic would have a deoxidant effect in mildly oxygenated pure copper, however the same effect has yet to be demonstrated for archaeological copper cast under primitive conditions. The assertion that the greater ductility of deoxidized castings necessarily outweighed the disadvantage of increased porosity also lacks support.

Although the degree of heterogeneity in arsenical copper alloys make it difficult to relate composition to mechanical properties

139

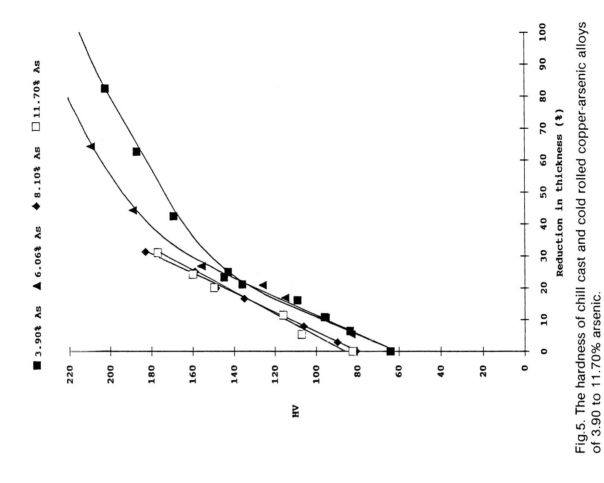

Fig.5. The hardness of chill cast and cold rolled copper-arsenic alloys of 3.90 to 11.70% arsenic.

Fig.4. The hardness of chill cast and cold rolled copper and copper-arsenic alloys up to 2.01% arsenic.

140

there is a compositional range on present evidence, broadly between 2 and 6% arsenic which represents the optimum in terms of ductility and work hardenability. It should be remembered, however, that casting practice is as likely to alter mechanical properties as minor compositional changes. It is perhaps also relevant that alloys in the upper half of this compositional range would have been visibly different from pure copper, being of lighter colour and perhaps displaying silver coloured surface layers.

Copper arsenic alloy of greater than 6 or 7% arsenic would have been very obviously different from pure copper but would have presented severe disadvantages in metalworking since it is very difficult to cold work without cracking. In addition casting such material might lead to arsenic loss. Although such losses might be insignificant in altering of the mechanical properties of the casting, the release of arsenious oxide vapour must have made such operations prone to hazard.

Acknowledgements.

The authors would like to thank Mr Roger Adams for his help with the production of the sand cast reference material and Dr G Gilmore (Universities Research Reactor Centre) for the neutron activation analysis of the samples reported in Table 1. The support of the Science & Engineering Research Council is also greatfully acknowledged.

References.

Allen, N.P. (1930). Experiments on the influence of gases on the soundness of copper ingots. *Journal of the Institute of Metals* 43(1), 81-124.

Budd, P. (in press). Eneolithic arsenical copper: heat treatment and the metallurgical interpretation of manufacturing processes. In: *Proc. Int. Symp. on Archaeometry, Heidelberg 1990*.

Budd, P. (forthcoming). A metallographic investigation of Eneolithic arsenical copper. PhD thesis, University of Bradford.

Chalmers, B. (1964). *Principles of Solidification*. Wiley, New York.

Charles, J.A. (1967). Early arsenical bronzes: a metallurgical view. *American Journal of Archaeology* 71, 21-26.

Charles, J.A. (1973). Heterogenity in metals. *Archaeometry* 15(1), 105-114.

Charles, J.A. (1974). Arsenic and old bronze. Excursion into the metallurgy of prehistory. Sir Robert Horne memorial lecture. *Chem.Ind.(London)* 12, 470-474.

Charles, J.A. (1979). From copper to iron: the origins of metallic materials. *Journal of Metals* 31(7), 8-13.

Faber-Morse, B & Gordon, R.B. (1986). Metallographic examination of pre-Columbian Mexican copper and silver artifacts from Mitla, Oaxaca (Mexico). *Archaeomaterials* 1, 57-67.

Hanson, D & Marryat, C.B. (1927). Investigation of the effects of impurities on copper. Part III: the effect of arsenic on copper. Part IV: the effect of arsenic plus oxygen on copper. *Journal of the Institute of Metals* 37, 121-168.

Hanson, D.; Marryat, C.B. & Ford, G.W. (1923). Investigation of the effects of impurities on copper. Part I: the effect of oxygen on copper. *Journal of the Institute of Metals* 34, 197-238.

McKerrell, H. & Tylecote, R.F. (1972). The working of copper-arsenic alloys in the Early Bronze Age and the effect on the determination of provenance. *Proc. Prehistoric Society* 38, 209-218.

Northover, J.P. (1989). Properties and use of copper-arsenic alloys. In: *Proc. of the Int. Symp. on Old World Archaeometallurgy, Heidelberg, 1987*. eds. A. Hauptmann, E. Pernicka & G.A. Wagner. Detschen Bergbau-Museums, Bochum.

Ottaway, B.S. (1982). *The earliest copper artifacts from the north Alpine region: their analysis and evaluation.* Schriften des seminars fur urgeschichte der Universitat Bern.

Ottaway, B.S. (1989). Interactions of some of the earliest copper using cultures in central Europe. In: *Proc. of the Int. Symp. on Old World Archaeometallurgy*, Heidelberg, 1987. eds. A. Hauptmann, E. Pernicka & G.A. Wagner. Detschen Bergbau-Museums, Bochum.

Selimkhanov, I.R. (1962). Spectral analysis of metal articles from archaeological monuments of the Caucasus. *Proc. of the Prehistoric Society* 28(4), 68-79.

Shalev, S. & Northover, J.P. (1987). Chalcolithic metal and metalworking from Shiqmim. In: *Shiqmim I: Studies concerning Chalcolithic societies in the northern Negev Desert, Israel (1982-4).* ed. T. Levy. British Archaeological Reports, International series 356.

Slater, E.A. (1972). Metallurgical aspects of Bronze Age technology. Unpublished PhD thesis, University of Cambridge.

Staniaszek, B.E.P. & Northover, J.P. (1982). The properties of leaded bronze alloys. In: *Proc 22nd Symp. on Archaeometry*, eds. A. Aspinall & S.E. Warren. Bradford, 1982.

Subramanian, P.R. & Laughlin, D.E. (1988). The As-Cu (arsenic-copper) system. *Bull. of Alloy Phase Diagrams* 9(5), 605-617.

THE SURFACE ARTEFACT ASSEMBLAGE FOR A PREHISTORIC COPPER MINE, AUSTRIA.

D. Gale

Department of Archaeological Sciences, University of Bradford, Bradford, U.K.

Abstract

This paper presents the analysis of a surface collection of stone implements made during the survey of a prehistoric copper mine, the Brandergang, in the Mitterberg region of Austria. Such recovery was only made possible by the disturbance of the vein area by a forestry track in the middle western slopes of the Einödberg above Einöden. The tool types fall into two main groups of activity, either mining or ore-dressing, as identified by their use-wear, namely battering, pounding and grinding.

Introduction

The Brandergang ore vein is located in the Salzach river valley, in the southern ore region or 'Südrevier' of the Mitterberg, Austria. The vein, one of three north-westerly trending copper veins, has been extensively worked prehistorically by continuous opencast pits, appearing as distinct gullies or overlapping pits, several metres deep. The considerable survival of these surface workings is owed to later limited extraction, consisting of horizontal adits interconnected by shafts, which resulted in minimal surface disturbance.

The ore vein has, fortuitously, been intersected by a zig-zagging forest track, cutting through the waste heaps lining the mining pits. This material, exposed in cuttings and re-deposited in the make-up of the track, contained numerous stone tool fragments, and a small number of pottery sherds and bone. The main distribution of finds was selected as the survey area which was located between the adit mouth of the Höchstollen at 950m and the 1150m contour. Outside the main survey area two wet processing sites were discovered exposed in the road cuttings to the hydro-electric power (HEP) construction

plant, above the level of the Arthurstollen, accounting for fourteen per cent of the collected assemblage. A number of stone tools were also recovered from the back fill of the HEP pipeline which cut prehistoric pits below Einöden, together with wooden artefacts including fragments of stemples and tapers.

The aim was, firstly, to determine the extent and nature of the prehistoric mining and ore-processing activity by mapping the surface finds and, secondly, to produce a topographical map of the surface workings. Details of the survey description and spatial analysis of the collected tools have been presented by Gale and Ottaway (1990, and in press).

The Brandergang ore vein consists mainly of chalcopyrite and some pyrite with rare occurrences of malachite and azurite (Pausweg, 1976). The host rocks are comprised of Palaeozoic green shales, weakly metamorphosed slates and phyllites.

The region has been investigated earlier this century by Kyrle, including limited excavation between the adits of the Unterer Höchstollen and the Arthurstollen (Kyrle, 1912 and 1913). In 1987, before the new HEP tunnel and pipeline were built, Prof. Eibner excavated at Höchbauer where he uncovered considerable evidence of domestic occupation (Eibner pers. com.).

Stone Implements

The stone implement assemblage presented here includes finds collected from the back filled area of the HEP pipeline and the two wet processing sites, both outside the main survey area. The total assemblage consists of 272 pieces dominated by stone tool fragments and splinters, exhibiting a high

Table 1. Stone tool assemblages. The artefact classes are divided into primary and secondary categories, ie. initial and residual uses.

Tool classes	Mining area		Wet processing sites	
	Primary	Secondary	Primary	Secondary
Tool fragments	125	57	20	15
Spherical pounders	35	1	4	
Modified stone hammers	32	7	4	1
Unmodified stone hammers	22	17	7	1
Cobble mortars	6		2	
Combination hammer and anvil stones	11	1		
Lower millstones	2		1	
Upper millstones	1			
Choppers		5		
Offset edge pounders		3		
Pecking hammers		1		

Table 2. State of completeness of primary and secondary tool classes. This is expressed by the number of measurable dimensions of the tool, ie length, breadth and depth, on a scale of 0 to 3, which have not been damaged as a result of the use of the tool.

Tool class	Dimensions			
	3	2	1	0
PRIMARY				
Modified stone hammers	1	5	10	20
Unmodified stone hammers			3	26
Spherical ore pounders	16	2	3	18
Cobble mortars	2	2	3	1
Combination hammer and anvil stones		1	4	6
Upper millstones	1			
Lower millstones			2	1
SECONDARY				
Modified stone hammers	1	5	1	1
Unmodified stone hammers	5	3	3	7
Spherical ore pounders	1			
Combination hammer and anvil stones	1			
Offset edge pounders	1		2	
Choppers	3	1		1
Pecking hammers	1			

144

degree of reduction (see table 1 and 2). These are mainly the product of heavy battering of stone hammers and ore-dressing stones. The tool categories below are defined by function, identified by use-wear, i.e. battering, pounding, grinding and cutting. The use-wear of the spherical pounders is much more distinct from the other heavier battering facets, so consequently fragments of this class will tend to be over-represented.

They consist of water-worn cobbles, usually of greenstone, employed for battering tools, along with coarser igneous rocks used for passive pounding and grinding tools. These have all been transported uphill from the Salzach river because no glacial erratics occur on the western slopes of the Einödberg. The bed rocks were not suitable for tools, although a few of the weakly metamorphosed slates have been found.

The intense use of the tools, demonstrated by the high degree of reduction (see table 2), is supported by the evidence of re-use (see table 1 and fig. 1). Three minor artefact classes have been identified which are entirely secondary, namely, offset edge pounders, choppers and pecking hammers.

The assemblage, as expressed in table 1, is sub-divided into primary and secondary assemblages, i.e. initial use and re-use. The secondary assemblage can be expressed in one of two ways. It can be said to represent the final activity, so that the true pattern should consist of only the final functions of the artefacts, which will of course include the primary state of tools that have not been re-employed (Gale and Ottaway, in press). In order to clarify the pattern of re-use, primary states have been excluded from the secondary assemblage in table 1. This has the advantage of including stone tool fragments as a secondary class for damaged primary tools, and expresses the number of tools which were discarded due to fracture and those that were discarded in a usable state (see Fig. 1).

Mining tools

MODIFIED STONE HAMMERS: This is a very fragmentary group, including two utilised stone tool fragments, exhibiting a high degree of re-use. There are no complete artefacts in this group although a few damaged examples, varying between 110 and 210 mm in length, may be considered to be complete. The hafting modification was kept to a minimum, usually either encompassing the midriff of the cobble with opposing notches, or, shallower grooving sometimes used in combination with notches, and, in a few cases, by a single patch of pecking to roughen the surface. These peck marks are distinct from the use-wear marks produced by pounding as they are more crater-like and without surface flaking.

All tools show intense heavy battering at both ends resulting in large spalling and fracture with very little of their original water-worn surface surviving, thereby confirming a simple hafting arrangement by a withy or rope. Many of the broken tools were re-employed as stone hammers, and those which fractured favourably, usually in the long axis, were remodified.

Shape as well as material seems to have played an important role in the selection of suitable rounded to sub-rounded cobbles. In addition to a predilection for elongated discoid cobbles a more pear-shaped profile suited this hafting method. Stones with natural 'waisting' and notches were also selected, and any integral edge protruberances were exploited by dual notching or with patches of pecking (Fig. 2c).

The type and extent of modification seems to be largely dependent on the rock type. Greenstones were particularly difficult to peck so these were notched by heavy battering at an oblique angle to produce hinged surface flaking. These notches are, on average, 37mm wide and 4mm deep. The deep midriff notches, up to 9mm in depth, are associated with the low metamorphic types (Fig. 2b). The pecked grooving was shallow, only around 3mm deep, and about 30mm wide. The groove types can be described as follows:

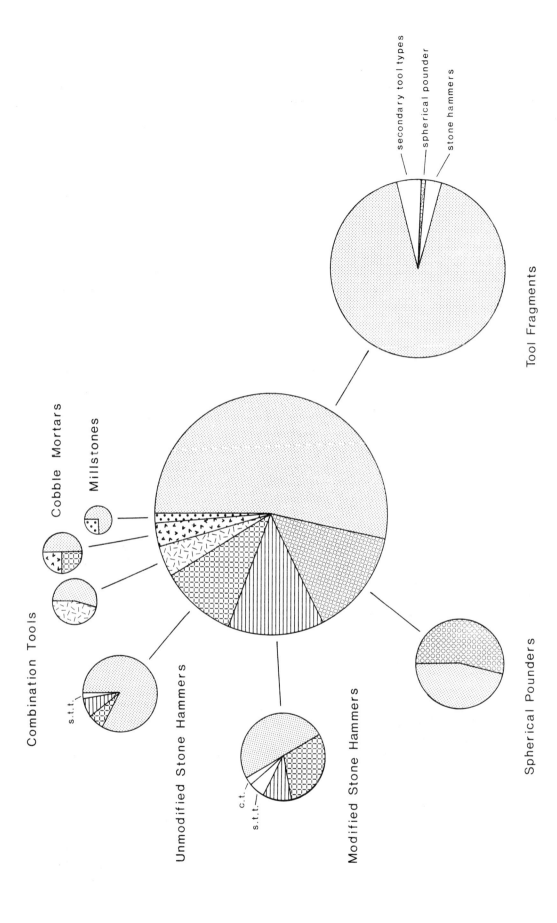

Combination Tools

Cobble Mortars

Millstones

s.t.t.

Unmodified Stone Hammers

c.t.
s.t.t.

Modified Stone Hammers

Spherical Pounders

secondary tool types
spherical pounder
stone hammers

Tool Fragments

Figure 1. Primary and secondary stone tool assemblages. The primary assemblage is represented by the central pie chart. The secondary assemblage, shown as a break down of the individual primary tool types, indicates the proportion of tools reduced to the tool fragment category and those re-employed or discarded in a usable state.

s.s.t. - secondary tool type.
c.t. - combination tool

146

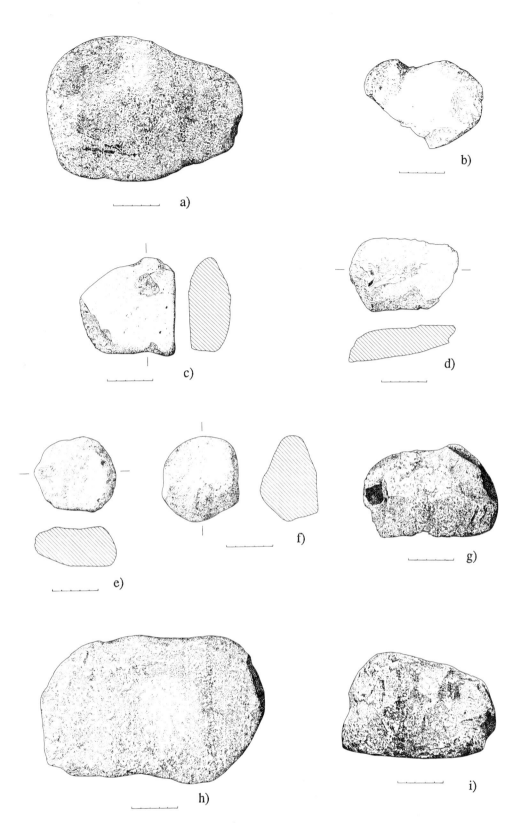

Figure 2.   a), b) and c) modified stone hammers
d) unmodified stone hammer
e) and f) spherical pounders
g) combination hammer and anvil stone
h) cobble mortar
i) upper millstone

147

a) incomplete - a groove which encompasses most of the cobble; b) C-form - a continuous grooving but incomplete over one face; c) face - a portion of grooving located on one face; and d) edge - a portion of grooving located on one edge (Fig. 2a).

UNMODIFIED STONE HAMMERS: These are characterised by evidence of heavy duty battering with extensive spalling, varying greatly in shape and size according to the reduction they have undergone from breakage and fracture (Fig. 2d). In order to differentiate these hammers from stone tool fragments they are defined as core tools. Although classified as hand-held hammers because they are all very fragmentary, the possibility of an earlier, hafted, phase cannot be ruled out. They include re-used stone tool fragments which have have already undergone a large reduction in size. Edge facets have a tendency to be less intensely battered although pecked facets have been recorded at the ends, also in conjunction with heavy battering, rather than at the edges.

Ore-Dressing Tools

SPHERICAL POUNDERS: This group, as the name signifies, consists of spherical to dicoid cobbles employed as ore pounders. Their shape and size variations reflect the constraints of a hand held crushing tool and there is a tendency for more spherical examples to be smaller in size. The long axes vary between 64 and 106mm with a minimum dimension of 33mm. Although only one re-used example has been identified other stone tool fragments may have been similarly employed, since these tools have well developed wear facets over their entire surface.

They have well developed broad and smooth edge pounded facets which become ridged in the plane of the tool when they are particularly broad (Fig. 2f). The facial facets exhibit more puntuated peck marks, typically with small batter flaking at the edge/face interface. Mature pounders exhibit pecked concavities, on average 35mm in diameter and 4mm in depth, which are most developed

on the facial facets and are then characterised by smoother peck marks (Fig. 2e). In five cases the shape of the stone has allowed ridging between pecked concavities to develop. The sphericity of the tool and the diameter of these facets for face and edge positions are not significantly related, while edge facets are generally less developed.

COBBLE MORTARS: This is a small group of particularly large, tabular, water-worn cobbles of coarse-grained rocks, up to 255mm in length, characterised by breakage in their depth axis. The crushing surface, generally developed on both faces, is identified by a broad, smooth surfaced, depression some 70mm in diameter and 8mm in depth (Fig. 2h). Only two mortars have been re-employed after breakage, for use as heavy duty stone hammers.

MILLSTONES: This artefact class is represented only by two fragments of lower millstones, together with a complete upper millstone recovered from the backfill of the HEP pipeline which was not included in the survey area.

The larger lower millstone fragment consists of a thin slab with a flat working surface. Its edge has been shaped by pecking so that it is lipped to its working surface, whereas the fragment from the Höchbauer wet processing site is lipped to the underside.

The upper millstone was found amongst recent back-flled top-soil at the level of the Alter Branderstollen adit. This type is characterised by a deep groove at the base, presumably for the fixing of a handle, and a smooth concave grinding facet (Fig. 2i). This example is very worn, the edge groove being partially destroyed by the grinding facet. It is unusual in that it has another groove over its top.

Combination Tools

HAMMER AND ANVIL STONES: These consist of fine-grained rocks, usually of greenstone, which are typically cubic in form and up to 170mm in length. One face of these water-worn cobbles has been battered resulting in a broad, flat, surface flaked facet, 65mm in diameter and with a surface reduction of about 5mm. The ends and edges have also been battered, the former having been generally more intensely worked (Fig. 2g). One example was a re-used stone hammer modified by edge notching and face pecking.

Secondary Tool Types

OFFSET EDGE POUNDERS: Represented by three re-used platy stone tool fragments with offset edge facets, fairly well developed, but not continuous, at between 30 and 65 degrees to the plane. The use-wear marks indicate a pounding tool somewhere in between battering and pounding.

PECKING HAMMERS: Only one artefact resembles a true pecking hammer, and this is a re-used stone tool fragment. In addition, two large stone hammer fragments, too big to be assigned to this class and damaged by spalling and fracture nevertheless exhibited well developed broad and ridged end facets.

CHOPPERS: An indistinct class of artefacts, referring to acutely edged stones, produced by chance from breakage whilst in service as another tool type. These would have been rather crude chopping tools characterised by small, hinged, use-wear flakes, the facets being generally undeveloped.

Discussion

There are considerable remains surviving due to limited modern extraction which, judging from the surface mining features, stone tool assemblage and ore type, indicate an Early to Middle Bronze Age exploitation. The artefact collection matches the ore-processing assemblage proposed for the eastern Alps by Eibner (1989) which demanded a high degree of beneficiation by pounding, grinding and wet processing. This survey has, clearly, identified an additional artefact class, namely the combination hammer and anvil stone.

The artefact assemblage is very fragmentary, exhibiting a high degree of reduction and re-use. Out of a total of 272 pieces, 139 (51 per cent) were assignable to specific tool classes. The primary assemblage consisted of 127 artefacts of which 20 per cent were re-used and 57 per cent were discarded in the form of stone tool fragments.

Three main tool categories can be recognised from the use-wear which reflect the main activites of ore extraction and beneficiation. Firstly, ore extraction is characterised by the existence of battering tools, stone hammers, modified and unmodified, and combination hammer and anvil stones for the reduction of the vein material. Secondly, there is the separation and reduction of the ore by pounding, represented by passive and active pounders, namely, spherical pounders and cobble mortars, and possibly including the offset edge pounders and pecking hammer. Finally, the ore was ground using two-part grinding mills consisting of a saddle-shaped grinding slab and an upper stone, hand-guided by a withy, handle which was attached in a similar fashion to the modified stone hammers.

For the activities of mining and ore-dressing located in the vein area, as identified by the tool types, nearest neighbour analysis has demonstrated that these activities are inseparable (Gale and Ottaway, in press).

The wet processing sites were located near running water by the exposure of horizons of fine sediment waste and not by their artefact types. Although the combination hammer and anvil stones were absent, heavy pounding tools were found. The latter may be explained by the proximity of the sites to the vein. Further analysis of the wet processing assemblage is hindered by the small sample size.

Acknowledgements

Thanks are due to my supervisor, Dr. B. S. Ottaway, Prof. C. Eibner, and Dr. Moosleitner for organising, advising and supporting the project, and I am also deeply indebted to

Michael Steinberger for all his painstaking work during the survey.

References

Eibner, C. (1989). Die Kupfergewinnung in den Ostalpen während der Urzeit, *Vortrage der 7, Niederbayerischen Archäolgentages*, 29-36.

Gale, D. and Ottaway, B. S. (1990). An Early Mining Site in the Mitterberg Ore Region of Austria, In: 'Early Mining in the British Isles', ed. P Crew & P. Crew. *Plas Tan y Bwlch Occasional Paper No. 1*, 36-38.

Gale, D., and Ottaway, B. S., (in press). Geophysical survey and surface artefact assemblage of prehistoric copper mining/working areas in Austria, In: *Proceedings of the XXVIIth International Symposium on Archaeometry Heidelberg 1990*. ed. E. Pernicka & G.A. Wagner.

Kyrle, S., 1912. Der zeitliche Stellung der prähistorischen Kupfergruben auf dem Mitterberge bei Bischofen. *Mitteil d. Anthropol. Gesellsch.* 42, Wein.

Kyrle, S., 1913. Der prähistorische Salzbergbau am Dürrnberg bei Hallein, *Jahrb. F. Alterumsk.* Bd.VII.

Pausweg, F., 1976. Die Bedeutung der Ur- und Frühgeschichtsforschung für die Lagerrstättenkunde am Beispiel des Kupfererzbergbaus Mitterberg bei Mühlbach am Hochkönig, Salzburg. *Festschrift für Richard Pittioni.* 125-129, Wein.

PREHISTORIC MINING FOR COPPER IN THE GREAT ORME, LLANDUDNO

David A Jenkins and C Andrew Lewis

Early Mines Research Group, c/o SAFS - Soil Science, UCNW, Bangor, Gwynedd LL57 2UW

Introduction

Mining techniques have developed rapidly over the past few centuries, and these developments have been well documented (eg. Hunt, 1887). There is some information on earlier Medieval (Agricola, 1556) and even Classical techniques (Pliny, eg. in Conophagos, 1980), but our knowledge of prehistoric times depends upon field evidence which is at present sparse. In recent decades there have been several important studies which have identified major mining operations that have supplied copper in the Bronze Age. Examples are sites in the Middle East (Rothenberg, 1972), middle Europe (Pittioni, 1951; Cernych, 1978; Jovanovic, 1979) and south west Europe (Rothenberg and Franco-Freijeiro, 1981). In the north west operations have been identified in south west Ireland (Jackson, 1968; O'Brien, 1987), although the dating of these has been disputed by some (Briggs, 1983). Although the pioneering work of Davies (eg. 1948) established evidence of ancient mining which he interpreted as Roman, until recently the consensus has been that the evidence for any Bronze Age mining which may have occurred in the UK had probably been obliterated by the intensive mining activities of the last few centuries (eg. Tylecote, 1986).

However, recent work in Wales has revealed evidence for Bronze Age mining at several sites and these are currently under investigation by members of the Early Mines Research Group. As yet, the evidence for mining at Mynydd Parys, Nantyreira and Cwmystwyth comes essentially from surface excavations (Timberlake, 1987; 1988) but that from the Great Orme is unique in that it derives from extensive underground workings. This paper presents a preliminary account of some of these underground features in relation to the geology of the Great Orme and also to the later historic workings by which they are intercepted and accessed. It will illustrate in general some of the practical difficulties presented by the investigation of possible prehistoric workings and the questions raised in their interpretation, as well as consider some of the methodologies that are now available to answer such questions.

The Great Orme Copper Mines

The Great Orme is an isolated precipitous hill some 680 feet high that overshadows the resort of Llandudno on the North Wales coast (Fig. 1). The Carboniferous limestone strata comprise rhythmic cycles including massive reef and rubbly limestones with interbedded mudstones (Warren et al., 1984). Regionally they dip gently to the north, but at the eastern end of the Orme they are folded into a basinal structure traversed by several high angle north-south faults. The limestones adjacent to these faults have been dolomitised and show a drusy nature with cavities up to a few m^3 lined by crystals of both dolomite and calcite, and it is in these rocks that the copper deposits occur. The primary ore takes the form of chalcopyrite [$CuFeS_2$] as crystals and occasional thin veins on, or in, the dolomite; such dolomite is often rotted to a soft granular state. In the upper parts of the mine the chalcopyrite has been extensively oxidised to green malachite [$Cu_2CO_3(OH)_2$] and rarer blue azurite [$Cu_3(CO_3)_2(OH)_2$], and these are accompanied by areas of black powdery manganese oxides [MnO_2] and traces of covellite [CuS]. Rare galena [PbS] has been found, and metallic copper [Cu] and cuprite [Cu_2O] were also recorded during active mining (Vivian, 1859) but have not been found since. Segregations of pitch-like asphaltine occur in some of the dolomites and secondary deposits, such as thin seams of malachite, jarosite, manganese oxides and rare azurite, are also concentrated in some of the mudstone bands.

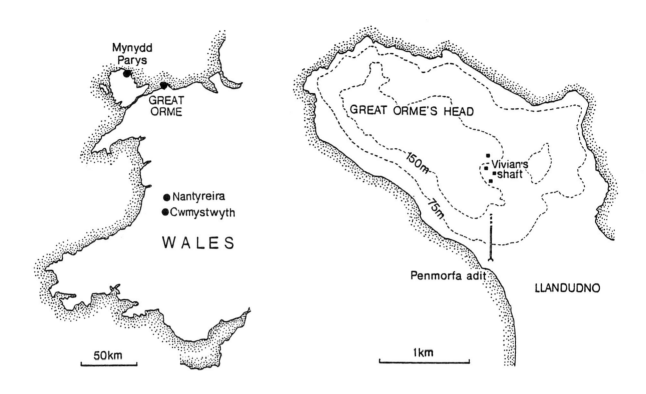

Figure 1. (a) Location of Bronze Age mining sites in Wales and (b) details of the Great Orme Mine.

The earliest records of mining for copper date back to 1692, and a patchy record of mining can be followed through the period of maximum output in the 1830s to the closure of the mines in the 1880s (Williams, 1979; Smith, 1989). A number of 'mines' were involved. In the area of interest, workings to the north-east were known as the 'Old Mine' and those to south-west as the 'New Mine'; they reached depths below sea level and were to share in the Penmorfa adit which was driven due north for 700m from just above sea level on the south shore, dewatering the workings above and allowing extraction of ore for a further 70m below sea level. To the east a separate mine system, the Ty Gwyn, was opened up under the edge of the Orme. In the Old Mine there are records of the nineteenth century miners breaking into earlier workings which contained calcite formations,

bones and stone tools: they acquired the arbitrary label of 'Roman'. In 1938-9, Oliver Davies carried out an excavation on the West Shore as part of his investigation of early mining for the British Association. He found evidence of habitation (shell layers, pottery sherds, charcoal) associated with mining (stone mauls) which he ascribed to a Roman date on the basis of the pottery and coins found in the mining area (Davies, 1948).

Although the mines in the Great Orme have now been subject to active exploration for some 30 years, it was only in the 1970s that access was gained to possible prehistoric workings. Their potential significance was first appreciated by James (1988) whose pioneering observations stimulated subsequent research. More recently, development of the surface area around

Maes-y-Facrell as a carpark has allowed, through the imaginative cooperation of the Council, access to further mine workings. This was achieved by the opening and securing of three other major shafts on the site, Vivian's, Owen's and Pyllau, of which the first two have given access to further prehistoric sites. Simultaneously with exploration underground, surface excavation around the Vivian shafthead has been carried out by the Gwynedd Archaeological Trust (GAT). This has exhumed, from beneath up to 5m of later mine waste, part of a deeply dissected early mining landscape. From these excavations abundant charcoal, shells and numerous bone fragments and stone tools have been recovered; the details of these excavations are to be reported elsewhere. This paper will concentrate on the underground workings accessed through Vivian's shaft, as depicted diagrammatically in Fig. 2.

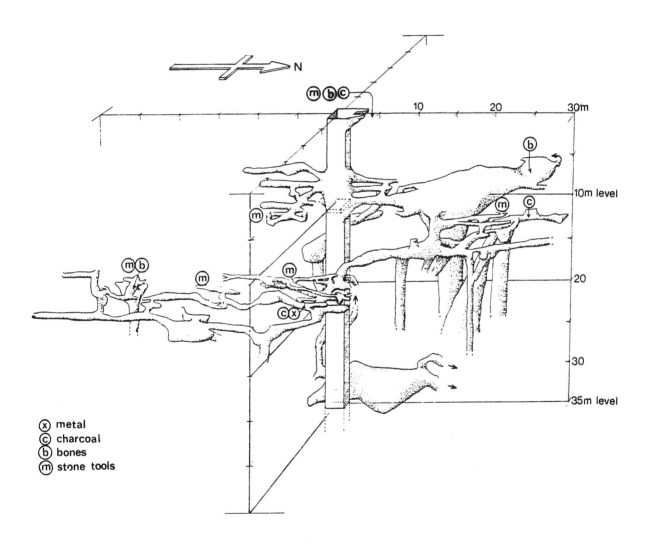

Figure 2: Simplified oblique view of the workings at the 10 m level in Vivian's Shaft.

Possible Criteria of ancient mining

During the course of the underground studies, a number of features have emerged as possible criteria of ancient mining. These derive from the passage morphologies and the associated waste products and artefacts, and may be interpreted in terms of the technology, such as fire-setting, that may have been employed. Considered individually these features may not always be convincing, but their cumulative effect is impressive (Fig. 3).

The morphology of mine passages and workings will depend on the nature and structure of the rock, and on the methods used in excavation. The passages and shafts which were driven through hard rock in the nineteenth century tend to be square cut and of 'comfortable' (ie. 1m x 1m) size, and give access to large ramifying stopes following the mineralised zones. Such workings are characterised by shot holes and by grooves made by picks of rectangular section, and in them have been found relics such as straw

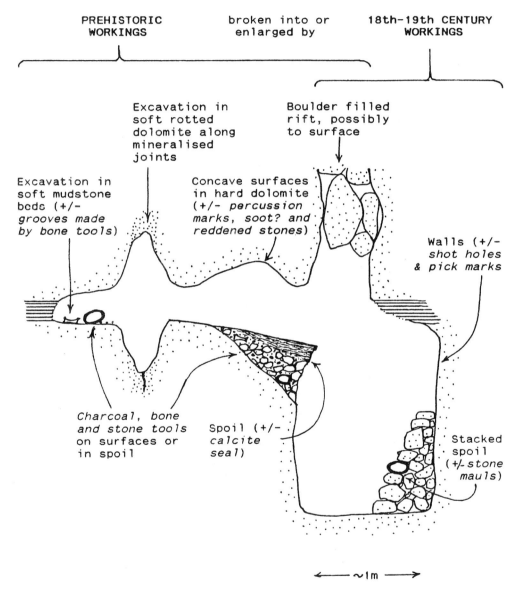

Figure 3: Diagrammatic view of features associated with Prehistoric mining.

fuses and their containers, powder horns, iron tools, waggons, windlasses, broken clay pipes, clogs and, strangely, the (ritualistic?) remains of two cats.

By contrast, other sections of the mine are very confined, comprising remarkably narrow (ie. 30-50cm) vertical or horizontal passages opened up along joints or bedding planes in dolomitised rock or along beds of mudstone; in several places they are only just wide enough to allow access. Their walls lack shot holes and pick marks and generally tend to be rounded in profile; occasionally some surfaces bear what appear to be percussion marks that could result from the impact of stone mauls, whilst rounded grooves have been observed in some of the soft mudstones. Often there is clear evidence that these early workings have been broken into by the later shafts, rises cross-cuts or passages described above.

Deposits of spoil within the underground workings comprise direct evidence of mining activity and thus a potential source of information. Their particle size distribution and shape reflects the techniques that were used. The qualitative impression is that they generally tend to be of smaller size (eg. 5-10cm), less angular and better sorted than are the products of iron picks and blasting in the eighteenth/nineteenth century, but quantitative data are required to confirm this. Rare shells of gastropods have been found in near-surface workings, and these raise the possibility of reconstructing prehistoric environments, although background information on such fauna on the Orme would be needed: a detailed search for other microfossils (pollen, phytoliths, etc.) that might have been carried in to the early workings also needs to be made. The deposits do contain charcoal, stone mauls, bone, and fragments of bronze as will be described below. However other artefacts, which might be expected amongst the products of such intensive labour, are rare - as indeed they are in more recent spoil; for example, no ceramic material has yet been recognised.

If fire-setting had been employed then the presence of charcoal, wood ash and reddened rocks might be expected. Fragments of charcoal are occasionally common and were one of the first clues to areas of early mining; they are to be found both *in situ*, on or in spoil, and also redeposited in water-lain or collapsed sediments at greater depths. The charcoal usually comprises fine particles less than 1 cm^3 in size, although occasional larger fragments of up to 3 cm^3 have been found (eg. site x, Figure 2). Those kindly examined by Dr M P Denne, of UCNW Bangor, have all been identified as oak and tend to show wide spaced regular growth rings suggestive of fast young growth; this could be compatible with coppicing. In the surface excavations alder has also been recorded. It is very unlikely that fragments sufficiently large to allow for dendrochronology will be found but, as described below, samples have been collected for C^{14} dating.

Wood ash does not survive physically, but would appear to be detectable chemically by a significant increase in water-extractable K$^+$ (but not of Na$^+$, Ca$^+$, or Mg^{2+}) levels (ie. 0.2 cf. 0.02 μg/g) and this enhancement may prove to be a useful criterion of wood ash where charcoal is absent. The effect of high temperature on the pure limestone would have been that of decarbonation (at $>800°C$) and probable slaking in the damp atmosphere. By contrast, from experiments, there is a distinct reddening (ie. to 'weak red' - 10R. 4/4) of the initially 'pale brown' (ie. 10YR. 6/3) ferriferous dolomite. Fragments of such reddened rock have been found underground in charcoal rich areas of the workings.

Smoke stains on the passage walls could be associated with underground fires. However, the field identification of such stains, and their differentiation from those resulting from manganese oxides, covellite, or a natural bitumen which occurs in the dolomitised rock, is not easy. An attempt to achieve this through Scanning Electron Microscopy shows promise. Artificial 'soots' show a distinctive fine grained (0.2-1.0 μm) particulate morphology in contrast to the microspherules of manganese oxides and the smooth surfaces of bitumen. With appropriate reference materials, it might well be possible to identify the components of the black stains on walls, and perhaps even identify the material used in any fire (or lamps?) if combined with organic microanalysis.

Stone mauls are another obvious feature of the early workings, some lying free in passages, other incorporated into deposits of spoil. Some 40 mauls have so far been discovered underground, weighing from 0.5 to 10 kg. Their anomalous well rounded shape within the angular rock debris stands out and is suggestive of beach deposits, and their fine grained dark grey green appearance is reminiscent of the microdiorite that forms the igneous intrusion of Penmaenmawr some 10 km to the west. No proper 'rilled' mauls have yet been found (cf. 'Group 2-4', Picken, 1988), but many show the usual signs of attrition at one end. On one such stone there is a sheath of prismatic calcite up to 1 cm thick around one end which has been fractured before the maul's inclusion in a bank of spoil, indicative of its considerable age.

In particular areas of the early workings bones are very common. These tend to be fragmentary, and those so far identified derive from pig and possibly cattle; the absence so far in the underground deposits of antler fragments, found experimentally to be so efficient as picks, is noticeable. However, many of the elongated bones appear to have been fashioned into a distinctive 'chisel/tool' of the order of 12 cm long, with one end sharpened to a point which has subsequently been rounded and polished, presumably by attrition through use in digging out the soft mudstone and dolomite. Indeed, in one instance grooves in mudstone can be matched with the tip of a tool found nearby. Most of these bone fragments are a distinctive blue-green colour, the stain penetrating a few millimetres below the surface. Analysis shows this to be due to copper (0.9%) and iron (0.5%), whilst others stained black also contain manganese (1.6%), but no mineral phases other than apatite have been detected by XRDA. At one site in the workings (Fig. 2) two smaller tabular fragments of bronze with a green oxidised surface, amounting to 0.6 g, have been found; their composition will be considered below.

A striking feature of obvious significance is the calcite which cements early deposits of spoil and, in some cases, seals them with layers of massive coarsely crystalline stalagmite up to 25 cm thick. Whilst there is no unique rate of calcite deposition, such thicknesses imply extended periods of time and hence age of the underlying deposits. Unfortunately, this calcite is probably too young to permit uranium-series dating, but banding is discernable within the calcite (eg. at 0.5 mm intervals) amounting in one sample 15 cm thick to some 100 successive layers. If such banding were to represent annual climatic cycles, then again they would give an indication of the interval between the successive phases of mining, but more data are again needed before quantitative interpretation would be possible. Another curious feature that has been noted in several of these stalagmitic deposits is a transition from the coarsely prismatic calcite to a thin surface layer of finer grained aragonite. On a geological timescale, the metastable aragonite inverts to the stable calcite, but the factors prompting this particular transition are not known. They could, for example, relate to a change in the chemical environment resulting from an intensification (nineteenth century?) of mining activity, or to a switch in extraction from malachite to chalcopyrite dominated ore, or even to a change in surface land use.

Dating of early mining

The first requirement is to establish the age of the early workings, for which there are several possibilities. The occurrence of charcoal on, or dispersed within, the spoil at a few localities has enabled sufficient material to be collected for C^{14} dating. So far, three dates have been obtained: (i) from material collected from spoil and submitted by James (1988) from some 27 m below ground in workings off the 'Roman' shaft; (ii) charcoal collected from surface workings by GAT; (iii) material sampled from workings off Vivian's shaft at the 20 m level (Fig. 2). These have given the following calibrated dates (Pearson and Stuiver, 1986):

156

(i)   2940 BP ± 80 (HAR-4845)
               cal BC 1410-920 (2σ range)
(ii)   3370 BP ± 80 (CAR-1184)
               cal BC 1885-1465
(iii)   3000 BP ± 50 (BM-2641)
               cal BC 1410-1070

All three dates relate to the early-middle Bronze Age comparable with those from Mynydd Parys, Nantyreira, Cwmystwyth and Mt. Gabriel (3550-3260 BP). The dates decrease with depth (ie. (i)=(iii)<(ii)) consistent with a progressive downward development of the early mine. To meet possible criticism of total reliance on charcoal, which could have derived from already ancient wood (Briggs, 1983), a further date (iv) has been obtained by GAT from collagen in the bones associated with surface workings at the Vivian shafthead. This agrees closely with the value for the related sample (ii):

(iv)   3290 BP ± 60 (BM-2645)
               cal BC 1740-1440

Other possible absolute and relative techniques to be considered include uranium series dating of stalagmitic calcites (although the upper limit of this technique is only 5000 years), residual magnetism of *in situ* fired dolomite, any equivalent of 'dendrochronology' in the stalagmites, and amino acid racemisation in the bones and their fluoride content. However, at this stage it seems reasonable to conclude that the workings for copper on the Great Orme date from the Early-Middle Bronze Age.

Ores and Mining Techniques

Two main types of ores would have been available to the early miners. The 'primary ores' (in the accepted geochemical sense) are dominated by chalcopyrite, mostly as crystals encrusting hard drusy dolomite, occasionally (in surviving exposures underground) as veins in rotted dolomite. The 'secondary ores' are dominated by malachite, as coatings on chalcopyrite crystals and veins in the hard dolomite, but also as thin veins in the mudstone. Analysis of a sample of the latter indicates 0.52% copper, from which the malachite was found to be concentrated readily by light crushing and panning. It is likely that the malachite would have been the ore first mined since it would have been visually obvious and accessible on the original surface outcrop. In the mudstone it would also have been mined with relative ease thus promoting the development of underground workings by the following of mineralised beds down dip into the valley side, so helping to explain the remarkable depths recorded. At a later date, and with improving technology, the chalcopyrite could also have been worked, especially in the softer rotted dolomite.

The nature of the workings must reflect the technology employed. The most striking feature is their small size and often tortuous nature which suggests that the workers were of small stature or perhaps even children. Yet these labrynthine workings reach remarkable depths (27 m) and/or distances (>100 m) from the surface which would demand considerable expertise. In this context the existence of natural cavities due to both limestone dissolution and dolomitisation could be an important consideration. The lithology is also relevant since the old workings are frequently seen to follow soft mudstone bands and dolomitised ore-bearing rock that is often also rotted and soft: these two materials would allow easy excavation even with bone tools, and early miners would have only needed to resort to other techniques (such as firesetting or the use of metal tools) where hard rock impeded access. The complete removal of the softer ore-bearing rock, often resulting in the characteristic smooth curved wall surfaces of harder rock seen in these old workings, could thus give a false impression of the labour involved.

To confirm this interpretation will require a detailed stratigraphic survey of the workings to locate and trace mudstone strata and rotted vein material in relation to early workings. This in turn needs to be based on an accurate underground survey which, given the three-dimensional complexity and tortuosity, will be a lengthy procedure. Such a survey might benefit from the aid of the newly improved locational devices, based on low frequency modulated magnetic fields, that are now available.

Evidence for other techniques includes the stone mauls and their presumed percussion marks, bone tools and the charcoal and reddened rock indicative of fire. As has been pointed out by many early writers, fire-setting is a practice that was followed well into the historic period (cf. Agricola, 1556) as has possibly been the use of stone mauls. Use of such criteria alone for prehistoric mining has therefore been suspect. However, at the Great Orme and elsewhere in Wales, they have now been placed in a dated prehistoric context, and also their efficacy demonstrated experimentally. Fire setting requires considerable volumes of wood and, on any large scale, would have hastened the deforestation of the Orme. The observation that the structure of the charcoal would be consistent with coppiced oak could therefore be relevant. Another interesting possibility to check would be the use of charcoal rather than wood underground; this would certainly reduce the labour involved and the difficulties of operating in confined space but may not prove to be so efficient. Other questions are posed by the major problems of ventilation at any depth underground and whether quenching of the heated rock was desirable or necessary; as yet there is also no indication as to the illumination used in these prehistoric workings.

Ore and product analysis

Some of the sulphide and carbonate ores have been analysed for their trace element contents which tend to provide a distinctive 'fingerprint' of the sources of such metals as copper (Jenkins, 1988). The analyses are presented in Table 1 and indicate that, in the context of Welsh chalcopyrites, those from the Great Orme can have relatively high values of Ag, As, Bi, Ge, Mo, Ni, Sn and low values of Pb and Zn: similarly the malachites have relatively high values of Bi, Co and Sn, but low levels of Mn, Pb, and Zn. The levels of As suggests that a search for the mineral olivenite (copper arsenate) on the Orme might

be successful, and this could have implications for the smelting of the ores (Pollard et al., this volume).

Against this background, it is interesting to consider the initial analysis of a sample (3 mg) of the small fragments of bronze found underground. The high level of Pb, which is notably sparse in the mineralisation of the Great Orme, and to a lesser extent that of Ni, suggest at this stage that the bronze is unlikely to have been the simple product of local copper ores, and that the metal came from elsewhere. However the high As, Ni and Pb, as well as Sn, contents are comparable to those of early Welsh bronzes (Northover, 1977). The high Fe content (0.2%; Craddock, personal communication) might also suggest the smelting of a slagged chalcopyrite (Craddock and Meeks, 1987) rather than malachite. A more detailed report on the composition of the bronze and its implications will be given elsewhere.

Conclusions

At present there are no validated prehistoric underground workings for copper in limestone of a comparable scale and nature in the UK or Europe (cf. Cernych, 1978; Jovanovich, 1979). Consequently there is no precedent or reference for the studies in the Great Orme Mines, and it has therefore been necessary to propose and test criteria through direct observations underground. Some useful guidelines relating to techniques might be drawn from comparisons with the mining for flint in the late Neolithic (Sieveking, 1979), but the major differences due to lithology would need to be borne in mind. At this stage in the investigation it may be concluded that copper ores had first been extracted by deep surface trenches along mineralised veins, followed by extensive penetration into the hillside through small passages along rotted dolomite veins and mudstone bands enriched in malachite. There is ample evidence for the use of fire setting below ground, but this may have been limited to areas where access necessitated the removal of hard rock; otherwise, mauls and bone tools may have proved quite

Table 1. Trace Element Contents of Samples from the Great Orme Mines.

Samples associated with workings at Vivian's Shaft.

Mal.1   Granular malachite vein, surface spoil.
Mal.2   Crystalline malachite vein, surface spoil.
Mal.3   Malachite seams separated from mudstone.
Chp.1   Chalcopyrite vein (c. 1 cm) in spoil, far eastern vein.
Chp.2   Chalcopyrite crystals encrusting dolomite, surface spoil.
Cu°   Metallic fragment (c. 1 g) found in spoil, far eastern vein.

Sample identity checked by XRDA, then pre-ignited/oxidised for 16 hours at 470°C. Values given in ppm ($\mu g/g$) in the initial ore mineral sample.

	(S)	Mal.1	Mal.2	Mal.3	Chp.1	Chp.2	Cu°
Ag	3.2	<	<	<	75	18	100
As	2400	<	<	<	3200	4200	>1%
Be	18	<	42	<	<	<	<
Bi	100	130	130	130	130	<	180
Co	7.5	100	<	320	56	18	130
Ge	32	<	<	32	100	130	75
Mn	10	75	24	<	42	240	<
Mo	18	<	<	<	240	18	<
Ni	3.2	180	42	1300	240	5.6	1800
Pb	32	<	<	<	<	320	c.32
Sb	320	<	<	<	<	<	00
Sn	32	130	56	32	56	24	750
Ti	100	<	<	1800	240	<	>1%
Y	10	42	420	<	<	<	<
Zn	750	<	750	<	<	<	<
							1300

Not detected in any sample (ie. < Lower sensitivity limit - S)

Au (<320); Cd (<420); Cr (<18); Ga (<3.2); In (<3.2);
Tl (<180); V (<100); Yb (<3.2); Zr (<18).

adequate. This would further suggest that the first ore worked was malachite (and azurite and any native copper), which would be consistent with the lack so far of any surface evidence for smelting involving slags. It is unlikely, however, that chalcopyrite would not have been worked in the later stages.

This activity has now been firmly dated to the Early-Middle Bronze Age. This makes it contemporaneous, to slightly younger, with mining at Mount Gabriel in Ireland and at Cwmystwyth, Mynydd Parys and Nantyreira in Wales. The mines in the Great Orme are distinguished, however, in having extensive, well preserved workings underground in limestone, and thus promise to be one of the most informative prehistoric mining sites. Much work needs to be done to fulfil this potential, both underground, at the surface, and in the laboratory. Below ground detailed survey is needed, such that workings can be related to stratigraphy and mineralisation, and details of the passage morphology and spoil recorded. Above ground the original mine entrances, together with any washeries, furnaces, habitation sites, etc., may yet be found preserved below the blanket of more recent spoil; their excavation could provide a greater range of data including, for example, ceramic material. In the laboratory, information is needed on the geochemistry of the ores and recent deposits, the nature of the prehistoric spoil and other artefacts. Although slow and expensive, such work will be of unique value.

References

Agricola, G. (1556). *De Re Metallica* (Reprint edited by H.C. Hoover & L.H. Hoover, Dover Publications Inc., New York, 1956).

Briggs, C.S. (1983). Copper mining at Mount Gabriel, Co. Cork: Bronze Age bonanza or post-famine fiasco? *Proceedings of the Prehistoric Society* 49, 317-333.

Cernych, E.N. (1978). Aibunar - a Balkan copper mine of the fourth millenium BC. *Proceedings of the Prehistoric Society* 44, 203-217.

Conophagos, C.E. (1980) *Le Lavrium antique et la technique greque de la production de l'agent.* Polytechnic of Athens, Athens.

Craddock, P.T. & Meeks, N.D. (1987). Iron in Ancient Copper. *Archaeometry* 29(2), 187-204.

Davies, O. (1948). The copper mines on the Great Orme's Head, Caaernarvonshire. *Archaeologia Cambrensis* 61-66.

Hunt, R. (1887). *A Historical Sketch of British Mining.* EP Reprint, 1978, Wakefield, Yorkshire.

Jackson, (1968). Bronze Age copper mine on Mount Gabriel, West County Cork, Ireland. *Archaeol. Austriaca*, 43, 92-103.

James, D. (1988). Prehistoric copper mining on the Great Orme Head, Llandudno, Gwynedd. In: *Aspects of Ancient Mining and Metallurgy*, ed. J. Ellis Jones, pp. 115-121. UCNW, Bangor.

Jenkins, D.A. (1988). Trace element analysis in the study of ancient metallurgy. In: *Aspects of Ancient Mining and Metallurgy*, ed. J. Ellis Jones, pp. 95-105. UCNW, Bangor.

Jovanovic, B. (1979). The technology of primary copper mining in South East Europe. *Proceedings of the Prehistoric Society*, 45, 103-110.

Northover, P. (1977). In: *Aspects of Early Metallurgy*, ed. A. Oddy, pp. 63-70. British Museum and Historical Metallurgical Society.

O'Brien, W.F. (1987). The dating of Mt. Gabriel type copper mines of west Cork. *Journal of the Cork Historical and Archaeological Society*, 92(251), 50-70.

Pearson, G.W. & Stuiver, M. (1986). High precision calibration of the radiocarbon time scale 500-2500 BC. *Radiocarbon*, 28(2B), 839-862.

Picken, J. (1988). Stone tools and early mining in Wales. *Archaeology in Wales*, 28, 18-21.

Pittioni, (1951). *Prehistoric mining in Austria: problems and facts.* In: 7th Annual Report, Institute of Archaeology, London, pp.16-43.

Pollard, A.M., Thomas, R.G. & Williams, P.A. (1990). Some experiments concerning the smelting of arsenical copper. This volume.

Rothenburg, B. (1972). *Timna.* Thames and Hudson, London.

Rothenburg, R. & Freijeiro, A.B. (1981). *Ancient mining and metallurgy in south west Spain.* Institute for Archaeo-Metallurgical Studies, London.

Sieveking, G. de G. (1979). Grimes Graves and prehistoric European flint mining. In:

Subterranean Britain, ed. H. Crawford, pp. 1-43.

Smith, D. (1989). *The Great Orme Copper Mines*. Llandudno, Gwynedd.

Timberlake, S. (1987). An archaeological investigation of early mine workings on Copa Hill, Cwmystwyth. *Archaeology in Wales*, 27, 18-20.

Timberlake, S. (1988). Excavations at Parys Mountain and Nantyreira. *Archaeology in Wales*, 28, 11-17.

Tylecote, R. (1986). *The Prehistory of Metal Working in the British Isles*. The Metal Society, London.

Vivian, W. (1959). On arborescent native copper in the Llandudno Mine near Great Ormeshead, North Wales. *Quarterly Journal of the Geological Society*, London, 15, 109-110.

Warren, P.T., Price, D., Nutt, M.J.C. & Smith, E.G. (1984). *Geology of the country around Rhyl and Denbigh*. British Geological Survey Memoir, NERC, HMSO, London.

Williams, C.J. (1979). *The Llandudno copper mines*. British Mining No.9 Monograph of the Northern Mines Research Society.

NORTHERN EUROPEAN METALWORKING TRADITIONS IN THE FIFTH AND SIXTH CENTURIES AD

C. Mortimer

Research Laboratory for Archaeology, 6 Keble Road, Oxford, OX1 3QJ

Introduction

Migration Period and medieval copper alloys are beginning to be the subject of scientific study, after a long time during which they were only subject to typological examination. Recent projects have looked at this material from various viewpoints, eg. long-term and with a widely dispersed geographical base (Capel, 1986 on pins, also currently Blades *et al.*, (this volume)), localised but without chronological commentary (Brownsword *et al.*, 1986 on Avon valley material), various types of artefact on a single site (Oddy, 1983 at Sutton Hoo). These have raised a number of theories about sources of copper alloy. It is generally accepted that much of the output of Anglo-Saxon metalworkers was made from recycled material and that, after the Roman period, brass was not produced in England until the eighth century AD (Gilmore, 1987).

This paper concerns fifth and sixth century Anglian English copper alloy jewellery, specifically excluding Saxon and Jutish material from the west and south of England. Research has concentrated on a single, common artefact form, the English cruciform brooch, found primarily in East Anglia. This allows a fuller examination of the archaeological and typological import and an understanding of the type of detail which may be extracted from the database.

In this initial discussion, I will only be considering information on the alloying elements and not the trace elements. With this data I shall try to determine a) whether recycling can be observed to be a significant feature of the metallurgical environment, b) whether brass was used and c) how typological features link with chemical compositions. The information can be compared with a selection of material from abroad. The chemical compositions are the result of analysis of around 500 drilled samples by atomic absorption and electron microprobe analysis. 48 samples analysed by both methods showed reasonable agreement (Mortimer, forthcoming).

Initial considerations

For any archaeometric study, it is essential to understand the archaeological attributes of the data. Our view will be distorted if we look at artefacts from different socio-economic environments. In this case, it would be unwise to unknowingly compare late fourth century Germanic material, brought in by mercenaries in the late Roman period, with heavily gilded, locally produced material of the mid-sixth century. A new typological study shows that some of the most common cruciform brooch forms continue in use over a long period and overlap in production periods should be kept in mind (Mortimer, forthcoming).

Technological differences between typological forms were minimised in this project since all the brooches were cast and the metalworkers appeared to have used quite consistent technology. In addition, there was little applied surface treatment which may have required particular alloy types, for example, silvering and tinning, at least until the latest forms. The broad geographical distribution does not appear to be an important factor since, when compositions were divided up according to groupings of sites, there appeared to be little significant difference between patterns of alloy use in individual areas within the main concentration.

Changes in patterns of alloy use

A considerable change in the pattern of alloy use can be observed between the earliest

162

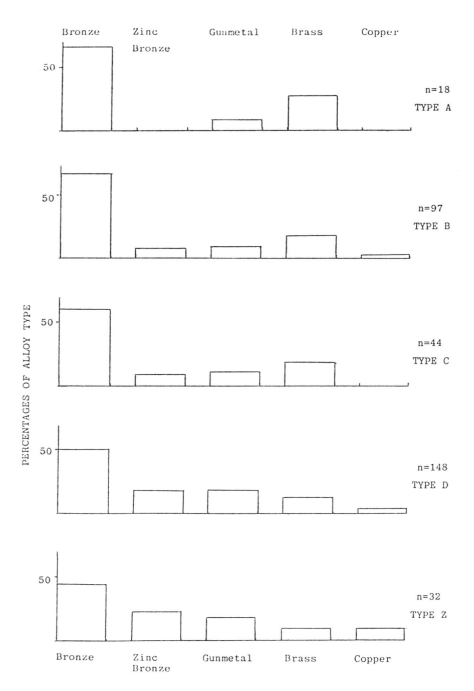

Figure. 1. Percentages of alloy types used in each English brooch type.

forms, type A, and that of the latest forms, type Z (Fig. 1). Bronze and brass are used in the early forms, but zinc bronzes and gunmetals are much more frequently found in the latest forms. These are archaeological types for which chronological overlap in extremely unlikely. The other forms show more subtle differences, although in general terms the alloy patterns could be seen to be intermediary between those of the earliest and latest brooch forms.The overall results conform to the picture expected if bronze and brass alloys were available for re-cycling. The increasing frequency in occurence of zinc-rich, low-tin bronzes demonstrates the continuing dominance of bronze in the alloy supply and the effects of recycling.

There are a number of significant differences between the patterns of alloy used in individual subtypes, particularly between brooches of the large, late type D forms, which have very distinctive typological features. This suggests that these subtypes could have been produced by specific workshops or, more generally, from environments with distinctive supplies.

The weight and size of individual brooches increases throughout the period and the practise of wearing more than one brooch does not significantly decline. So, it seems clear that the copper alloy industry in England did not suffer from a lack of basic materials. There is no clear method of determining the amount of metal in circulation and its speed of turnover, although models have been proposed (Caple, 1986). Hence it is difficult to estimate what proportion of fresh metal might have been available. Bronze is likely to have been produced throughout this time, since it occurs persistently through each artefact type, but brass does not seem to have been made in any quantity, even in the latest period of production.

It should be mentioned that lead was found to be present in all cruciform brooches, normally between 2 and 6%. There appears to be little correlation between lead content and chemical or archaeological types, with the exception that bronzes with high lead levels (more than 6%) are found only occasionally, in types B, C and D.

Contemporary material abroad

Cruciform brooches are one of the artefact forms used widely to comment on connections between England and other regions on the North Sea Coast and to give dates to these theoretical structures. The earliest English cruciform brooch styles are very close to the north German and Danish styles. However, by the mid-fifth century, it is a simple matter to distinguish visually between brooches found in the two areas and further constructional differences between products of these areas are now being discovered. In the late fifth and early sixth centuries, Scandinavian cultural contacts are thought to have become important for Anglian

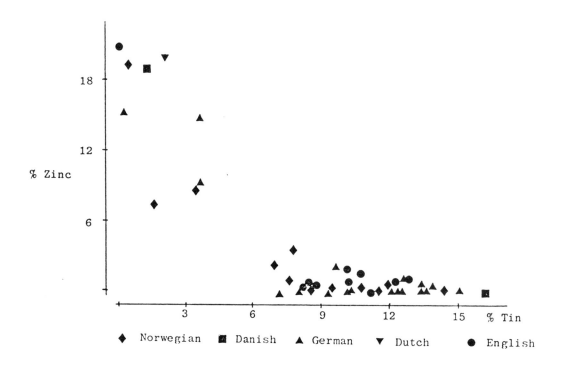

Figure 2. Early cruciform brooch compositions

England (Hines, 1984). The comparison of chemical data from continental and Scandinavian cruciform brooch styles may add further information to illuminate the nature of these contacts.

Late fourth century and early fifth century examples from foreign sites are mainly either bronze or brass, as in England (Fig. 2). Other copper alloy types become frequent later in all areas (Fig. 3). Only the latest Norwegian brooches appear to differ from the trend generally observable in the late fifth/early sixth century, that of a continuum between brass-like and bronze-like compositions.

Looking at bronzes in more detail (Fig. 4), by the latest period for which comparative material is available, ie. the early to mid sixth century, two different groups of bronze compositions clearly emerge. One group is that with rather high tin and low zinc content which was common in the earliest phases and the other is rather low in tin and higher in zinc. The exceptions to this rule are mainly English or Dutch.

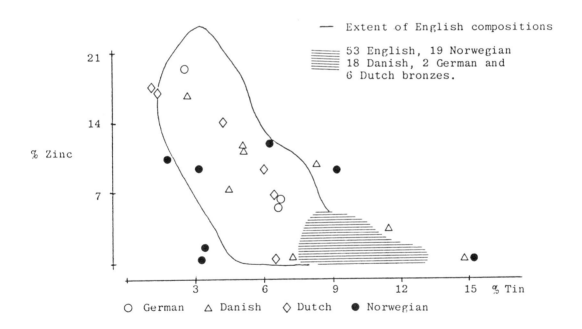

Figure 3. Later Cruciform brooch compositions.

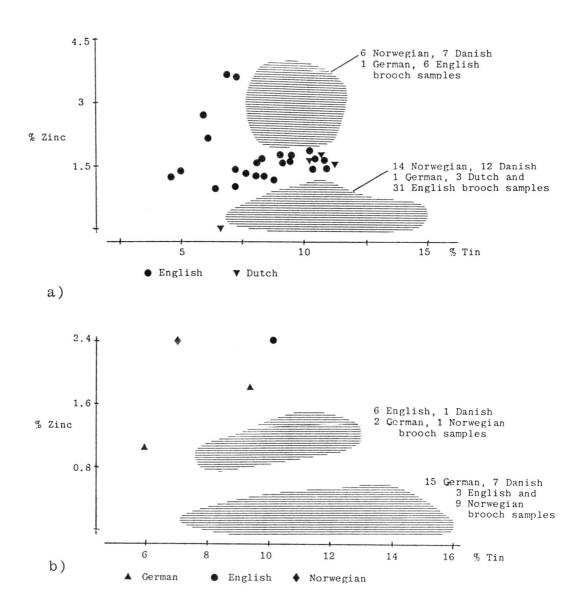

Figure 4. Bronze alloys used for cruciform brooches, a) early brooches, b) later brooches.

Conclusions

There are a number of important conclusions to be drawn. The patterns of alloy production in the metalworking environment which produced the earliest cruciform brooches are similar throughout all regions of production. The copper alloys of northern Germany and England (which are through to have the closest archaeological links at this time) are chemically as similar to each other as they are to material from other areas. In the late fifth century, each of the regions that produced cruciform brooches suffered what we would see as a decline in the standard of metal used to cast brooches. In the latest phases of production, the types of alloy and their proportions in the overall database are similar throughout the distribution area, with the possible exception of Norway. When considered in detail, the bronzes are divided into two groups, based on zinc content, although English and perhaps Dutch bronze compositions stand out by virtue of a continuum of zinc concentrations.

166

Some possible explanations for these patterns

Reasons for the move towards impure alloy types could include the partial or complete severance of the normal metal supply routes. In discussions concerning Anglian England archaeologists have attached considerable importance to typological links but there are very few items which were imported from, rather than 'influenced' by, other cultural areas. The emphasis is on development of shared typological themes, not on traded artefacts (Hines, 1984). Of course, lack of evidence for trade in artefacts does not necessarily preclude the existence of an ingot or ore trading system at this time although, again, there is sparse archaeological evidence for such a trade.

During the fifth and sixth centuries AD, supplies of gold and silver were relatively plentiful in the south-east of England and they were used for casting jewellery. In the rest of England, precious metals were used sparingly, mainly as thin coatings and there were several other types of treatment used at the time to colour metal surfaces (Mortimer et al., 1986). Breaks in communication and trade have also been noted in Scandianvia at this time. In these circumstances, recycling would be an important economic factor. It has been suggested that remelting 'old-fashioned' brooches is the reason for the relatively rarity of great square-headed brooches in Kent (Leigh, 1980). Perhaps the same is true for the early cruciform brooches of which there remain few.

Were the Anglo-Saxons metallurgically inept or technically adaptive?

It could be suggested that the increasingly impure alloys were the result of a decrease in technical knowledge. Methods of casting do not appear to change in England during this period, but the decoration becomes much more complex and detailed. Although the backs of the late, decorated brooches look clumsy, high standards of guilding, tinning and jewel setting are evident on the front. These surface treatments may be partly related to an attempt to disguise the slightly more coppery colour of the base metal. It seems that, even if the English copper alloy metalworkers were largely powerless over composition, they produced fine artefacts nonetheless and could ameliorate some of the more undesirable effects.

In Norway and Sweden, casting techniques do change significantly in the late fifth century (Germany and Denmark appear to drop out of production, by this time). In the earliest periods of cruciform brooch manufacture, the Scandinavian manufacturing style is similar to those elsewhere. Later, it seems likely that a lack of metal and the technical requirements of artefact styles with high relief forced the development of the thin casting technique in Norway. However, the stylistic development of the Norwegian brooches was by no means as complex as that in England and the production becomes repetitive. Perhaps in this case, development of technical skills and development of artistic imagination are incompatible and metalworkers from either side of the North Sea went in opposite directions.

Acknowledgements

This research was carried out at the Research Laboratory for Archaeology, Oxford. The analyses were performed by Dr P. Northover & Dr C. Salter, using the facilities at the Department of Materials, Oxford.

References

Blades, N., Bayley, J. & Walsh, J.N. (1990). The analysis of ancient copper alloys using ICPS. (this volume).

Brownsword, R., Ciuffini, T. & Carey, R. (1986). Metallurgical analyses of Anglo-Saxon jewellery from the Avon valley. *West Midlands Archaeology*, XXIX, pp101-112.

Caple, C. (1986). An Analytical Appraisal of Copper Alloy Pin Production 400-1600 AD. Unpublished PhD thesis, University of Bradford.

Gilmore, G. (1987). Metal analyses of the Northumbrian stycas. In: Coinage in Ninth Century Northumbria, ed. M. Metcalf, pp 159-172. Tenth Oxford Symposium on Coinage and Monetary History, *British Archaeological Reports* no. 180.

Hines, J. (1984). The Scandinavian character of Anglian England in the pre-Viking period. *British Archaeological Reports* 124.

Leigh, D. (1980). The Square-Headed Brooches of Sixth Century Kent. Unpublished PhD thesis, University College, Cardiff.

Mortimer, C., Pollard, A.M. & Scull, C. (1986). X-ray fluorescence analysis of some Anglo-Saxon copper alloy finds from Watchfield, Oxfordshire. *Journal of Historical Metallurgy* 20, 1, pp 36-42.

Mortimer, C. (forthcoming). Some Aspects of Early Medieval Copper Alloy Technology, as Illustrated by a Study of the Anglian Cruciform Brooch. D.Phil., Oxford.

Oddy, W.A. (1983). Bronze alloys in Dark Age Europe. In: *The Sutton Hoo Ship Burial*, Volume 3, part II, pp.945-961. British Museums Publications.

SOME EXPERIMENTS CONCERNING THE SMELTING OF ARSENICAL COPPER

A.M. Pollard, R.G. Thomas and P.A. Williams

School of Chemistry and Applied Chemistry, University of Wales College of Cardiff, P.O. Box 912, Cardiff CF1 3TB

Abstract

This paper describes a series of smelting experiments to produce arsenical copper. The results suggest that early practice at the onset of extractive metallurgy could have involved the exploitation of secondary copper minerals with temperatures obtainable in simple bowl of bonfire type furnaces. At these temperatures the level of arsenic and other trace elements obtained in the smelt can be shown to be kinetically controlled by time, temperature and the particular reducing conditions of the furnace. The results suggest that the arsenic content of early bronze age metalwork is an entirely fortuitous occurrence resulting from the accidental inclusion of secondary arsenates in the smelt.

Introduction

Although several workers have considered the origins of the arsenic content in Early Bronze Age artefacts (Allan, 1970; Craddock, 1980; Tylecote et al., 1977) the bulk of the work until now has been based upon the supposed exploitation of arsenic-bearing minerals of the primary zone of base metal ore bodies. However, experiments have failed to reproduce all of the alloys known in antiquity (Pollard et al., 1990). By the way of comparison, relatively little consideration has been devoted to the study of secondary copper(II) arsenates as an ore source. In view of the existence of suites of secondary arsenate minerals such as cornwallite, cornubite and especially olivenite with other secondary copper(II) minerals such as malachite in the British Isles (Dewey & Eastwood, 1925; Greg & Lettsom, 1858; Heddle, 1923; Magalhaes et al., 1988; Symes and Embrey, 1987) it was decided to examine the possibility that the production of 'arsenical coppers' was the result of smelting appropriate mixture of secondary copper(II) minerals.

Experimental

Synthetic olivenite was prepared using the methods described by Guillemin (1956), dried in vacuo over silica gel, and its purity checked by X-ray powder diffraction using $CuK\alpha$ radiation. The infrared spectrum of the authentic mineral was recorded in KBr pellets using a Perkin Elmer 783 spectrophotometer. Preparations of the synthetic compound were then checked against this reference spectrum.

Synthetic malachite was prepared by the addition of $CuCl_2.2H_2O$ to a solution containing excess sodium carbonate. The product was isolated after the recrystallisation of the transiently amorphous blue phase had occurred. The precipitate was then filtered using a glass sinter, washed with water, then acetone, and dried in vacuo over silica gel.

The powders thus prepared were ground and mixed together in various proportions to a total weight of 1g, placed in a vitresol crucible, covered with charcoal and heated in an electric furnace at temperatures of between 600 and 1000°C. Preliminary experiments established that the layer of charcoal was sufficient to maintain reducing conditions and to prevent serious loss of arsenic during reduction.

Experiments involving natural olivenite and malachite were carried out using a mineral specimen collected from Wheal Edward, Cornwall, supplied by the Geology Department of the National Museum of Wales, and using malachite collected from the same region. The minerals were crushed and hand picked prior to mixing and smelting in the manner described (*vide supra*).

The metal obtained was isolated from the charcoal by flotation, dried in an oven at 105°C and remelted in the furnace at 1000°C. The samples were then mounted, polished and prepared for metallographic examination by etching with alcoholic iron(III) chloride solution. The copper content of the metal was measured using the iodometric titration method of Vogel (1958). Arsenic concentrations were measured either by the method Liu & Chen (1982) or, when higher concentrations of arsenate were encountered, by the bismuth nitrate method (Williams, 1979). The concentration of antimony and silver in the natural ores, and in the alloys produced by reduction, were measured by atomic absorption spectroscopy using a Varian AA-275 spectrophotometer.

Results and Discussion

At temperatures greater than 950°C the smelt charge was rapidly reduced to its constituent metals. The rate of absorption of gaseous arsenic into the copper produced was such that the production of a liquid phase, at the particular temperature chosen, was achieved within 10 minutes of placing the crucible in the furnace. Once this had occurred, the rate of absorption of arsenic into the now liquid alloy proceeded at an increased rate. The final arsenic recovery in the smelted alloy was never less than 80% by weight of the furnace charge, and is comparable with that obtained by previous workers (Tylecote, 1980). Metallographic examination of the alloys revealed that they contained considerable amounts of the eutectic composition (20% As by weight). Unless the furnace charge was kept below 4% arsenic then the alloy composition consistently lies above the range of the Early Bronze Age Type A metal as assigned by Northover (1980), (see Table 1).

Table 1. Measured Arsenic Concentration in Smelt Charges and Alloys Produced at 1000°C.

%age Arsenic in Ore Charge	No. of Replicates	%age Arsenic in Metal
2	4	0.05-0.5
4	4	3.2-3.7
6	4	5.8-5.9
8	4	7.8-7.9
10	2	9.7-9.8
15	4	14.7-14.9
20	3	19.6-19.7
33	4	26.4-28.2

Reduction of the temperature at which the smelting is carried out has a dramatic effect on the composition of the alloy produced. From Figure 1 it can be seen that, provided the temperature of the furnace is such that the alloy never becomes molten, the final arsenic concentration is given by the limit of solid substitution and is thus independent of charge composition. In practice, the limit achieved is somewhat less than this due to the rate of loss of arsenic from the crucible. In our experiments we were unable to achieve a final concentration in the alloy of greater than 4% (Table 2). This result is a natural consequence of phase relationships.

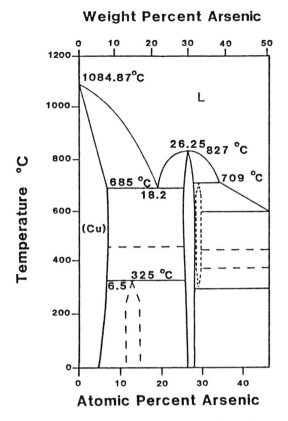

Weight Percent Arsenic

Figure 1. Binary Phase diagram for the copper-arsenic system (after Subramanian & Laughlin, 1988).

Table 2. Measured Arsenic Concentration for Smelt Charges and of Alloys Produced at Below 1000°C.

%age Arsenic in Ore Charge	No. of Replicates	%age Arsenic in Metal	Temperature (°C)
2	3	n.d.-0.1	750
2	4	n.d.-0.3	800
2	3	n.d.-0.4	900
5	3	0.07-0.9	800
5	4	0.05-0.9	850
5	4	0.05-1.1	900
10	4	0.5-2.0	800
10	3	0.2-0.8	900
20	4	0.5-3.1	800
20	4	0.5-2.2	900
33	4	0.2-1.5	700
33	4	0.2-2.1	750
33	4	0.5-2.8	800
33	3	0.7-2.9	850
33	4	0.5-3.5	900

The results obtained by the smelting of natural malachite and olivenite are exactly the same as for the smelting of the synthetic compounds. However, their use allowed us to examine the effect of smelting on other trace elements indicative of Type A metal, notably antimony, contained within the ore charge. Microscopic examination of the mineral specimens did not reveal the presence of any discrete antimony minerals, but the malachite and olivenite were found to contain 0.03 and 0.01% Sb, respectively, by atomic absorption spectroscopy. Under the conditions obtaining in a charcoal-filled crucible, antimony is said to be reduced in the same manner as arsenic (Tylecote et al., 1977).

Measurements of the antimony content from furnace charges containing 6% olivenite were made after reduction times of between 2 and 15 mins at 850°C (Table 3). From the results of these experiments it is difficult to draw any conclusions as to the possible relationship between the antimony:arsenic ratios in the ore and in the metal, beyond the observation that both elements diffuse into the impure copper at different rates and that the final composition of the alloy with respect to these elements is dependent upon time, temperature and the particular reducing conditions of the furnace.

Table 3. Percentage Antimony in Recovered Metal, from Smelt Charges of 95% Malachite, 5% Olivenite at 850°C and 1000°C, after Various Smelting Times.

Time (mins)	%age Sb (850°C)	%age Sb (1000°C)
3	n.d.	n.d.
5	n.d.	0.002
7	0.005	n.d.
8	n.d.	0.010
10	0.010	0.090
12	0.011	0.013
14	0.012	0.015
16	0.010	0.014

These experiments show that metal of the same compositional Type A (with respect to As) as that described by Northover (1980) for the early period of the Early Bronze Age can be produced by the reduction of secondary copper(II) minerals containing arsenate in any proportion up to 33 mol% (a value which corresponds to pure olivenite), provided that the temperature of the furnace is kept below that at which the smelted alloy becomes molten. The temperatures at which this might have been achieved correspond to those attainable in bonfire or shallow pit type furnaces as described by Craddock & Meeks (1987).

172

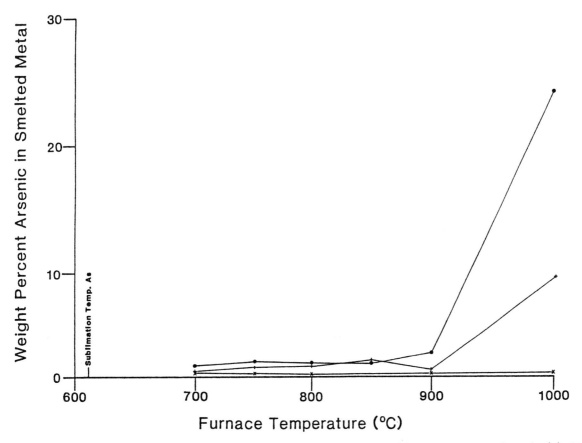

Figure 2. Variation of average arsenic content of alloy produced with temperature of smelt; (x): 2% As in charge; (+): 10% As in charge; (o): 33% As in charge.

Acknowledgements

The Science-based Archaeology Committee of SERC are thanked for their provision of a studentship to R.G.T.

References

Allan, J.C. (1970). Considerations on the antiquity of mining in the Iberian Peninsula. In: *Royal Anthropological Society Occasional Paper 27*, pp. 23-27, London.

Craddock, P.T. (1980). Deliberate alloying in the Atlantic Bronze Age. In: *The Origins of Metallurgy in Atlantic Europe*, Ed. M. Ryan, pp. 369-385, Stationary Office, Dublin.

Craddock, P.T. & Meeks, N.D. (1987). Iron in ancient copper. *Archaeometry 29*, 187-204.

Dewey, H. & Eastwood, T. (1925). Copper resources of the Midlands, Wales, Lake District and Isle of Man. *Special Report of the Geological Survey of Great Britain*, Vol. 30, HMSO, London.

Embrey, P.G. & Symes, R.F. (1987). *Minerals of Devon and Cornwall*, British Museum (Natural History), London.

Greg, R.P. & Lettsom, W.G. (1858). *A Manual of the Mineralogy of Great Britain and Ireland*. Van Voorst, London.

Guillemin, C. (1956). Contribution a la mineralogie des arsenates, phosphates et vanadates de cuivre. Part 1. Arsenates de cuivre. *Bull. Soc. franç. Minér. Crist.*, 79, 7-95.

Heddle, H.F. (1923). *Mineralogy of Scotland*, Henderson and Sons, St Andrews.

Liu, F. & Chen, D. (1982). *Anal. Abs.*, 42, 28130.

Northover, J.P. (1980). The analysis of Welsh Bronze Age metalwork. In Appendix to Savory, H.N. (1980). *Guide Catalogue of the Bronze Age Collections*, pp. 229-243. National Museum of Wales, Cardiff.

Magalhães, M.C.F., Pedrosa de Jesus, J.D. & Williams, P.A. (1988). The chemistry of formation of some secondary arsenates of Cu(II), Zn(II) and Pb(II). *Mineralogical Magazine* 52, 679-690.

Pollard, A.M., Thomas, R.G. & Williams, P.A. (1990). Experimental smelting of arsenical copper ores: implications for Early Bronze Age copper production, *Proceedings of the Early Mining Workshop, Maentwrog*. Ed. P. Crew & S. Crew, Plas Tan y Bwlch (Occasional Paper No.1), pp.72-74.

Subramanian, P.R. & Laughlin, D.E. (1988). The As-Cu (arsenic-copper) system. *Bulletin of Alloy Phase Diagrams* 9, 605-618.

Tylecote, R.F., Ghaznavi, H.A. & Boydell, P.J. (1977). Partitioning of trace elements during smelting of copper. *J. Archaeol. Sci.* 4, 305-333.

Tylecote, R.F. (1980). Summary of results of experimental work on early copper smelting. *In Aspects of Early Metallurgy Occasional Paper*, No. 17, Ed. W.A. Oddy, pp. 5-12, London.

Vogel, A.I. (1958). *A Textbook of Quantitative Inorganic Analysis*, Second Edition, Longmans, London.

Williams, W.J. (1979). *Handbook of Anion Determination*, Butterworths, London.

THE VARIATION OF INCLUSION MORPHOLOGY AND COMPOSITION IN FERROUS ARTEFACTS

David E. Starley

Department of Archaeological Sciences, University of Bradford, Bradford, West Yorkshire, BD7 1DP, England

SUMMARY: Two studies examine the relationship between iron alloys, and their residual slag inclusions, in high quality Dark Age and Medieval artefacts. An attempt is made to relate this to the changing technology of the period.

Introduction

Given the value of iron as a material and the probable importance of the iron production and working industries, the scientific examination of iron from archaeological sites has been remarkably limited. This has been largely due to problems inherent to the material: The heterogeneous nature of early iron, its numerous production stages, wide range of ore sources and corroded state of many artefacts, make it an unattractive subject for archaeometric analysis. Recent studies have, however, considered the potential of the analysis of slag inclusions within the metal.

Iron produced by the bloomery smelting process invariably contains large numbers of slag inclusions, retained from both the original bloom production and from subsequent smithing operations. The compositional relationship between ore, metal, bulk slag and inclusions has been investigated for archaeological, ethnographic and experimental material (Todd, 1984) and attempts have been made to provenance the ore used to produce Iron Age currency bars (Hedges and Salter, 1979). This study explores the relationship between the composition of the metal and the shape and composition of the slag inclusions in some of the highest quality products of the Dark Age and Medieval Iron Industry.

Case 1: Hamwih knife blades

The great skill of Dark Age metalsmiths has been attested by the metallographic examination of iron artefacts such as knives and other edged tools and weapons. These high quality items often benefit from composite structures, whereby a hardened steel edge was combined with a more resilient back of low carbon iron, sometimes containing a significant phosphorus content. This implies that the Dark Age smith had access to a range of ferrous alloys.

Three possible explanations were proposed:

1. Low and high carbon regions of inhomeneous blooms were identified and worked separately before being recombined in the artefact. In this case similar inclusion populations would be expected in all materials.

2. The smiths obtained low carbon iron and subsequently carburised this to produce steel. If this were the case then phosphoric iron, which is notoriously difficult to carburise should be seen as exotic material.

3. Prepared steel, produced either from a different ore source or by a variation in smelting technique, was manufactured separately or imported. Inclusions within the steel would therefore be expected to show not only lower oxide concentrations (due to the more reducing conditions of steel production) but perhaps also different impurities carried over from an exotic ore source.

Method

A systematic three way study of the prepared knife blade sections from Hamwih (Saxon Southampton) was undertaken as follows:

1. Metal Composition

Bulk metal analysis was not carried out, however, metallographic examination of the etched samples allowed the following regions within the knives to be identified by their characteristic structures:

a) Low carbon (ferritic) iron
b) Phosphoric iron
c) Steel (generally on the cutting edge)

2. Inclusion morphology

An optical microscope connected to a visual display unit (V.D.U.) allowed measurement of the size, and calculation of the aspect ratios (defined as maximum length/maximum width along perpendicular axis) of 563 inclusions, which lay along a series of random traverses. These were then classified according to the following scheme:

Aspect ratio

a) Spheroidal = 1
b) Sub-round   = greater than 1, less than 2
c) Elongated   = 2 or greater, less than 5
d) Stringer   = 5 or greater
e) Occasional inclusions of irregular form were disregarded from further analysis.

3. Inclusion composition

Analysis for major and minor elements (above 0.5 weight %) was undertaken on a further 275 slag inclusions, from the same samples, using a scanning electron microscope with energy dispersive X-ray analyser operating at a beam current of 18.5 kV and a spectrum collection time of 50 seconds.

The compositional data were used to calculate mineral phases within the inclusions using a simplification of Kresten and Serning's normative calculation (Kresten & Serning 1983), suggested by McDonnell (1986). In this procedure it is assumed that fayalite crystalises out first; its "iron oxide" component (in fact comprising the oxides of iron, manganese and magnesium), combines with silica in the ratio 70:30 until one component becomes exhausted. If excess "iron oxide" is present this is assumed to form an iron oxide phase. If excess silica remains after the formation of fayalite it is assumed to combine with the oxides of sodium, potassium, phosphorus and calcium to form a glassy component.

Using the results of these calculations ternary diagrams were plotted to show the relative proportions of glassy, iron oxide and iron silicate phases, within inclusions, for each of the metal types identified metallographically. For statistical purposes the following classification of inclusion composition type was made;

Contains

a) Iron Oxide.   >90% Fe, Mn & Mg oxides.
b) Iron Silicate >50% iron silicate phase.
c) Glassy   >50% glassy phase.
d) Unclassified (Not included in further analysis)

The resultant grouped data were examined using the chi-square test for association.

Conclusions

Surprisingly, although steel showed a lower overall proportion of slag, the size and shape of the inclusions were not significantly different from those in the ferritic or phosphoric irons. Furthermore, no association between slag inclusion morphology & inclusion composition was evident within the regions. By contrast a limited number of inclusions examined on weld lines between the regions of different metal type, had high aspect ratios and were particularly rich in silicates, poor in glassy phase.

The value of morphological studies could not be demonstrated in this first test case. This may have been due to difficulties in adequately classifying the true shape of the inclusions from a random two dimensional section. However, when comparing the metal types identified metallographically with inclusion composition, strong links were found. Inclusions within steel matrices were strongly biased towards the glassy phase and were notably low in manganese. Thus separate origins for the steel and ferritic iron are suggested, involving at least a variation in smelting conditions and possibly different ore sources.

Case 2: Medieval armour

On-going work applies a more sophisticated form of the analytical and metallographic procedure, outlined above, to the examination Medieval plate armour. Previous work by Campbell (1935), Brewer (1981) and Williams (1987) has demonstrated the use of a range of ferrous alloys, and particularly the availability of high quality steel. The often very precise dating and provenancing of armour also makes it well suited for a pilot study of ferrous technology. In this project armour is being investigated as part of a major study of iron and steel production in fourteenth and fifteenth century Europe, where great expansion in the production of iron coincides with the introduction of the blast furnace.

The main advance in analytical approach lies in the use of wavelength dispersive electron probe microanalysis (E.P.M.A.) of the metal matrix. This offers two advantages over a purely metallographic interpretation; firstly, a more objective distinction can be made between ferritic and phosphoric iron and secondly, it enables the study of the partitioning of elements between slag inclusions and their metal matrix. Gordon (1984) has suggested the use of phosphorus partition coefficients derived in this way, to distinguish between bloomery iron and that produced as cast iron in a blast furnace and subsequently decarburised in the finery. However, because of the very low phosphorus content of most steel artefacts, a more comprehensive range of elements is being investigated.

Difficulties have been experienced in obtaining material of undeniably charcoal blast furnace/finery origin. However, preliminary results have indicated very low impurity concentrations in the iron and steel (typically totalling less than 0.2%). This would suggest that armourers continued to use material from the traditional bloomery furnace, with its' less highly reducing operating conditions, into the period in which the blast furnace is known to have been in production.

Acknowledgments

The contents of this paper include work carried out under the supervision of Dr. Gerry McDonnell and Dr. Barbara Ottaway with co-operation from English Heritage (Hamwih knives) and The Royal Armouries (Medieval armour). Current research is funded by The Science and Engineering Research Council.

References

Brewer, C.W. (1981) Metallographic examination of Medieval and post Medieval armour. *Journal of the Historical Metallurgy Society* 15(1), 1-8.

Campbell, W. (1935). On the structure of armour, ancient and modern. *Metals and Alloys* 6, 267-272.

Gordon, R.B. (1984). The quality of wrought iron evaluated by microprobe analysis" In: *Microbeam Analysis*, ed. A.D. Romig, Jr. & J.I. Goldstein, San Fransisco Press.

Hedges, R.E.M. & Salter, C.J. (1979). Source determination of iron currency bars through the analysis of slag inclusions. *Archaeometry* 21, 165-175.

Kresten, P. & Serning, I. (1983). *The calculation of normative constituents from the chemical analysis of ancient slags.* Jernkontorets Forskning H25, Stockholm.

McDonnell, J.G. (1986) The classification of early ironworking slags. Unpublished PhD Thesis, Aston University.

Todd,J.A. (1984) The relationship between ore, slag and metal composition in pre-industrial iron. In: *Microbeam Analysis*, ed. A.D. Romig, Jr. & J.I. Goldstein, San Fransisco Press.

Tylecote, R.F. (1985) The early history of the iron blast furnace in Europe; a case of East-West contact. In: *Medieval Iron in Society*. Jernkontorets Forskning H34, pp.158-173. Stockholm.

Williams, A.R. (1987) The manufacture of armour in Germany: The metallurgical evidence from specimens in German Museums, 14th-17th Centurles. *Waffen und Kostümkunde*. 2, 90-106.

NEW EVIDENCE FOR EARLY PREHISTORIC MINING IN WALES - PROBLEMS AND POTENTIALS

S.A. Timberlake

Early Mines Research Group, 12 York Street, Cambridge CB1 2PY, UK.

Introduction

The validity of Chalcolithic-Bronze Age copper mining within continental Europe is now generally accepted and numerous sites have been thoroughly investigated over the last few decades eg. Rudna Glava and Chinflon (Jovanović, 1980; Rothenberg & Blanco Freijeiro, 1980); Aibunar (Černych, 1978); and the Mitterberg (Gstrein & Lippert, 1987). Characteristic artefacts associated with such sites include stone hammer mining tools often associated with the evidence of fire-setting within the mines themselves.

However, the question of provenance for supplies of copper required during the British Bronze Age has never been satisfactorily or conclusively resolved, particularly for the earlier part of the period, and thus recent excavations of possible mining sites in Wales and Ireland herald a new and interesting dimension to this investigation.

Radiocarbon sampling of burnt wood associated with stone hammers and mining debris from recently excavated Welsh sites (Cwmystwyth, Parys Mountain, Nantyreira, and the Great Orme) have now all given dates within the Early-Middle Bronze Age period c.1900-1150 Cal BC.

The purpose of this paper is to briefly review the *surface* excavation and survey work carried out to date, and to discuss the difficulties of interpretation and dating of sites poor in typological dateable artefacts and with particular problems of preservation. Lastly to assess the potential for further excavation and a programme of scientific research - to include palaeoenvironmental work and a metal provenancing study based on the analysis of ores from identified or possible early mines within the region.

A Review of the Excavations

The earliest references to discoveries of primitive stone tools and charcoal during the re-opening of old mine workings in Wales appear within contemporary mining accounts and topographical literature of the eighteenth century (Lewis Morris, 1744; Sykes, 1796). By the middle to the latter years of the nineteenth century such accounts became increasingly frequent, as mining and prospection activity intensified, and a professional interest in antiquarian matters was seen to be both fashionable and pragmatic (Stanley, 1850 and 1873; Hunt, 1884), and sometimes even a route to promoting faded mining ventures (Spargo, 1870).

However, there was little or no archaeological interest shown in any of these sites until at least the 1930's when Oliver Davies, following from his major study of Roman mining in Europe (Davies, 1935), re-examined a number of early mines in Wales as part of the study of a British Association for the Advancement of Science Committee which was set up to look into this matter (Davies, 1937 and 1938). Foremost amongst those sites he looked at were Cwmystwyth, Parys Mountain, Nantyreira and the Great Orme's Head Copper Mine (Fig. 1). At all of the localities he carried out small trial excavations on areas of old mine spoil, recovering stone hammers, querns or hand mortar stones, and occasionally charcoal. In conclusion, he considered that the remains indicated mining of the Roman or 'Old Celtic' period (Davies 1939 and 1947).

The recent phase of examination of early Welsh mine sites began during the early 1980's with the work of Duncan James, who spent several years investigating a complex series of underground galleries on the Great Orme near Llandudno. Large masses of

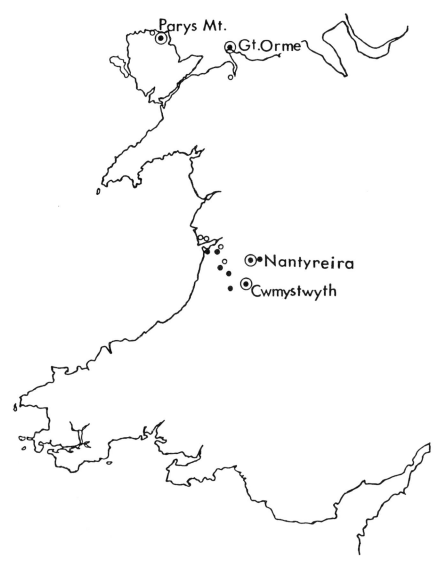

Figure 1. Map of Wales showing some of the mine sites with field evidence (solid circles) or documentary record (open circles) for finds with stone hammers.

ancient pack fill (waste rock or 'deads' piled underground) containing bone and stone tools and charcoal, heavily cemented over by calcite, were discovered driven through by later mineworkings. A radiocarbon date of 2940 ± 80 BP (HAR 4845) was obtained from charcoal within a small area of pack at a depth of 27 metres from surface (James, 1988). Accepting the limitations of a single date this did nevertheless suggest a Bronze Age working for the mine.

Independently of this, surveys of both the author and John Pickin (on different occasions) within the mid-Wales mining area,

had located a number of the sites for which there were documentary records for finds of stone hammers. At quite a few of the mines evidence for the hammers and primitive workings were still to be found. Of the sites so far identified with observed or recorded evidence for stone tools, 18 are located within the mid-Wales area - of these 14 have produced finds within recent years (Pickin, 1988). Where sites have been examined in greater detail, there appears to be a general association between shallow opencast workings, copper mineralization, and broken pebble tools. Nevertheless, lead and zinc mineralization is almost always present, if not

180

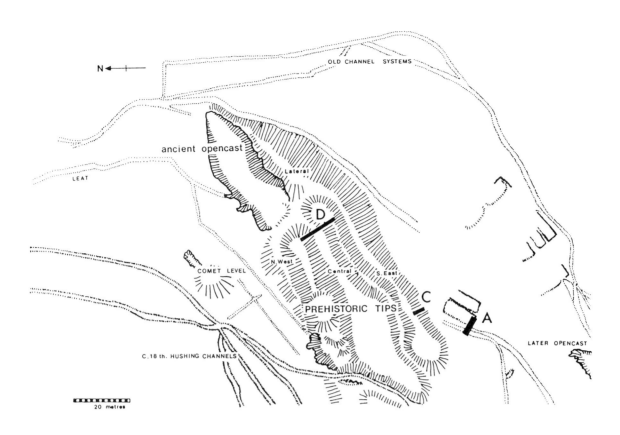

Figure 2.   Environs of old opencast on Copa Hill, Cwmystwyth.

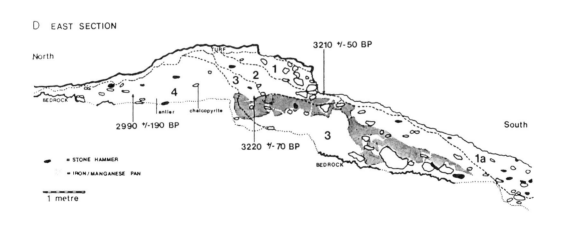

Figure 3.   Section through prehistoric tip (Site D) cut in 1986. Radiocarbon dates indicated.
Copa Hill, Cwmystwyth.

181

Table 1.   Radiocarbon dates for early Welsh Mines

Sample		BP Date	Possible calibrated age range(s) from Pearson and Stuiver, 1986 (unless otherwise stated) Cal BC	
			1° error 68% confidence	2° error 95% confidence
Cwymstwyth Q-3076 Q-3077 Q-3078	roundwood, ash, hazel, oak, charcoal	3220 ± 70 BP 2990 ± 190 BP 3210 ± 50 BP	1540 to 1425 1440 to 960 1525 to 1430	1685 to 1370 1685 to 810 1590 to 1410
Parys Mountain BM-2584	branchwood of some age: oak, charcoal	3550 ± 50 BP	1960 to 1875 or 1835 to 1820 or 1795 to 1785	2040 to 1750
BM-2585 BM-2586		3490 ± 50 BP 3500 ± 50 BP	1885 to 1745 1895 to 1745	1950 to 1690 1970 to 1695
Nantyreira BM-2585	branchwood of some age: oak and birch, charcoal	3390 ± 80 BP	1865 to 1845 or 1770 to 1610	1900 to 1520
BM-2583		3500 ± 50 BP	1855 to 1845 or 1765 to 1675	1880 to 1610
Great Orme HAR-4845	charcoal (unidentified)	2490 ± 80 BP		1470 to 975* *(at 4° error, after Clarke, 1975)
CAR-1184[1]	branchwood: alder (charcoal)	3370 ± 80 BP		
BM-2641	short-lived oak (charcoal)	3000 ± 50 BP	1375 to 1345 or 1320 to 1210 or 1180 to 1165	1410 to 1070
BM-2645[2]	bone	3290 ± 60 BP	1675 to 1515	1740 to 1440

[1] & [2] = samples from gallery at surface excavation.

182

predominant, at most of the localities, and at one or two of these there has been no record of copper at all, at least within the historical period.

Much needed impetus to the investigations of the Welsh sites was afforded by the pioneering work of Jackson (1968) and latterly O'Brien (1987) at Mt. Gabriel primitive mines in Co. Cork, SW Ireland. Although the geological setting for these mines (strike-bound copper rich sedimentary red-bed deposits within Devonian sandstones) is quite different from that of the Welsh examples (hydrothermal vein deposits), there are nevertheless many parallels both in the technology of working and the tool assemblages found. Twenty-five primitive drift adits worked by fire-setting and stone hammers have now been located on the eastern slopes of Mt. Gabriel. These have, by and large, escaped disturbance from later mining and prospecting activity in the area, some of them concealed in hollows under several metres of peat bog. One of these adits, Mine No. 3, has been completely excavated and has since produced 14 radiocarbon dates c.1700-1500 Cal.BC from wooden artefacts, charcoal, and peat found in secure stratigraphic contexts within the tips and also in the mine itself.

In Wales excavations carried out by the author on Copa Hill, Cwmystwyth in 1986 (Timberlake, 1987) yielded three C^{14} samples dated to c.1500 Cal. BC. These were obtained from burnt wood (Fraxinus, Quercus and Corylus) associated with the numerous split hammerstones, discarded lead ore, and some antler fragments within a section through sealed mine spoil in one of the ancient tips (Trench D, Fig. 3). The site is fairly unique within the Welsh context in that such a large area of undisturbed early tip still survives (and from which many hundreds of hammerstones have eroded, see Fig. 6), as does a large, shallow, opencast (45 m x 17 m) at the top of the lode (Fig. 2). In purely archaeological terms this must represent one of the most promising sites for investigation within mainland Britain.

Two other O. Davies sites: Parys Mountain near Amlwch in Anglesey and Nantyreira on the eastern slopes of Plynlimon in mid-Wales, were examined by the newly formed Early Mines Research Group during August 1988. At both sites early tips identified by Davies in the 1930's were re-examined, and again burnt wood associated with partly modified and un-modified split hammerstones were recovered (Timberlake, 1988a). A total of three C^{14} samples from Parys Mountain and two C^{14} samples from Nantyreira were obtained. These gave Bronze Age dates also (Table 1), apparently somewhat earlier than those from Cwmystwyth but with a margin of error that could not be considered significant. At both these sites there was some evidence of later spoil tip sealing (and preserving) the early horizons, although the surrounding disturbance from later mining activity mitigates against the likelihood of many other remains being found *in situ*.

Surface excavations on the Great Orme by Gwynned Archaeological Trust during 1988 (Dutton, 1988) and 1989 have revealed an interesting mining landscape of back-filled vein open-cuts working malachite and possibly chalcopyrite within a rotted dolomite limestone. Numerous bone tools, antler points, stone tools and charcoal fragments have been recovered. Coeval Bronze Age C^{14} dates, corresponding well to the dates turning up from the other Welsh sites, have now been obtained from *both* bone and charcoal samples sealed in sediments inside of a small gallery near to the surface (Table 1). Complementing this investigation, exploration of the underground mine on the Orme by Andy Lewis and Dr. David Jenkins has revealed further early workings, and recently some major new discoveries have been made (Lewis, 1988; Jenkins & Lewis (this volume)).

At Cwmystwyth a second season of work carried out during August 1989 and involving excavation within the opencast itself, has turned up evidence which now adds a new dimension to the interpretation of the site. At the front of the opencast, beneath an eroded hushing dam, up to two metres or more thickness of deposits have been found, including several waterlain 'standstill' horizons

Figure 4.   Entrance to fire-set gallery against NW face of opencast on Copa Hill (1989). Note concavity formed by stone hammer pounding marks in shale of roof. Drawing by B. Craddock.

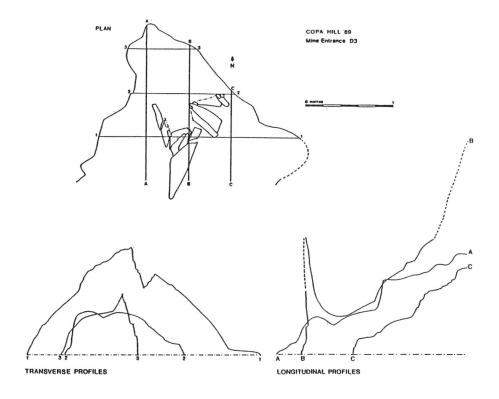

Figure 5.   Plan and profiles of mine entrance (same).

184

of peat and silt overlaying layers of primary mine spoil on the steep-dipping south face of the quarry. Against the opposing north wall, a fire-set gallery or drift complete with evidence of hammerstone pounding marks in the roof, has been uncovered at about 2 metres depth (Fig. 4). The entrance to this was found sealed by *in situ* (naturally eroded?) shale, and capped by spoil from an 18th to 19th century prospection shaft - hence a virgin, undisturbed working beneath. Excavation of this feature (Figure 5), which appears could be the start of an underground mine, will be carried out during a further season's digging here in 1990.

Problems of the excavation and interpretation:

Whilst the scientific study for the evidence for early metal mining has now clearly begun in this country, we are nevertheless faced with the dilemma that we still know very little indeed about it. Apart from C^{14} dates and an association with stone mauls there is little that helps key these sites into the conventional archaeological context. The workings are exceedingly primitive, and are fairly similar. However, despite the rather more intensive investigations of the last year or so, they have failed to produce any recognisable and culturally identifiable artefacts, habitation sites, or processing areas.

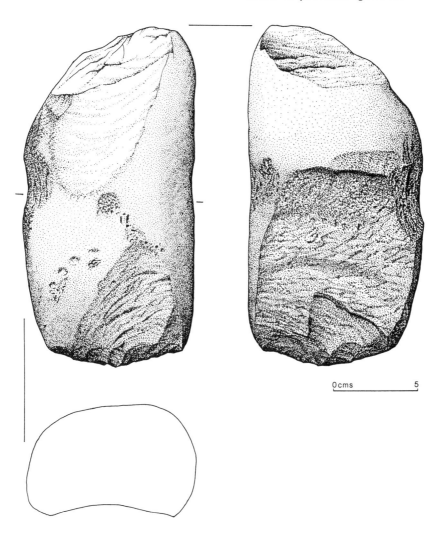

Figure 6.   Stone hammer of local gritstone. NB. notched wasting as a modification for hafting. Surface find on prehistoric tips Copa Hill, Cwmystwyth (1989).

occasionally occur, even within normally acidic mine waste (eg. antler at Cwmystwyth), and this fact should always be borne in mind. The mobility and ubiquity of manganese and iron hydroxides within these tips is similarly liable to conceal the evidence for artefacts, and the weight and hardness of overlying rock might crush poorly preserved material. Waterlogged conditions however, particularly where this occurs inside of the mineworkings, may on occasions lead to exceptional preservation of wooden and organic material - as has been noted at Mount Gabriel (O'Brien, pers. comm.).

The converse of the acid mine tip example is that the preservation of bone within a limestone environment, such as on the Great Orme, tends to be very good. Any comparison therefore of tool assemblages between one mine and another, if preservational factors are not taken into account, could be very misleading.

iii) Habitation, finished processing, or smelting areas have not so far been found in association with any of the sites examined in the British Isles. The reasons for this are not entirely clear. It could just be due to the insufficiency, and so far, small scale nature of our investigations, or perhaps to the problem of peat cover or later mining activity erasing or concealing the evidence altogether. Alternatively, the evidence may not have been there in the first place. Perhaps these activities were never carried out on site?

Many of the localities are fairly isolated, both from immediate sources of water for processing purposes, and also from suitable or existing temporary or permanent habitation sites. Clearly in a number of cases firewood and stone tools would have had to have been brought up to the sites on a fairly regular basis for mining. Examples might be Cwmystwyth and Parys Mountain for instance. One must also consider the possibility that mining could have been seasonal work carried out by a (primarily) farming community during the slack season.

Both the volume and rate of production in prehistory (at such typically low-grade deposits by 19th century standards) was probably small, although we must entertain the possibility that mining may have been carried out intermittently over a longer period of time, and therefore a small scale enterprise *could* have produced a reasonable tonnage of ore. A useful exercise here would be to look at one example of a recent quantification calculation (Timberlake, 1990) based on estimates for ore grade and total tonnage of waste produced on Copa Hill, Cwmystwyth. The figures quoted are not intended to be accurate statistics but rather as a rough outline to help us reassess our assumptions concerning such sites:

Measurements of the area and depth of the prehistoric tip(s), with a correction for material removed by hushing and natural erosion, suggest that at least 3 200 cubic metres of waste were removed in antiquity. Samples taken from the tip had an average density of 1.5 gm/cc and therefore the tonnage represented could be of the order of 4 800 metric tonnes at least. Assuming, conservatively speaking, that the vein carried only small pockets of chalcopyrite (at 20-30% copper) and that good ore consisted of no more than 1% of the total volume extracted, then at a minimum of 30-40% recovery rate after hand-picking, crushing, and smelting, we might be looking at a final production of about 2-4 tons of copper metal. Whilst it is possible that this figure is an underestimate of the true amount, it does nevertheless illustrate the need for caution in interpreting from an historic standpoint what was or was not a viable ore deposit in prehistory.

The significance of all this to the discussion of why we are not finding evidence for smelting and habitation on site should not be lost. Community mining may well have produced little more than one or two baskets of hand-picked cobbed ore per day. An amount that might easily be transported down to a processing/smelting site associated with a permanent or temporary habitation, as suggested at Aibunar (Černych, 1978). Whilst some Bronze Age mining such as at Mitterberg in Austria (Pittioni, 1951) and Chinflon (Rothenberg & Blanco Freijeiro, 1980) in south west Spain have produced evidence for smelting either on or close to the mining site, it would seem prudent to take cognisance of the above observations before we either presume all the evidence has been destroyed, or else proffer theories for slag-free smelting techniques.

One of the criteria upon which Briggs (1988) dismisses the evidence for Bronze Age mining at Cwmystwyth is the lack of evidence for Bronze Age settlement in the area. The fact that little fieldwalking has ever been carried out in the vicinity and that a large part of the area is now obscured by thick blanket peat appears not to have been entertained. Indeed, recent examination of the area has located a possible Bronze Age cairn on the hillside immediately opposite the mine. Nevertheless, we must be wary here of the use of any such arguments - the presence or absence of Bronze Age monuments within the vicinity need not *necessarily* support or refute the antiquity of mining sites.

iv) Another purely practical problem faces the investigator of early mining sites: much of the crucial evidence is likely to be preserved underground. At sites such as the Great Orme the underground workings are accessible but nevertheless one here faces the risk of contamination from later sources. At Cwmystwyth the presumed underground workings are almost certainly preserved intact but are very difficult of access. Large amounts of overburden filling these workings will require removal, involving deep excavations, shoring, and the possible use of mechanical diggers.

Because large amounts of sterile waste need to be removed one cannot always adhere to the same procedure and rigours of recording and excavation carried out at more conventional archaeological sites. Inevitably this will require a greater reliance on vertical sectional recording.

v) It is an important part of our study to address the problems and limitations of absolute dating techniques, such as radiocarbon, and to answer some of the criticisms that have been levelled against it - particularly as related to the dating of mining sites.

Whatever criticisms have been raised concerning the interpretation of the dates, the contexts for the C^{14} samples both from Mt. Gabriel mines and the Welsh sites are generally good (O'Brien, 1987, Timberlake, 1987 and 1988a). Similarly, the broad agreement in age range between almost 25 radiocarbon dates from similar types of sites in Britain and Ireland is almost impossible to explain away by coincidence, particularly if claims are to be seriously entertained that they are all being improperly sampled or mistakenly interpreted. At this point it is perhaps pertinent to address the criticisms of Briggs re:

Mt. Gabriel (Briggs, 1983); the Great Orme (Briggs, 1988); and Cwmystwyth (Briggs, 1988); that the early C^{14} dates could represent the use of fossil fuel sources.

Recent excavations of Mine No.3 infill on Mt. Gabriel have produced over 8 000 fragments of waterlogged oak and hazel roundwood, some exhibiting metal tooling marks, from which at least four C^{14} dates have recently been obtained (O'Brien, 1987). In Wales coincident C^{14} dates have recently been obtained from both bone tool and charcoal samples from identical locations within a small mine gallery on the Great Orme - clearly the bone here could not have come from any fossil fuel source. At Cwmystwyth the C^{14} dates were all obtained from branchwood (ash, oak and hazel), which was prevalent throughout the tips. It is difficult to conceive of any extraction of fossil brushwood from peat beds on such a vast scale that could produce nearly 5 000 tonnes of fire-set mine waste, even if such fossil fuel existed locally (Timberlake, 1988b). The fossil wood source referred to be Briggs (1988) - a thin birch rootlet horizon - appears too insubstantial to be seriously considered as a source of fuel for fire-setting. It should also be noted that tracks from peat cuttings referred to by Briggs clearly post-date the mine workings, as do peat-drying racks which have been shown by recent excavation to be post-mining horizon (Timberlake, 1990). Whilst it is important to be aware of the dangers of C^{14} sampling intrusive or re-used fossil material, and to be ever alert to this possibility, the balance of evidence clearly outweighs this likelihood here, and indeed refutes it in the cases mentioned. In fact the same blanket argument could be used to discredit C^{14} dating throughout prehistory - indeed a cremation beneath a cain in the Welsh uplands (in the absence of artefacts) might equally be claimed to be nineteenth century on the same account!

The real limitations of radiocarbon for dating primitive mine sites is probably in the expectation that it could be used to produce an *accurate* chronology of working amongst mine sites of a broadly similar age. In the case of the examples listed in Table 1 the answer is probably not. Radiocarbon is a blunt, if useful tool and to obtain the sort of resolution that we are looking for we will almost certainly have to look to other methods.

vi) Although, as we have seen, the old adage that 'mining destroys the evidence for its own origins' is not *entirely* true, we can but note that in some cases, at least, this does appear to be so. Almost certainly, at some sites in Wales, the evidence for early mining has been all but destroyed.

Why however, we should ask ourselves, is there no evidence at all of primitive stone-hammer mining elsewhere in some of the more obvious copper producing areas of Great Britain such as Cornwall and Devon; Ecton, Staffordshire; Snowdonia and the Lake District?

vii) There is certainly a danger, particularly as regards the Welsh primitive mine sites, of making assumptions both about the metal being sought and the ores being worked. The association with copper appears to be significant in most cases and naturally one would assume that in the Bronze Age man was interested in copper rather than anything else. Nevertheless some of the sites have significant amounts of lead, silver, and possibly even gold and one can't eliminate these from the reckoning, particularly since we know that the oxidised or gossan areas of the veins, which were probably the first to be exploited in antiquity,

188

would have had greatly enriched values of gold and silver. Furthermore, we are told that by the Acton Park phase of the Middle Bronze Age lead was being intentionally added to Welsh bronze (Burgess & Northover, in Tylecote, 1986).

The problem over which ores were being worked, and when, has important implications for our understanding of the development of metallurgy in this country. Whilst it appears that malachite could well have been a viable ore at the Great Orme, it is difficult to imagine how the early miners could have worked anything other than chalcopyrite at Cwmystwyth and on Parys Mountain, although again one must be wary of jumping to conclusions over negative evidence (such as of an ore which has now been removed). The problem will not be an easy one to resolve.

viii) At the moment we appear to have a number of Bronze Age mines but we are as yet quite unable to relate these ore sources to metal types in use during the British Bronze Age. Little in the way of metal analyses has ever been carried out on ores from known early mine sites. Such work might help to establish the relative importance of particular sites during the Early Bronze Age.

Application of the varied scientific techniques to the problem of primitive mines:

It would seem clear that answers to some of the enigmatic problems posed by these sites will only be found through a multi-disciplinary approach and coordinated programme of scientific research. The application of new and existing techniques both at the pre-excavation, excavation, and post-excavation stages of the work should yield dividends in terms of new information about these sites - some of which have proved very difficult to interpret through the conventional archaeological method.

The following techniques and their appropriate application are suggested for future work:

(a)   C^{14} dating.

Reliance on this technique will continue to be important. As a blunt dating tool it should be used to try and obtain approximate *terminus post quem* and *terminus ante quem* dates for early mining activity rather than almost impossible attempts to chronologically date Bronze Age horizons themselves. There seems little or no point in 'overkill' on the number of C^{14} dates obtained, most workers now accept the prehistoric origins of these sites and this resource should be used more selectively in the future. Bone should be sampled in preference to wood, and wooden artefacts or tooled wood in preference to fire-wood. This will help to eliminate criticisms over the mistaken sampling of fossil fuel.

(b)   Archaeomagnetic dating.

As with radiocarbon the accuracy of this technique very much depends on the part of the 'calibration curve' one is looking at (in this case the curve or spiral is the track of polar wandering). The advantage with this is that, assuming you have eliminated sampling error, one is dating a single definitive event in terms of the contemporary geomagnetic field.

Primitive mine sites may provide two contexts in which archaeomagnetic sampling can be applied. Firstly, an *in situ* fire-set rockface if little or no subsequent erosion or working of it has occurred. The suitability of this may well depend upon the original temperature reached (almost certainly below the Curie Point in this case). A smelting hearth would probably supply a rather better date for instance. Secondly, a sequence of waterlain sands, clays, or silts particularly if this represents a still-deposited sediment with a sufficient number of magnetic grains. The thick sequence of sediments infilling the opencast at Cwmystwyth may provide a good example of this.

(c) Soil techniques.

In situ sampling of phosphate levels in the absence of bone being preserved in the acidic tips. Similarly, sampling of potassium levels. High levels of this are a good indicator of burnt wood ash from fire-set waste, difficult to identify by any other means (D. Jenkins, pers. comm.).

(d) Careful excavation techniques to record, and in some cases remove, films of very poorly preserved material such as bone or pottery. Likewise removal of waterlogged wood or other organic artefacts. Provision for the conservation of waterlogged wood, if found, is essential.

(e) Geophysical techniques and survey.

The use of shallow seismic to determine the depth and profile of completely infilled opencast workings. This is clearly an important consideration prior to any major excavation within an opencast area such as at Cwmystwyth, both on logistical and safety grounds.

A resistivity survey to pick up buried structures beneath peat or turf cover on upland sites. Similarly a magnetometer survey would probably detect areas of burning and/or smelting within the vicinity, whilst a magnetic susceptibility meter might prove useful in either locating or confirming areas of fire-set rockface and waste during the course of an excavation.

Finally, underground surveying may be enhanced by the use of a 'mole-phone' (electric field recording) enabling the accurate positioning of an underground surveyor from the surface. This technique has been used with some success in the survey of cave systems and recently of prehistoric workings beneath the Great Orme.

(f) Palaeoenvironmental work.

The need for palynological survey in the areas around early mine sites has already been identified (Briggs, 1988; Timberlake, 1988b). In the absence of settlements etc. this could be used to detect human influence on the local vegetational history during prehistory - either clearance for agricultural or for fire-wood (felling or coppicing) for fire-setting purposes at the mines. Such a study has been recently initiated at Cwmystwyth sampling the local pollen record from the blanket peat bog above the mine. The results are awaited. Radiocarbon dating of the pollen cores should form an integral part of this work.

Environmental analyses (fauna and flora) of bulk samples from buried peats/organic horizons within mine sediments could reveal much about climate, environment of deposition, seasonal working, or else the abandonment history of the mine.

(g) Experimental archaeology.

Some work has already been carried out in this field (Pickin & Timberlake, 1988) - both experimental work on fire-setting and also the use of stone hammers. Since most of the evidence for the use of fire-setting and stone tools has come from material that has already been discarded from the mines, perhaps the only way left to us to really understand the methods employed is to reconstruct the tools and fire-setting operations, and to

190

start experimenting ourselves. Unlikely practices will soon be eliminated this way and practical solutions found to improve on efficiency, whether it is in the hafting of hammers or increasing yields in fire-setting. Quantification exercises achieved through this means are invaluable tools in realistically interpreting these sites.

(h) Metal provenancing studies.

Detailed work has been carried out by Northover (1980) on analyses of Bronze Age metalwork from Wales, although little in the way of any serious attempt has been made to relate this to ore sources known in antiquity. Of fundamental importance then would be to sample likely ores from mines of known or presumed Bronze Age date and to try and fit these analyses into the known impurity patterns for Welsh Bronzes. Variables in these patterns such as produced by the limitations of the smelting practice are easier to account for if one known a range of possible ores and ore sources in the first place. Lead isotope analysis of copper ores and artefacts (many of the mid-Wales coppers are lead rich) may be a very useful tool in this work. This sort of study has been carried out previously on Aegean ores (Gale, 1978).

Extremely important information for our study such as: the operating lives of Welsh mines, main centres of production, distribution networks and relative importance of Welsh copper sources during the Early Bronze Age may only ever be established through this kind of work.

Conclusions:

Prehistoric primitive mines present their own peculiar and rather unique problem of interpretation for archaeologists. In the absence of many artefacts or structures with which to fit into the conventional archaeological framework (or known material culture) or the Bronze Age, and with problems of indifferent preservation and difficulties of excavation, it is proposed here that a rather more comprehensive multi-disciplinary approach to the investigation be carried out. Many of the existing techniques suggested, plus others of a rather more innovative experimental kind, might be appropriate to this work. Clearly a reasonably strong funding base and the support of relevant institutions for this work is needed.

The importance and age of these sites should no longer really be disputed. What is now needed is a concerted effort to tackle the many unanswered questions that remain. The source(s) of copper during the British Bronze Age is still one of the great vexating questions which face prehistorians today. Almost certainly these primitive mine sites are the beginning of the answer.

Acknowledgements:

I would like to express my gratitude to the following for helpful advice, support, and discussion concerning this work: David Jenkins, Peter Crew, Duncan James, Paul and Brenda Craddock, Billy O'Brien, Danny Dutton, Andy Lewis, David Bick, Tim Mighall, Janet Ambers, James Thorburn, Phil Andrews and John Pickin. I would also like to acknowledge the support of the National Museum of Wales, British Museum, Society of Antiquaries of London, Board of Celtic Studies, Cambrian Archaeological Association, and the Historical Metallurgy Society in helping fund the excavation work at Cwmystwyth.

References

Briggs, C.S. (1983). Copper mining at Mount Gabriel, Co. Cork, Bronze Age bonanza or post-famine fiasco? in *Proc. Prehist. Soc.* 49, 317-333.

Briggs, C.S. (1988). The location and recognition of metal ores in pre-Roman and

Roman Britain and their contemporary exploitation, in *Aspects of Ancient Mining and Metallurgy: Acta of a British School at Athens Centenary Conference at Bangor, 1986*, U.C.N.W. Bangor, ed. J. Ellis Jones, 106-114.

Burgess, C. & Northover, P. (1976). The Welsh Bronze Age and its position in Brithis Bronze Age metallurgy, in press.

Cernych, E.N. (1978). Aibunar-a Balkan copper mine of the fourth millennium BC, in *Proc. Prehist. Soc.* 44, 203-217.

Craddock, P.T. (1980). ed. *Scientific Studies in Early Mining and Extractive Metallurgy*, British Museum Occasional Paper No.20, B.M. Research Laboratory, London.

Jovanovič, B. (1980). Primary copper mining and the production of copper (Rudna Glava), In: *Scientific Studies in Early Mining and Extractive Metallurgy*, British Museum Occasional Paper No.20, pp. 31-40. British Museum Research Laboratory, London.

Rothenberg, B. & Blanco Freijeiro, A. (1980). Ancient copper mining and smelting at Chinflon (Huelva, SW Spain). In: *Scientific Studies in Early Mining and Extractive Metallurgy*, British Museum Occasional Paper No. 20, pp. 41-62. British Museum Research Laboratory, London.

Davies, O. (1935). *Roman Mines in Europe*, Oxford University Press, Oxford.

Davies, O. (1937). Mining sites in Wales, *British Assoc. Ann. Reports*, Section H(4), Nottingham.

Davies, O. (1938). Mining sites in Wales, *British Assoc. Ann. Reports*, Section H(1), Cambridge.

Davies, O. (1939). Excavations on Parys Mountain, *Trans. Anglesey Antiq. Soc.* 40-42.

Davies, O. (1947). Cwmystwyth Mines, *Arch. Cambrensis* 99, 57-63.

Dutton, L.A. (1988). Great Orme Copper Mine, *Arch. in Wales* 28, p.46.

Gale, N.H. (1978). Lead isotopes and Aegean metallurgy, *Proc. of the 2nd International Scientific Congress on Thera and the Aegean World*, 529-545.

Gstrein, P. & Lippert, A. (1987). Untersuchung bronzezeitlicher Pingen am Hochmoos bei Bischofshofen, Salzburg, *Arch. Austriaca* 71, 89-100.

Hunt, R. (1884). *British Mining*, Book 1, 36-40, Crosby Lockwood.

Jackson, J.S. (1968). Bronze Age copper mines on Mt. Gabriel, West Co. Cork, *Arch. Austriaca* 43, 92-103.

James, D. (1988). Prehistoric copper mining on the Great Ormes Head, Llandudno, Gwynedd. In: *Aspects of Ancient Mining and Metallurgy: Acta of a British School at Athens Centenary Conference at Bangor, 1986*, ed. J. Ellis Jones, pp.115-121. U.C.N.W., Bangor.

Lewis, A. (1988). Great Orme Copper Mines, *Arch. in Wales* 28, p.46.

Lewis Morris (1988). An account of lead and silver mines in Cwmmwd y Perveth, National Library of Wales Manuscript, p.46.

Northover, P. (1980). The analysis of Welsh Bronze Age metalwork, an appendix to: *Catalogue of the Welsh Bronze Age Collections*, ed. H.N. Savory, pp. 229-241. National Museum of Wales, Cardiff.

O'Brien, W.F. (1987). The dating of the Mt. Gabriel type copper mines of West Cork, in *J. Cork Hist. and Arch. Soc.* 92(251), 50-70.

Pickin, J. & Timberlake, S. (1988). Stone hammers and fire-setting: A preliminary experiment at Cwmystwyth Mine, Dyfed, *Bull. Peak District Mines Hist. Soc.* 10(3), 165-167.

Spargo, T. (1870). *The Mines of Wales, Their Present Position and Prospects*, 1-79, London (republished 1975; Simon Hughes, Y Lolfa, Talybont, Dyfed).

Stanley, W.O. (1850). Note on Great Orme and Parys Mountain Copper Mines, *Arch J.* 7, 68-69.

Stanley, W.O. (1873). Notes on vestiges of Roman workings for copper in Anglesey, *Arch J.* 30, 59-62.

Sykes, Sir. C. (1796). *Journal of a Tour in Wales*, National Library of Wales M/s 2258C, p.63.

Timberlake, S. (1987). An archaeological investigation of early mine workings on Copa Hill, Cwmystwyth, in *Arch. in Wales* 27, 18-20.

Timberlake, S. (1988a). Excavations at Parys Mountain and nantyreira, *Arch. in Wales* 28, 11-17.

Timberlake, S. (1988b). Bronze Age mining at Cwmystwyth: the radiocarbon dates, *Arch. in Wales* 28, p.50.

Timberlake, S. (1990). Excavations and fieldwork on Copa Hill, Cwmystwyth, 1989. In: *Early Metal Mining in the British Isles: Plas Tan y Bwlch Occasional Paper No.1 - The Proceedings of the Early Mining Workshop at Plas Tan-y-Bwlch, Snowdonia National Park Study Centre, 17-19 Nov. 1989.* Ed. P.S. Crew, Plas Tan-y-Bwlch, 1990, pp.22-29.

Tylecote, R.F. (1986). Table of 'chronology of the Copper and Bronze Ages in Britain and Ireland' from: Burgess, C. & Northover, P. in *The Welsh Bronze Age and its position in British Metallurgy* (in press), in *The Prehistory of Metallurgy in the British Isles*, Table 13, p. 27, The Institute of Metals, London.

RECENT HIGHLIGHTS IN ARCHAEOMETALLURGY

R F Tylecote

Institute of Archaeology, University of London, London WC1H 0PY.

This report draws attention to recent work and is culled from conferences, meetings and excavations on the subjects which have taken place during the last few years. It deals with work from China and India on iron, early copper in Europe, tin in the Mediterranean, Roman iron in Britain, underwater archaeology in France and Portugal, and conferences in Beaune, Paris (St. Germain), and Northern Italy.

Crucible Smelting

In 1975 Richard Harrison (Harrison *et al.*, 1975) reported the finding of 12 cm diameter pots near Madrid dated to the Beaker period (Chalcolithic) which had internal deposits of slag and which were made from material different from the ordinary ware of the period - more grit and porosity. There was some copper waste and tools in nearby sites.

In January 1989, at St. Germain, S. Rovira reported more smelting crucibles from Almizaraque, a Chalcolithic site near Almeria (Fernandez-Miranda *et al.*, 1989). It was clear that the highest temperatures had been inside, while modern crucibles were normally used with external heating. No tuyeres were found but many early crucibles were heated from the top. The products were high in arsenic and low in iron (0.08% Fe), and it is claimed that these were produced directly from high arsenic oxidised ores.

In a recent thesis, part of which had been published in World Archaeology, Bennett (1989) refers to reaction vessels in which copper has been smelted and probably poured to make cast ingots. Similar finds are coming from the Feinan region of Jordan (Hauptmann, 1989). The problem is that if these 'reaction vessels' have been heated internally by tuyeres inside or above, they are really furnaces rather than crucibles which, by

definition, are heated externally. It is quite clear, however, that in the Chalcolithic period and later, simple furnaces which were little more than crucibles were being used to smelt copper. Very often the ores contained little iron and were not self-fluxing so that viscous slags of low iron content from which copper could not easily be separated were being produced. But in the same period some smelters were producing fayalite-type slags from which copper could separate more efficiently.

Bronze Age Melting

The demonstration at Beaune in 1988 by Philippe Andrieux showed how a crucible full of bronze could be melted with two elbow tuyeres and bag bellows worked by one man, and poured with the aid of a charred stick pushed into a hole in the handle of the crucible (Fig. 1). This confirmed the presumed manner of use of a lot of Mediterranean crucibles.

Recent Finds of Tin Ingots and Iron Bars in the Mediterranean

Recent work by the French underwater unit, Direction des Recherches Archéologiques Sous-Marines, has located and excavated a wreck off the island of Bagaud, one of the Hyères group (Lond, 1985). The importance of this find to the British archaeometallurgical scene is the finding of plano-convex tin ingots similar in shape but not in size to those from the south-west of Britain. We have, of course, previous finds of tin ingots from this area such as those from Port Vendres, but these, like the ones from Sardinia, are purse-shaped.

The ship's cargo, dated by its pottery to 125-75 BC, consisted of a layer of concreted iron bars (billets) lying on top of stamped plano-

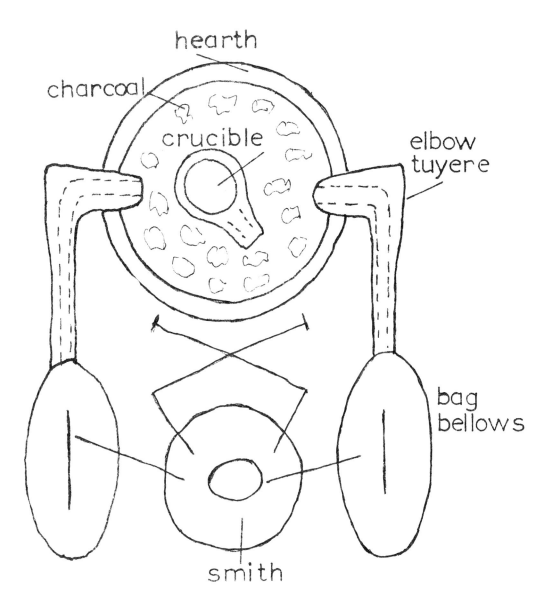

Figure 1.   A Bronze Age melting furnace worked by Ph. Andrieux at Beaune in April, 1988; one man could blow bellows and tip crucible into mould.

convex tin ingots. The ship was only 10-15 m long but was heavily loaded with a cargo of 7-10 tonnes of iron bars and 2-3 tonnes of tin ingots. There were about 100 tin ingots averaging about 25 kg each which makes them a good deal larger than the British 10 kg ones. They contain some stamped markings in Greek such as a coin-like figure of Hermes, and a rectangular cartouche - ARISTOKR on their flat sides. The legend written around the head of Hermes appears to contain the words translated as HYPO KELTON which could be translated as 'low Celts' according to the researcher Luc Long. He thinks that the low Celts (= southern Celts) referred to could be those mentioned by Caesar as the Celts of Britain or Gaul but for several reasons prefers those of Iberia (Celtiberians). It is Andalusia that is preferred by Luc Long as the source of this is closer to know Greek trading posts and in this he has the support of Strabo. The exact area would be between the Rio Gardiana and the Guadalquiver including the Sierra Morena in which Rio Tinto lies.

While most of the Mediterranean tin appears to have come from NW Spain (Galicia) it is possible that the area of the Guadiana was responsible for this tin and Penhallurick in his book (1986; 96) does show deposits just north of the Guadiana.

Romano-British Lead Ingots off the Coast of Brittany

The French Under-Water Unit, Recherches Archéologiques Sous-Marines, based on Marseilles, have for some seasons been carrying out work off the north coast of Brittany at Ploumanac'h. This has been reported by M. L'Hour (1985). They have been finding numerous lead ingots of plano-convex and rectangular shape. These weighed 22 tonnes altogether. The latter are inscribed CIV BR and ICENI on the sides suggesting a British Origin. These inscriptions are more primitive than the usual Romano-British ones of the late 1st and early 2nd centuries AD and are believed to be 2nd to 4th century. The inscriptions have been written on the sides and on the bottom (as cast) and appear to have been scrawled on the inside of the mould with the finger. In some cases, not unnaturally, the letters are written back to front. The lead isotope ratios grouped well into one group which is believed to be British. While the letters BR can be interpreted as BRIGANTES and therefore connected with the lead-producing areas of Derbyshire and North Yorkshire, the word ICENI comes as a bit of a surprise as this tribe is not thought to have operated as far north as the lead-producing region on the east coast. Perhaps this shows a new role for the powerful ICENI, that of middleman.

The fact that lead was being taken to Brittany like coals to Newcastle comes as no surprise to the French, as Algerian wine is being imported into that country today.

The fewer plano-convex ingots, weighing 6 tonnes, ie. about 80 kg each, had large rounded holes in some of them that looked as though they may have been caused by cavitation corrosion - there was a 5 knot current running over the wreck - or by the more normal galvanic corrosion caused by large charcoal inclusions.

New Type of Iron Smelting Furnace in Roman-Britain

At Laxton Hall, Northants, a large type of iron smelting furnace for Britain, but similar to some found in the Burgenland, was found in 1986 (Jackson & Tylecote, 1988). Primitive bellows cannot penetrate a furnace more than 50 cm diameter (25 cm each side). Yet these furnaces exceed 100 cm. Some suggestions have been put forward such as the deep penetration of the tuyeres towards the centre, which seems to be a waste of such a furnace and makes the slagging of the furnace lining difficult to explain. Alternatively the large furnace could be housing for a number of small furnaces situated just inside the periphery. Each tuyere would produce a bloom and the large furnace would be thermally more efficient than a number of small ones (Figures 2a and 2b).

Corrosion of Iron; the Mass Effect?

A well-known 4th century AD Delhi pillar has some parallels in China dating to the 14th century (Tylecote, 1984a and 1984b), but which are made of cast iron. Why does this type of object so well resist corrosion? In both Peking and Delhi, humidity is generally below the critical 60% R.H. for iron in the day time, and the mass effect prevents the pillars from cooling down to below the dew point at night. The humidity on the Yellow River at Lanzhou is probably higher and this may explain the somewhat poorer condition of the Lanzhou pillars. There are also some wrought iron pillars in central India, the state of which has not been reported. But it is probable that in spite of the higher humidity of the more tropical climate the mass effect protects them. Here is an opportunity for some experiments. It would be possible with the aid of attached thermocouples to find whether the surface temperature fell below the dew point at night.

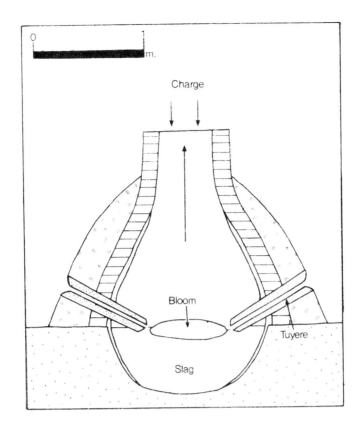

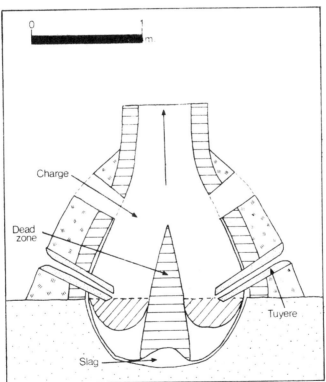

Figure 2.   Alternative reconstructions of Romano-British iron smelting furnaces found at Laxton Hall, Northants (after Jackson and Tylecote, 1988).

An 18th Century Spanish Wreck off the Portuguese Coast

In February, 1786 the Spanish vessel 'San Pedro de Alcantara' bound for Cadiz from South America was wrecked in a storm on the headland of Peniche 75 km north of Lisbon. It carried about 440 people and some of the bodies were buried in a small cemetery on top of the cliffs marked by a memorial cross. As the ship was wrecked in shallow waters salvage has been going on ever since. Recently, controlled underwater excavation on the wreck has been directed by Jean-Yves Blot of the Museum of Peniche and on the cemetery by his wife, Maria Luisa de B H Pinheiro Blot, of the Archaeological Museum in Lisbon.

We know from the finds, and its manifest which was located in the Hague, that the ship was on its way from the port of Callao in Peru which it left in 1784. When wrecked it was carrying 600 tonnes of copper 'bars' and some pigs, as well as cartons of silver coins and some gold. The copper was probably loaded at Talcahuano, a port near Concepcion, and is therefore most probably Chilean. It is possible that this shipment was a once-only event designed to overcome the shortage which was developing in Spain due to the war with Britain cutting off their previous imports from Sweden. The copper 'bars' measure about 65 x 25 x 10 cm thick and weigh over 100 kg. They have 'ears' on either side on the top (casting side) surface and a large shrinkage cavity in the centre (Figure 3a). The fact that Chile was capable of making such large ingots of copper in the 18th century seems incredible. It demonstrates the then scale of the metallurgical industry under Spanish and Portuguese control at that time and shows the enormous increase in size that had developed since that described by Alonso Barba in his book 'Arte de los Metales' of 1640.

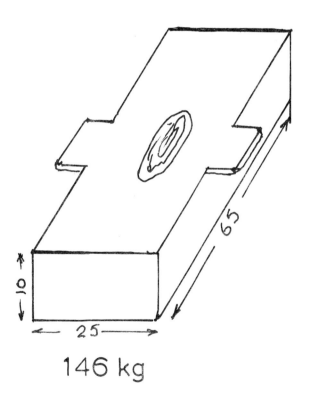

146 kg

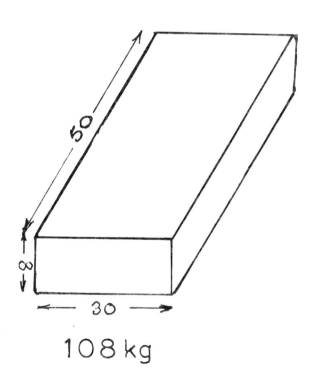

108 kg

Figure 3.   Copper ingots from 18th century Chile. (a) as found off the coast of Portugal in 1786. (b) as described by Basil Hall in 1824. Dimensions in cm.

In the period 1761-1775 the king of Spain received 10 000 quintals (460 tonnes) of copper from Peru while other customers got 65 000 q. It is possible that tin also came from the same area (Bolivia still being one of the world's tin producers); this all went to make guns in the foundries of Spain.

The Dutch archives in the Hague have a drawing of the method of loading the ship. The copper consisted of 6 930 bars and 13 'salmons' which was tightly packed together with the silver coins and probably some bars of tin.

Similar ingots were being made in Chile in the 1820s using blast furnaces run by water-powered bellows and dried wood as fuel (Hall, 1824) (Figure 3b). This would seem to follow German tradition rather than the British one of using reverberatory furnaces. The rich oxide ores would produce little slag. Analyses of the ingots found at Peniche are being made in Lisbon.

Hammer Mills in the Val Camonica

The Iron Committee of the Union Internationale des Sciences Préhistoriques et Protohistoriques met in the Val Camonica in Northern Italy in the autumn of 1988. The adequate rainfall of this southern Alpine region has been harnessed to supply the water-power requirements of early Italian industry, and some of it is still being used today. At Bienno there is a working hammer-forge making edge tools, especially shovels and spades. The water wheels, supplied with more than sufficient water from leats from the hills above, were of the Bulgarian Samokov type (Georgiev, 1971) in which water was directed on to the iron wheels through a vertical pipe (Figure 4). Thus the wheels were more like impulse turbines rather than the normal water wheel which is turned by the weight of water in the buckets. These wheels were no more than 2 m diameter and ran at a speed of 30 rpm. Instead of cams, 4 short bars were placed longitudinally on the shaft operating the tails of the helves. The frames were of stone and the rest of the hammer, hursts and bearings were normal. Although

clearly of an old traditional design these units were quite capable of doing the job which they were doing efficiently, and it would have been difficult to find modern equipment capable of replacing them. One wonders how widely this type of water wheel was used on the north side of the Mediterranean and what its origins were.

Italian Blast Furnace at Capalbio (Pescia Fiorentina), Tuscany

This is a typical example of a Bergamesc blast furnace in which air was blown in from the front by means of the hydraulic machine known as a *trompe*, and a slag and metal tapped also from the front but below the tuyere (Tylecote, 1987; 342) (Figure 5). It would appear to have been developed from the *stuckofenon* used for the production of solid bloomery iron rather than cast iron.

At Capalbio the Elban ore was roasted in lime kilns, either to make it more easily broken or to remove sulphur. Most of the cast iron was sent to forges for conversion to wrought iron.

Such a unique furnace whose future was uncertain is clearly a desirable object. The first excavation of this site was directed by Crossley (Crossley & Trinder, 1983) and carried out by a team from the UK with both British and international funding (ICCROM). It was done on the casting floor of this modified Bergamesque blast furnace which consumed roasted Elban ore and charcoal. Besides pig beds in the casting floor the dig revealed the line of the overhead conduit carrying air from the *trompe* to the tuyere in front of the furnace. The air supply was changed several times and the dig revealed later (post 1843) underfloor passages for the conduction of air to the tuyeres at the side or rear of the furnace. The *trompe* system was replaced by a cylinder blower worked off a water wheel around 1843. The furnace was blown out in 1864.

The water exiting from the *trompe* had fed a large quenching tank at the right of the casting floor.

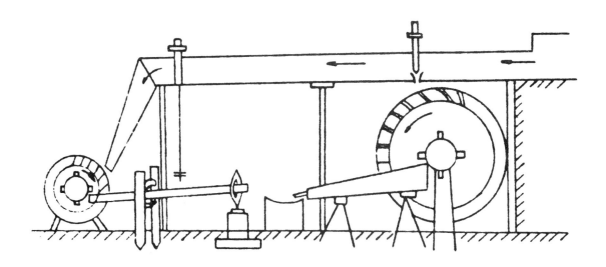

Figure 4.   Type of hammer mill found in bulgaria and called the Samokov. The water wheel (impulse type) is the same as that used today in Northern Italy in the Val Camonica (after Georgiev, 1971).

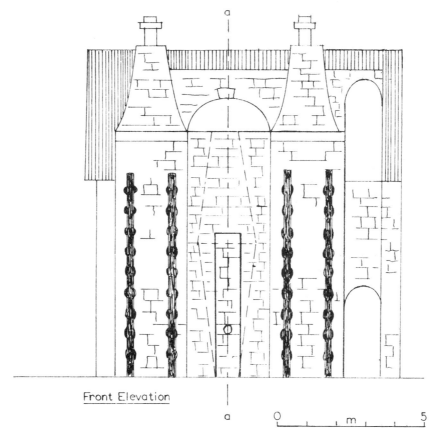

Front Elevation

Figure 5.   Type of Bergamesc blast furnace standing today at Capalbio (Pescia Fiorentina) in Tuscany (after Crossley and Trinder, 1983)

The buildings are extensive and even contain the remains of a vertically shafted water wheel now modified to drive an olive mill. It is hoped that details of the lower part of the *trompe* will be found during further work.

General Assessment

Research on archaeometallurgy has two main aims:

1. To provide information on the techniques used in early times to produce metals and to work on them, and

2. To help archaeologists to decide where the metal objects were made so that sources of metal and trade routes can be found.

The first aim can be achieved by fieldwork and laboratory experimentation, and much successful work has been done on this aspect. The second question, however, is not so easily answered. By themselves elemental chemical analyses of objects cannot help much, but when allied with lead isotope provenancing, they have begun to provide the sort of answer the archaeologist has been looking for.

Problems still await solution, but no-one can be dissatisfied with the pace of progress in this field at this time.

Acknowledgement

I am indebted to Dr E Newell for drawing my attention to the Journals of Basil Hall.

References

Bennet, A. (1989). The contribution of metallurgical studies to South-East Asian archaeology. *World Archaeology*, 20(3), 329-51.

Crossley, D.W. & Trinder, B. (1983). *The Ferriera at Pescia Fioentina, Tuscany*.

Fernandez-Miranda, F., Delibes, G., Fernando-Posse, D., Martin, C. & Rovira, S. (1989). paper given at the Conference on the Discovery of Metal, St. Germain, January 1989.

Georgiev, G.A. (1971). Die alte Eisengewinnings Industrie in Bulgarien. *Geology* 20(4/5), 597-608.

Hall, B. (1824). Extract from a journal written on the Coasts of Chile, Peru and Mexico in the years 1820, 21 and 22. *Edinburgh* 2, 14-5.

Harrison, R., Quero, S. & Priego, C. (1975). Beaker metallurgy in Spain. *Antiquity* 49(196), 273-8.

Hauptmann, A. (1989). The earliest periods of copper metallurgy in Feinan, Jordan. In: *Old World Archaeometallurgy*, ed. A. Hauptmann, E. Pernicka & G.A. Wagner, pp. 119-39. Proc. Int. Symp. on Old World Archaeometallurgy, Heidelberg, 1987, Bochum, 1989.

L'Hour, M. (1985). Un site sous-marin sur la Cote de Armorique: L'épave de Ploumanac'h. *Rev. Arch. de l'Ouest* 2, 1-19.

Jackson, D. & Tylecote, R.F. (1988). Two new Romano-British iron working sites in Northants. *Britannia* 19, 275-98.

Long, L. (1985). L'épave antique Bagaud 2. *Proc. VI Congresso International de Arch. Submarina, Cartagena, 1982*, pp. 93-8.

Penhallurick, R.d. (1986). *Tin in Antiquity*. Institute of Metals, London, p. 97.

Tylecote, R.F. (1984a). Early metallurgy in India. *Metals and Mat. Technol.* July, 343-50.

Tylecote, R.F. (1984b). The long-term corrosion of cast iron. Paper given to the *Third International Conference on Chinese Science, Beijing, August 1984*.

Tylecote, F.R. (1987). *The Early History of Metallurgy in Europe*. Longmans, London.

GEOLOGY AND METALLURGY IN THE INTERPRETATION OF ANCIENT MINING AND SMELTING

Lynn Willies

Peak District Mining Museum, Matlock Bath, Derbyshire, DE4 3NR.

Introduction

Although there is an increasing number of studies of small, usually very early mines, notably in Spain, Israel, Austria, Yugoslavia, and more recently in Wales and Ireland, the problem of tackling the archaeology of large scale mineral deposits is more unusual. Mining archaeologists - those prepared to work deep underground - are slowly increasing in number, but even so investigations by the writer and co-workers at Rio Tinto in Spain, in Israel, and at three sites in Rajasthan, India, has been a substantial part of recent research. Sites tackled have mainly been due for destruction, and work has concentrated on archaeological survey. There is no lack of need for full excavation, but given the almost total lack of field data currently available, and the difficulties of protracted excavation underground, simple survey is currently probably the best means of recovering data in face of a quite stupendous rate of destruction. We have been fortunate both in gaining financial support and permission from international mining companies, and in consulting or working with their geologists and mining engineers. We have benefited too from working alongside metallurgical archaeologists of the Institute of Archaeology at London University, and the British Museum. Our working together at the very least has prevented some considerable errors in interpretation, and has revealed a great deal about ancient large scale mining methods between some 3000 and 1200 years ago.

Modern Mining Methods

Mining operators nowadays mainly prefer very large deposits, of the order of (sometimes many) hundreds of millions of tonnes of ore, which with modern beneficiation technology can be of extremely low grade. Compared with the relatively modest damage done to surface archaeology by agriculturalists and developers, mining operations are extremely thorough. Not only is the deposit (and archaeology) blasted and displaced by some of the world's most powerful equipment, with up to 750,000 tonnes of material per day being shifted on larger sites, but much of it is also ground down to particles of a few microns diameter. The only archaeological survivals of this are usually tools caught as "tramp iron" on the magnets, but of course these are entirely out of context, and only occasionally have any wood attached which might be dated.

A Typical Natural History of Mining

However, although destructive, even the modern mining process can yield data, particularly in the early stages of development. To help appreciate the possibilities, and the problems, it is useful to consider a "natural history", which seems fairly typical of many large scale, anciently worked sites.

Ancient Mining

A: Exploitation of outcrop, shallow trenching, and "coyote-holing" following rich ore.

B: Systematic exploitation by opencast methods, and deep under-ground methods, including tunnel and human drainage systems.

C: Reworking of shallow deposits.

Medieval and Post-medieval Mining

A: Repetition of the above in both old and new mining areas (and also much small scale).

B: Driving of short and long levels, and use of human, animal, and water-power drainage.

Industrial age

A: Drainage levels, steam pumping and winding, plus earlier methods.

B: Systematic stoping and transport methods capable of extracting large quantities of lower grade ore.

C: There is likely to be several phases of the above, following economic or technological developments.

Modern mining

A: Deep mining, almost totally removing old workings or going beneath them. This latter may cause collapse of upper levels.

B: Opencast to depths of several hundred metres, with total destruction, both at the maximum depth reached, and over a much wider area needed for overburden removal or disposal.

Clearly if an ancient site has subsequently been untouched, then the archaeology will remain relatively undisturbed. Unfortunately on sites of any but local significance, this is rarely so. Fortunately, in all but the latest methods, some remains will normally survive of older periods, and not all is necessarily lost today if the site can be archaeologically monitored, especially if this is done during the exploration phase of development. This might be anything up to fifteen years, even without postponement of a project. Generally a modern mining site will go through four main stages in its development; Exploration and geological assessment, development, exploitation, and finally, reclamation.

Geologists, Metallurgists and Ancient Mines

Exploration geologists in the early phases of investigation utilise very similar skills to those of the field archaeologist. They are interested in all signs of previous working (which might easily cause both economic and physical hazards), including the accessible workings, the outcrops, and associated features such as beneficiation areas and slag heaps. In India for example, no major common metal deposit has yet been found without there being some evidence of ancient workings: Hindustan Zinc now use them as an exploration method. The use of inclined drills in exploration may reveal the presence of former workings, and at Rajpura Dariba, the drill core contained wood of c1800 years BP, from a depth of nearly 250 metres. Once a likely prospect is revealed, then drilling will be adapted to a close-spaced pattern, which, after analysis of the results, will reveal the presence, distribution, and concentrations of metals, and incidentally, possibly of former workings.

Again at Rajpura Dariba, though surface workings seemed from our initial survey to be for copper (which is very conspicuous in its oxidised form), at depth, copper zones were unaffected by ancient mining, and the ancient operation is now seen as having most likely been for argentiferous lead ores, or near surface for very rich oxidised silver carbonate ore. This conclusion was confirmed by metallurgical survey of slag heaps, which found ancient lead smelting and refining for silver on a great scale, but not copper. Since the ore in ancient workings has generally been most effectively removed, or may not be conspicuous, this prevented at least initial errors of interpretation. Previously it had been thought that silver in India had largely been imported, but the combined results at Rajpura Dariba, and also at Rampura Agucha, show that large domestic supplies were available 2500 years ago, when the first silver "punched coins" appeared there. Interestingly, though the radiocarbon results from Rajpura Dariba were published in mining publications, they attracted little interest from the archaeological world.

At Zawar (Rajasthan, India), which is in the development and exploitation stage, initial surveys by Craddock and the writer in 1982 produced a remarkable paradox. Whereas enormous waste-heaps of used zinc retorts were thirteenth century AD or later, which itself was remarkable enough, since the metal was not prepared in the west until the eighteenth century AD, the large scale mines nearby turned out to have associated radio-carbon-dates of between 1800 and 2200 years BP. This gave considerable impetus to both mining and metallurgical investigations on subsequent visits. It now appears that the deeper mines are mainly ancient, and that larger open-cuts near surface were the source of the later ores. These have been reworked, and careful examination shows the signs of earlier working in many places. It thus cannot be presumed that the shallowest workings as found today are necessarily the oldest surviving.

The Zawar samples are still being examined, geologically and metallurgically, as well as by conventional methods, so conclusions can only be tentative. However it does seem that lead was also extracted at Zawar in ancient times. Some older, fairly small lead slag heaps show considerable zinc content, those younger, but still ancient, have much less, which suggests the zinc was being more carefully separated: This may mark a transition in the brass-making process, in which the easily used, shallow deposits of oxidised ore of zinc - calamine (smithsonite) were successfully replaced by the deeper, and much more abundant, sulphide ore, zinc blende (sphalerite). It is hoped that charcoal samples from the different mineral zones in the mine will throw light on the dating of this major metallurgical development, which seems otherwise unobtainable at present.

At Zawar too, it has been possible to show that the form of the workings, which might relate mainly to the morphology of the deposit or to the technique of mining, are most likely largely controlled by the former.

A feature noted there was how higher (broadly earlier) workings spread across the deposit laterally, whilst the deeper workings followed down by narrow "finger workings", which were, we initially thought, necessary to isolate each working from the others because of water. However, drill core analysis during exploitation shows a similar pattern of finger deposits in adjacent, previously unworked, areas.

Benefits from the mining process

Most of these stem from the access which would not otherwise be possible in deep mines. The Zawar Mines for instance yielded a great deal of evidence of firesetting, used to break down the rock. Very little is known about this process, and most of that is from relatively recent times, notably in sources such as Agricola (Hoover & Hoover, 1950), or in more recent descriptions from mines in Sweden or Norway, where the technique remained in use into the 1890s. These show huge piles of timber set alight against the wall of the workings. The characteristic features of fireset mining are round-arched openings and chambers (see Figure 1), with a great deal of ash, charcoal, and rock debris. These are obviously very suitable for dating purposes. After they had been dewatered, the deep workings at Zawar (and Dariba, below), showed that firesetting was carried out far below the water-table, which might otherwise be assumed impossible. However, as well as the anticipated ability to break down walls and roof, it is clear also that techniques had been evolved to sink shafts and declines, and to "bench" horizontally in the floor. The deepest working seemed to have a much higher charcoal to ash ratio than the shallower, possibly due to the oozing of water under pressure, although an alternative hypothesis is that charcoal was used in the fire to increase the temperature of the seat of the fire to assist downcutting. The fires also seem to have been much smaller than later sources indicate, with several near to each other, which may indicate a continual cycle of firing, breaking and clearing debris, and firing again, one after the other.

Figure 1. Typical Firesett Working in the Zawarmala Mine: note pillar and rounded arch form. This area survives, but is no longer accessible due to modern working methods. (LW. 113/8).

Figure 2. Well preserved basket found at the base of the silver deposit in the ancient opencast at Rampura Agucha. Over a dozen such baskets were found in a section some five metres long, exposed by modern opencasting. (LW. 122/15).

At Rampura Agucha, which is undergoing development, despite a major programme of drilling, and the removal of a half-million tonne milling sample, the ancient presence of a major silver-ore body was unsuspected by the mine geologists. The possibility was made clear by large finds of silver smelting debris by Paul Craddock. The deposit here is about 1.5 km long by 120 m wide, to depths of 300 metres at least, and is one of the world's largest zinc reserves, although no zinc seems to have been exploited in ancient times. The deposit is marked by a shallow depression, and because the ore occurs in what, in India, are usually barren rocks, it was not found until a mines inspector reported slag. Archaeological survey carried out during the opening of the modern pit, following reports of the finding of ancient underground workings, found evidence of a vast open pit, circa 2500 years old, to depths of some ten metres or so. This, it appears now, was the base of a silver carbonate "bonanza", more or less at the ancient water table, produced by oxidation of the original ore deposit, and concentration of silver by downward movement of solutions to the contact zone between oxidising and reducing conditions (see Figure 2). Comparable bonanza deposits were found in both America and Australia late in the last century. Without our monitoring at an early stage of development, the evidence of such a body would have been entirely lost.

A similar silver deposit probably occurred at the East Lode of the Rajpura Dariba Mine, also circa 2500 years old. Here there is another large opencast of ancient origin, which after draining of water by the modern mine, revealed a large timber revetment holding back the waste-tips (See Figure 3). This appears to have been necessary to allow opencast mining to continue where the lode hades (inclines steeply) under the ancient tips. The modern working had previously used the water in the ancient opencast for the beneficiation plant, at a rate of 600 cubic metres a day for a year without the level notably diminishing: the implication of course is that the ancient mining required an even higher rate of drainage, using human or animal power, since the mine is on a huge plateau with the water table only some 10 m or so below surface. The bulk of the water today finds its way into the mine, and much must have done so in the past. It was on the East Lode that the 250 m deep underground workings were found. The survival, both at surface and underground, of so much large timber suggests the climate was no drier in the past than today.

The deepest workings so far entered underground at Rajpura Dariba are at a depth of around 100 m, and are up to 3000 years old. These can in parts be explored from the modern workings, but will eventually be mainly destroyed, and of course will again flood when mining and pumping ceases. The workings were in ancient times reached by a least one large shaft, sunk on the hade through rather softer rock next to the ore bearing bed. This had a 3.4 x 2.8 m crossection, and is not visible at surface. Large, possibly preformed, timbers of a type found in a working nearby, which were up to 5m long by 300 mm diameter, were probably used to make stagings. In another ancient working, the remains of a dam were found, placed in a fireset cavity in a wall, some 20 metres above the bottom, fronted with two large carefully worked timbers. From the dam a three metre long wooden ladder, still in position reached to another dam, and it is clear that water was bailed from dam to dam either to surface, or to a drawing shaft (see Figure 4).

At Rio Tinto in Spain, the ancient workings by the pre-Roman Iberians, and in the Roman period have almost, but not quite entirely, been removed. Mining engineers and geologists have however left fragmentary accounts of what they found during reworking after 1873, and information is contained also in their notes and plans. There is also the evidence provided by some six million tonnes of slags. Metallurgical research has established that over two thirds of these slags result from silver smelting and the remainder from copper (until recently they were believed all to be from copper smelting). Geological investigation of very low grade ore remaining under the now almost entirely moved rich, and anciently mined, oxidation zone has

Figure 3. The timber revetment, used to hold back the ancient tips at Rajpura Dariba. There were at least four such lifts, each about four metres high. It was exposed when modern working drained a water filled ancient pit. (LW. 120/25).

Figure 4. A dam used for lifting water, at a depth of about 80 metres on the South Lode of the Rajpura Dariba Mine. The steel ladders and ropes are modern. Note the characteristic curve of the rock face left by firesetting. (LW. 109/16).

established the areas where silver and copper were respectively enriched, whilst boreholing has shown that only six million tonnes of slag were produced, rather than earlier estimates of sixteen to thirty million tonnes. Archaeological examination has shown up to three horizons (probably due to successively lower water tables) were silver enriched, rather than the one originally postulated. This has subsequently been confirmed by analysis of bore hole and test pit samples. Taken together this allows a match between the ore originally available and the resultant slag, removing the need previously to postulate the import of ore to the site from elsewhere.

The Rio Tinto evidence has been gathered at almost the last opportunity, and the demonstration of the existence of the surviving workings has led to the probability of their preservation, despite enormous economic pressures on the Company. However in Israel, a current project has the possibility of investigating workings which are largely untouched since exploitation. The Wadi Amram Mines near to Eilath and the Timna Valley are in sandstones which were very easy to exploit, and which although low grade, were also easily beneficiated. Results so far show workings of three or possibly four types, representing at least two quite distinct periods of working. Ramifying passages, with a ruling section of about 0.7 m high and wide, but with both wider sections and chambers appear to date to the 1st and 2nd century AD. There was also probably a much later larger scale reworking with pillar and stall, and possibly a subsequent robbing of pillars which led to the collapse of a large area, which is likely, from the pottery so far found, to be Islamic, 7th or 8th century. Artefacts appear common (despite partial burial under sand which has "bitted" from the walls and roof) and as well as pot, include very well preserved cloth, string, basketwork, leather, and probable human coprolites. However, unlike on modern mine sites, there has been no previous geological survey, there is no basis of large scale sections and plans, and surface mapping is based on aerial photography. Access, though not excessively difficult is not easy either, and away from entrances archaeological excavation will be very difficult. Because of sediments brought in by flash-floods, except in the unlikely event of reworking, it will never be possible to investigate the bottom of the mine. Paradoxically, results in some respects will thus, at least at present, be less complete than where modern mining destruction does take place.

Conclusions

In India, it has been possible to examine workings both at surface and underground, during both the exploration stages, and actual exploitation, with the result that archaeological information is available to place against data collected by modern geologists, mining engineers, and archaeo-metallurgists. In isolation, conclusions drawn by any of the separate parties would have been far more prone to error, if not of commission, then certainly of ommision. Because destruction by modern mining is an unfortunate reallity, it is necessary to work with mining companies in a multi-disciplinary way to recover and reconstruct as much of the past as possible.

Acknowledgements

Thanks are due to my colleagues from Peak District Mining Museum who have participated in the Indian, Spanish, and now the Israeli investigations. I am especially endebted to Professor Beno Rothenberg of the Huelva Archaeo-Metallurgical Project, Institute of Archaeology, University of London, for support for work in Spain and Israel, and to Dr Paul Craddock and the British Museum for Spain and India, to RTZ and RTM who have given access and support for the work in Spain; to Hindustan Zinc, to the British Museum, and the British Academy, who have supported the work in India.

References

Hoover H.C. & Hoover, L.H. (1950). *Trans. of Agricola, Georgius De Re Metallica*, 1556. Dover Pubs. New York.

Craddock, Paul; Gurjar, L.K. & Hegde, K.T.M. (1983). Zinc Production in Medieval India. *World Archaeology* 15:2, 211-17.

Craddock, Paul. (1984) How Zinc was Smelted in Ancient India. *New Scientist*, 29 March 1984, 23.

Willies, Lynn. (1984) Ancient Lead and Zinc Mining in Rajasthan, India. *World Archaeology* 16:2, 222-33.

Willies, Lynn. (1987) Ancient Zinc-Lead-Silver Mining in Rajasthan, India - Interim Report. *Bull. Peak District Mines Historical Society* 10:2, 81-124.

Willies, Lynn. (1989) Ancient Mines in Rajasthan. *Mining Magazine*, January 1989, 31-35.

A report on the work at Rio Tinto will be published, edited by Beno Rothenburg, shortly, by the Institute of Archaeology, London University, whilst the final report of the work in Rajasthan is in preparation, to be edited by Paul Craddock. An interim report on Wadi Amram will be available in the summer 1991 issue of the Bulletin of the Peak District Mines Historical Society.

USE OF SPECTROSCOPIC METHODS IN THE METALLURGICAL STUDY OF BRONZE AGE COPPER ALLOYS FROM THE HASSE COLLECTION

H. Wouters, L. Butaye, F. Adams, and P. Van Espen.

University of Antwerp (UIA), Department of Chemistry, Universiteitsplein 1, B-2610 Wilrijk, Belgium

Introduction

The occurrence of tin-rich surfaces on ancient bronze objects seems to be more widespread than was previously thought. Such objects do not always display the expected silvery coloured surface, but often show various shades of green, which makes them very difficult to distinguish visually from the copper corrosion product. Other objects have a characteristic black coloured surface with a lasting lustrous appearance. These certainly cannot originate from copper sulphides as mineral alteration products.

The results of an initial examination with energy-dispersive X-ray fluorescence of Belgian Bronze Age tools and weapons belonging to the Hasse Collection (Museum Vleeshuis, Antwerp) show tin-enriched surfaces on about 30 of the 180 analysed bronze specimens. Meeks (1986) cited that high tin contents on the surface can arise either from simple tinning, from cassiterite reduction via the cementation process, from tin sweat during casting or finally from selective corrosion. These possibilities have different microstructural features. Relying solely on optical microscopical investigation it is very difficult to make a distinction between the different processes.

For the determination of corrosion layers, techniques capable of investigating very thin surface layers of materials are prerequisite. Electron probe micro-analysis (EPMA) is often used for such problems but the detection limit reached by this method and its inability to detect low Z elements, constitute a restriction for detailed corrosion research. With secondary ion mass spectrometry (SIMS) the analytical signal originates from the utmost surface layer, however the technique is not only an additional method for surface characterisation, due to the much lower detection limit (ppm to ppb levels), the analytical capabilities are greatly enhanced. The present research attempts to illustrate the capability of applied surface analysis with SIMS as a refinement of the results obtained by EPMA and metallographic studies, and aims to contribute to our understanding of the formation of tin-rich surfaces on Bronze Age copper alloys. We will limit the discussion which follows to two characteristic examples.

Description of the artefacts and sample preparation

The first example of a high tin bronze from the Hasse collection, is the Middle Bronze Age rapier (reg. no. 2533) which was bought, in broken condition, by George Hasse, as a find from the Scheldt-bed near Ghent. It shows a lustrous, compact, green coloured patina containing distinct greyish coloured parts. Non-destructive energy-dispersive X-ray fluorescence analysis of the surface indicates 50% of tin, 3.5% of arsenic and a trace amount of lead, nickel, zinc, antimony and iron. For further investigation a sample was removed from the blade, across the line of the fracture. It was mounted in a copper ring and ground and polished by the standard technique, finishing with 1 μm diamond paste.

A second example of tin enrichment on the surface of an artefact from the same collection, is the Late Bronze Age socketed axe (reg. no. 2326) found at Jambes, on the border of the river, while excavating the foundations of a house. The object shows a very smooth and uniform lustrous black patina. Non-destructive energy-dispersive X-ray fluorescence analysis of the material indicates 45% of tin and trace concentrations of lead, arsenic, nickel, antimony, zinc and iron. For further investigation, a 'V'-shaped sample was removed from the socket end, mounted in a copper ring and polished by the standard technique, finishing with 1 μm diamond paste.

Analytical procedure

The removed and polished samples from the objects were analysed by EPMA, a JEOL 733 Superprobe used at a primary electron energy of 25 KeV. Energy dispersive analysis was used for the determination of the elemental composition and quantitative analysis was performed with software based on standardless ZAF corrections (Raeymaekers, 1986). A linescan was made across the cross-sectioned samples, whereby every 20 ìm an X-ray energy spectrum was collected over 100 seconds.

The metallographic studies were carried out using a Reichert MeF2 reflection optical microscope. The use of polarized light made it possible to observe the different features of the structures in more or less real colours. The samples were examined both unetched and after etching in alcoholic ferric chloride.

Lateral distributions in the patina were obtained by applying a CAMECA IMS 3f micro-analyser. Although the capabilities of SIMS are very promising, the problem of surface charging during analysis of the insulating patina needs to be overcome. Experimental work indicated that by using an O⁻ primary beam on a graphite coated bronze sample, mounted in a graphite-rich resin, the surface charging was partially eliminated. The measurement parameters for the SIMS mass spectra and the surface mappings of the corroded copper alloys had to be optimized. A full account of this work will be described in a forthcoming publication (Wouters et al., in press). The ion optics of SIMS provide direct microscopic imaging by retaining a one-to-one correspondence between the point of origin of an ion emitted from the surface and its position in the final detected image. The resulting ion micrograph represents X-Y elemental distributions with a lateral resolution of about 1 μm. The ion images were recorded with a high sensitivity camera

(Dage-MTI Inc., 66-SIT).

Results and discussion

The microstructure of the body metal of the sword-blade corresponds with a heavily-worked, annealed and final cold-worked 11% tin bronze in which the eutectoid has been broken up and resides at the grain boundaries. On the surface the δ phase of the eutectoid is sometimes changed to the solid ε compound. The presence of the eutectoid together with the heavily worked structure make it unlikely that the tin-rich surface of the object originated from tin-sweat. The surface layer must hence be due either to a cassiterite-reduced eutectoid or a 450°C heat-treated tinned surface. The ε compound in the tin-rich layer can never arise from cassiterite reduction, thus its presence is a strong indication of heat-treated tinning.

The possibility of selective corrosion must also be taken into account. In this case the tin rich layer may be the result of a preferential disintegration of the α phase instead of the tin-rich δ component. This so-called corrosion inversion is rather an effect of liquids in the burial environment (Werner, 1967). Which elements might have been invoked in the corrosion phenomenon was studied by application of SIMS. The ion micrographs show the distribution of copper as a major element, which is oxidized overall, except in a partial uppermost surface where copper sulphate occurs. The tin-rich surface is completely oxidized and arsenic and antimony are diffused towards this surface layer. Silver is only found in a few crystals in the upper corrosion layer. Chlorine resides at the superficial surface layer, the grain boundaries and in a distinct layer at the border of the corrosion products and the metal. This behaviour of chlorine lends proof to the hypothesis that the corrosion of copper is stimulated by the presence of the chlorine anion (Organ, 1963). The elements sodium, potassium and calcium are distributed exclusively over the entire corrosion layer, whereas silicon and aluminium are only found in a smaller part of the surface region. From these trace analyses it is shown that the compact and lustrous patina of oxidized tin occurs as a deep green colour due to the natural corrosion with salts. The answer to the question of whether the tin-rich surface arises from deliberate tinning or selective corrosion, rests in favour of the hypothesis of a tinned object overheated to about 450°C.

The remarkable increased hardness of the socketed axe is explained by the observation under reflected, polarized light, of beta phase martensitic needles in the α phase grains. This feature could only be obtained by rapidly quenching of the material. From an EPMA study it was seen that tin was highly enriched towards the surface together with an increase in phosphorous, iron and silicon. More detailed element distributions were recorded with the ion microscope. The tin-rich surface appears to be completely oxidized. Only at places different from this surface layer do copper and tin seem to be a homogeneous alloy. The concentration of sulphur on the surface is low, so that the possibility of black coloured copper sulphide corrosion must be of little importance (Duncan *et al.*, 1987). At very distinct places near the surface intergranular corrosion of the alloy has taken place. These observations support the statement that the object was tinned almost immediately after production, rather than surface enrichment arising from selective corrosion of copper which would leave the tin-rich δ compound unchanged. The black crystals of the corrosion layer seems to be romarchite which was first found in tin patinas by Organ (1971). Further analysis by X-ray diffraction will be carried out to confirm this hypothesis.

Conclusions

In order to consider all possible mechanisms for tin-enriched surfaces it is important to combine optical microscopy with elemental analytical techniques such as EPMA and the more sensitive SIMS. The examples show the convenience of measuring ion micrographs for relatively rapid qualitative indication of phases and inclusions, which are indispensable for investigations of the tin-enriched surfaces found on Bronze Age copper alloys.

Acknowledgements

P.V.E. is a fellow the the Belgian National Science Foundation (NFWO). We sincerely appreciate the cooperation of Ms. J. Lambrechts, Conservator of the Vleeshuis Museum, for placing the collection at our disposal for this study; E. Warmenbol is thanked for his erudite advice.

References

Duncan, S.J. & Ganiaris, H. (1987). Some sulphide corrosion products on copper alloys and lead alloys from London waterfront sites. In: *Recent Advances in the Conservation and Analysis of Artifacts*, ed. J. Black, pp. 109-18. Summer Schools Press, University of London, Institute of Archaeology, London.

Meeks, N.D. (1986). Tin-rich surfaces on Bronze - some experimental and archaeological considerations. *Archaeometry* 28, 133-62.

Organ, R.M. (1963). Aspects of Bronze patina and its treatment. *Studies in Conservation* 8, 1-9.

Organ, R.M. (1971). Romarchite and Hydroromarchite, two new stannous minerals. *The Canadian Mineralogist* 10, 916.

Raeymaekers, B. (1986). Characterisation of particles by automated electron probe microanalysis, PhD. Thesis, University of Antwerp (UIA), Belgium.

Werner, O. von (1967). Contribution to the phenomenology of corrosion and patina formation of antique copper alloys. *Prakt. Metallogr.* 4, 3-15.

Wouters, H.J., Butaye, L.A., Adams, F.C. & Espen, P.E. van (in press). Application of SIMS in patina studies on Bronze Age copper alloys.

OPTICAL DATING: PROGRESS AT OXFORD

M.J. Aitken, B.W. Smith, E.J. Rhodes, N.A. Spooner, D. Questiaux, S. Stokes, C. McElroy & S.H. Li

Research Laboratory for Archaeology, 6 Keble Road, Oxford, OX1 3QJ.

Extended Abstract

In this companion technique to thermoluminescence (TL) the signal used for dating is optically-stimulated luminescence (OSL); in both techniques the luminescence results from the release of electrons trapped at crystal defects in mineral such as quartz, feldspar and zircon. The electrons are held at such defects since capture there in antiquity, the build-up of trapped electrons with time being due to the weak flux of nuclear radiation (alpha, beta and gamma) from potassium-40, thorium and potassium present in sample and burial soil as trace impurities together with a minor contribution from cosmic rays. With OSL electrons are released only from traps which are light-sensitive ('bleachable') and this gives the technique a strong advantage in application to the dating of unburnt sediment. In this application the event dated is the last exposure of the mineral grains to daylight and as long as the exposure is sufficient (upwards of 10 minutes) the OSL signal is set to zero. On the other hand in the TL signal there is a contribution from hard-to-bleach traps as well and consequently there is an appreciable residual TL signal at deposition; uncertainty about the level of this residual is a drawback to the use of TL in this application, particularly for young samples and samples which have had only short daylight exposure at deposition. As far as radiation dosimetry is concerned the two techniques have the same requirements and limitations.

Optical dating was demonstrated as feasible by Huntley et al. (1985) using green light from an argon ion laser and subsequently various wavelengths from a krypton laser (Godfrey-Smith et al., 1988). Work in Estonia reported at the 1987 Cambridge TL meeting (Hutt et al., 1988) showed not only that a xenon arc lamp could be used but also, surprisingly, that a stable dating signal could be obtained using infrared wavelengths (860 and 930 nm); later Hutt & Jaek (1989) reported use of an infrared laser diode. In UK, in addition to an argon laser and a xenon lamp there has been use of an array of infrared diodes (LED); both at Durham and Oxford low cost modules have been developed which enable standard TL sets to be adapted for OSL use (Poolton & Bailiff, 1989; Spooner et al., 1990).

At the 1987 Glasgow Science and Archaeology meeting we reported successful dating of quartz grains from two windblown deposits containing palaeolithic artefacts (at Hengistbury Head, UK and at Chapeyron Rouge, Morocco). Subsequently we have continued this quartz-based dating in a wide variety of contexts within the range 100 to 100 000 years, in some of which the sediments were waterlain; as reported by Smith et al. (1990) the agreement with age estimates based on other techniques is good. We are now extending to feldspar and zircon grains as well as to polymineral fine-grains from loess deposits in France, Italy and China and from river and ditch silt on UK archaeological sites.

The majority of the quartz-based dates have been obtained using an argon ion laser as by Huntley et al. (1985). The laser is also used for the other minerals mentioned above but for feldspar grains and polymineral fine-grains an infrared diode array has also been used; there is good agreement with laser results confirming that deep enough traps can be reached with the 880 nm infrared used, despite its low quantum energy. A special test of both infrared and laser methods has been made using alluvium bracketed between two Roman levels found at the Wellington

(Hereford) gravel quarry site; there is satisfactory dating agreement, as reported by Spooner *et al.* (1990).

References

Godfrey-Smith, D.I., Huntley, D.J. & Chen, W.-H. (1988). Optical dating studies of quartz and feldspar sediment extracts. *Quaternary Science Reviews*, 7, 373-380.

Huntley, D.J., Godfrey-Smith, D.I. & Thewalt, M.L.W. (1985). Optical dating of sediments. *Nature* 313, 105-107.

Hutt, G., Jaek, J. & Tchonka, J. (1988). Optical dating: K-feldspars optical response stimulation spectrum. *Quaternary Science Reviews*, 7, 381-386.

Hutt, G. & Jaek, J. (1989). Infrared stimulated photoluminescence dating of sediments. *Ancient TL*, 7(3), 48-53.

Poolton, N.R.J. & Bailiff, I.K. (1989). The use of LEDs as an excitation source for photoluminescence dating of sediments. *Ancient TL*, 7(1), 18-20.

Smith, B.W., Rhodes, E.J., Stokes, S., Spooner, N.A. & Aitken, M.J. (1990). Optical dating of sediments: initial quartz results from Oxford. *Archaeometry* 32(1), 15-29.

Spooner, N.A., Aitken, M.J., Smith, B.W., Franks, M. & McElroy, C. (1990). Archaeological dating by infrared-stimulated luminescence using a diode array. Radiation Protection Dosimetry (in press).

RED FIRED SANDSTONE: A GOOD RECORDER OF ANCIENT GEOMAGNETIC FIELD?

D. Atkinson & J. Shaw

Geomagnetism Laboratory, Institute of Prehistoric Sciences and Archaeology, Oliver Lodge, University of Liverpool, P.O. Box 147, L69 3BX.

Abstract

The magnetic properties of samples from an ancient red sandstone kiln have been investigated to determine their possible archaeological applications. The magnetic mineralogy and domain states have been defined. Also the problem of magnetic refraction or the presence of a magnetic fabric have been investigated. The results suggest that fired sandstone is useful for archaeomagnetic directional dating, but the determination of firing temperature or the intensity of the ambient magnetic field at the time of firing are unobtainable due to the thermochemical nature of the sandstone's magnetisation.

Introduction

Continued developments in archaeomagnetism have led to the assembly of a reference curve of geomagnetic directional secular variation for Britain extending back to 1000 BC (Clark *et al.*, 1988). The curve is now routinely used to date fired archaeological artefacts that are in situ. It also provides information about the geomagnetic field (Tarling, 1989). Most of the data for the curve are derived from baked clays; other materials such as burnt soil and brick have been utilised to a limited extent. It is the aim of this study to investigate the magnetic properties of baked red sandstone, collected from a hearth in Chester, to determine its value to archaeomagnetism.

Sampling

The sandstone samples investigated came from a single site in Chester. The sandstone is probably the local Triassic sandstone which is predominantly composed of quartz grains with a haematite cement. Samples were removed from a small curved wall considered by the site archaeologists to be part of a kiln or hearth. The wall was constructed from large, roughly worked blocks of sandstone with soft cement between them. The inner side of the wall was blackened with ashes. Blocks from the top of the wall were oriented using a suncompass, mounted on a small tripod, and a theodolite.

In the laboratory the blocks were reoriented in the horizontal plane and set with plaster of paris for subsampling. Four blocks were subsampled using a standard 25 mm diameter palaeomagnetic drill. The distribution of subsamples in individual blocks allowed the correlation of magnetic properties with distance from the inside of the kiln (ie the heat source).

Thermal demagnetisation behaviour

Incremental thermal demagnetisation was undertaken on between four and ten samples from each of the blocks using a Magnetic Measurements thermal demagnetiser. The direction and intensity of remanence was measured after each heating stage using a Molspin spinner magnetometer.

The demagnetisation behaviour of subsamples was similar between blocks and showed systematic variations of blocking temperature spectra and magnetic vector components when plotted against distance from the heated surface of the blocks.

The variation of blocking temperature spectra against distance is illustrated by subsamples from block 2 (Figure 1). The temperature of the spectra peaks is seen to reduce and the peaks broaden with increasing distance from the heated face.

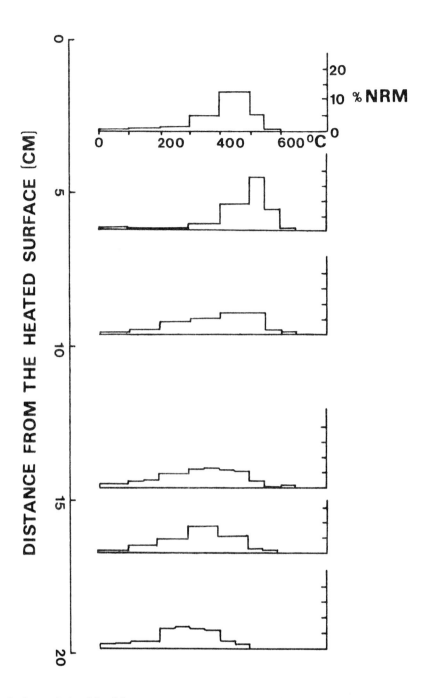

Figure 1. Variation of the blocking temperature spectra with distance from the heated surface, obtained from incremental thermal demagnetisation.

The vector composition of the natural remanent magnetisations (NRMs) is dependent on the distance from the heated surface. A low intensity component was removed from all subsamples after heating to 200-300°C. All subsamples showed an intense northerly/positive component of which 95% was removed by heating to 470-550°C.

For subsamples nearest to the heated surface this was the only definable component (Figure 2). For the subsamples taken furthest from the heated surface a second higher temperature component was also observed. The very low intensity of this component did not allow an accurate determination of its direction.

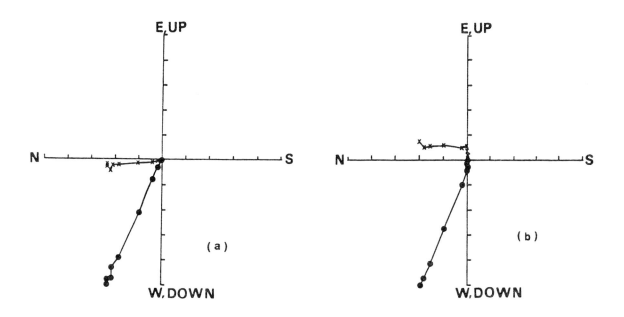

Figure 2. Orthogonal vector plots of the NRMs of subsamples (a) nearest to the heated surface and (b) furthest from the heated surface. Crosses indicate the horizontal projection and circles the vertical projection.

Rock magnetic properties

In common with the demagnetisation behaviour, the magnetic mineralogy is a function of distance from the heated surface. The variation of NRM intensity and bulk susceptibility against distance are shown in Figure 3. Isothermal remanence acquisition (IRM) curves with a maximum applied field of 3.6 T indicate a rapid change in mineralogy. Subsamples closest to the heated surface show simple behaviour saturating at 250-300 mT. With increasing distance from the heated surface this component is very much reduced in intensity and a much higher coercivity component is observed to saturate toward

4 T (Figure 4). Uncalibrated hysteresis measurements in a maximum field of 1 T show similar characteristics, with only samples closest to the heated surface being saturated. The ratio of saturation remanence to saturation magnetisation (Mrs/Ms) obtained from saturated loops fall between 0.29 and 0.31.

Thermomagnetic behaviour of subsamples obtained from heating in air is of two types. Those subsamples closest to the heated surface have single Curie temperatures around 580°C and those further than 7 or 8 cm from the heated surface have Curie temperatures in the region of 650-720°C.

219

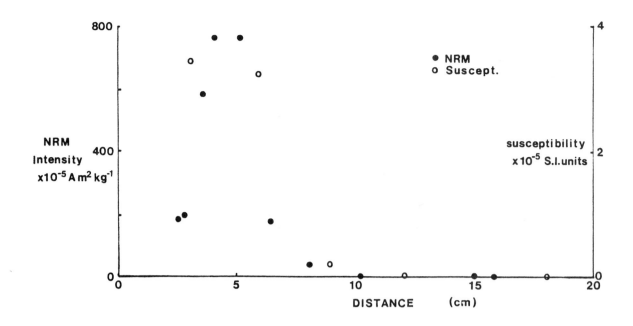

Figure 3. Variation of the NRM intensities and room temperature susceptibilities with distance from the heated surface.

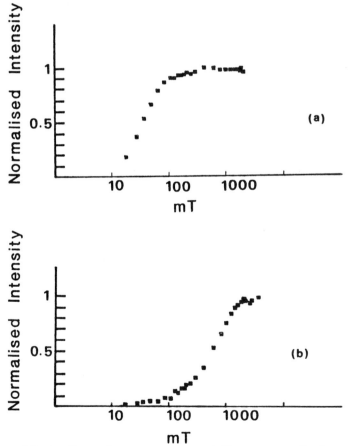

Figure 4. Comparison of the isothermal remanence acquisition curves of subsamples (a) closest and (b) furthest from the heated surface of the blocks.

High temperature susceptibility measurements were undertaken using a Bartington MS2 bridge. The temperature variation of susceptibility of subsamples up to about 9 cm from the heated surface of the blocks was similar (Figure 5). The Curie temperatures determined from the descending arms of the peaks of the curves are between 500 and 560°C. The temperature variation of susceptibility of subsamples from further than 9 cm from the heated surface of the blocks was not distinguishable from the noise level of the bridge. Anisotropy of magnetic susceptibility (AMS) measurements, using a Molspin minisep anisotropy delineator, indicate that the blocks have a magnetic fabric. The ratio of maximum to minimum susceptibility being similar within and between blocks. The ratio varies between 1.014 and 1.065. These values are slightly larger than the actual ratios because the length to width ratio of the subsamples was 1, not the ideal value of 0.9 calculated by Hounslow et al. (1988). The direction of the maximum and minimum axes are the same within blocks but vary between blocks. These AMS measurements indicate a very small degree of alignment of the magnetic grains.

Discussion of magnetic properties

The most striking feature of the magnetic properties is their dependence upon distance from the inner side of the wall. This distance dependence, which is consistent between blocks, strongly suggests that the magnetic properties are the result of firing the structure.

Assessing the reliability and applicability of the magnetic record obtained from fired red sandstone, both specific to this study and for the general case, requires an understanding of demagnetisation behaviour, magnetic mineralogy, domain states, and possible refraction effects.

The variation of the magnetic vector components of subsamples can be explained by the variation of temperature within blocks. Before heating the sandstone blocks would have held a remanence arising from the growth of haematite when cementation occurred. With heating or firing, the magnetic minerals may alter and their magnetisations will be partially or completely reset toward the ambient field direction depending on

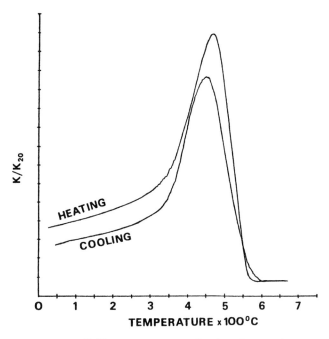

Figure 5. High temperature susceptibility measurements showing an increase in susceptibility up to the Hopkinson peak.

whether the temperature of heating exceeds the maximum blocking temperature of the minerals. Subsamples closest to the heated surface have only a low temperature viscous component, and a steep northerly component, whereas samples further from the heated surface have an additional component at higher unblocking temperatures.

Firing has had a major effect on the rock magnetic character of the sandstone. Subsamples nearest to the heated surface have high susceptibilities and NRM intensities. These subsamples are saturated by 250-300 mT during IRM acquisition, and have Curie temperatures in the range 500-580°C. This is indicative of a single phase of titanium poor magnetite. Mrs/Ms ratios obtained from hysteresis loops suggest grain sizes predominantly in the single domain region (Thompson & Oldfield, 1986).

Subsamples taken furthest from the heated surface have very low susceptibility and

remanent intensity and are not quite saturated by 3.6 T. Curie temperatures for these samples are in the range 650-700°C. The rock magnetic parameters for these samples suggest that the predominant magnetic mineral is haematite.

The mineralogy of intermediate samples shows a rapid gradation from the magnetite rich surface to the haematite rich distal zone. This gradation was only observable in IRM acquisition, the haematite signal being swamped by magnetite in other analyses. The mean ratio of Mrs/Ms was taken as representative of the magnetite in the blocks. It was used with the SIRM intensities of he magnetite component observed during IRM acquisition, and the saturation magnetisation value for magnetite (92 Am^2kg^{-1}, O'Reilly,1984) to calculate the magnetite content present at increasing distance from the heated surface (see Figure 6).

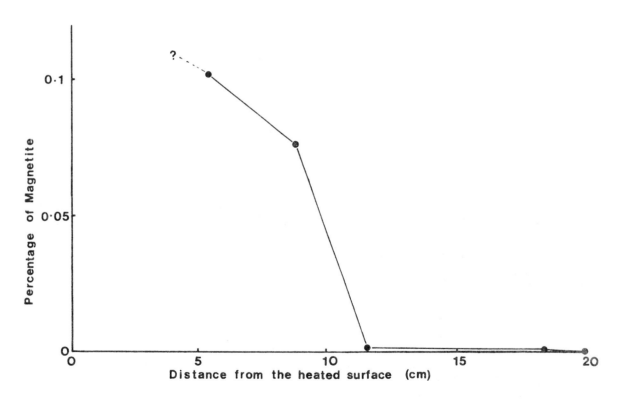

Figure 6. Variation of the magnetite content of the rock with distance from the heated surface (shown as the percentage by mass of magnetite).

The primary nature of haematite in red sandstone and the concentration of magnetite at the heated surface suggest that magnetite is the result of heating or firing. The magnetite is probably derived from reduction of haematite in a hot carbon monoxide atmosphere produced by fuel, the ashes of which are preserved on the kiln wall. The rapid reduction of magnetite with distance may be due to a combination of duration and temperature of heating and the porosity of the rock.

With an understanding of the mineralogy it is clear that remanence is not a simple thermoremanent magnetisation (TRM). The growth of magnetite particles, by the reduction of the primary haematite, combined with the temperature and duration of firing, results in a more complex thermochemical remanence (TCRM). Because of this complexity a description of the variation in the blocking temperature spectra cannot be based solely on thermal activation theory and hence no temperature information can be easily obtained.

Deflection of the direction of remanence from that of the ambient field can arise from shape anisotropy (refraction) or the presence of a magnetic fabric. Both aspects have been investigated.

The problem of shape anisotropy leading to magnetic refraction has been investigated by several workers (eg Aitken & Hawley, 1971; Stacey and Banerjee, 1974; Abrahamsen, 1986). The approach followed in this study is that of Abrahamsen (1986) which is essentially the same as that of Stacey and Banerjee, (1974). The angular deviation from the ambient field direction for an elongated body is derived in the general form:

$$Deviation = v - \arctan\left(\frac{\tan v}{1 + K_a}\right) \quad (1)$$

where v is the angle of the magnetisation to a perpendicular to the surface of the body and

K_a is the apparent susceptibility in S.I. units ie. the magnetisation at the blocking temperature. The maximum deviation occurs where the angle of magnetisation is in the range $45 < v > 52$ (Abrahamsen, 1986). Values for the maximum apparent susceptibilities of the red sandstone have been obtained from the Hopkinson peaks observed during high temperature susceptibility measurements. The maximum angular deviation of the magnetisation has been calculated by considering a block as a series of vertical slices and then using equation (1) as an approximation. The maximum deviation occurs close to the heated surface but is only 0.54 degrees.

The presence of a weak magnetic fabric in the sandstone is indicated by AMS measurements. The average percentage anisotropy of subsamples (the percentage difference between the maximum and minimum susceptibilities) is 3.8%. The presence of a magnetic fabric will influence the accurate recording of the ambient field, but the deflection of magnetisation has been shown to be negligible when the AMS is less than 5% (Hrouda, 1982).

Conclusions

Usually archaeomagnetic work is undertaken to determine the age of last firing of an in situ artefact using geomagnetic directional secular variation. Also information about the intensity of the geomagnetic field at the time of firing and the firing temperature can be obtained.

Firing red sandstone in a reducing environment clearly alters its magnetic signature. This has both advantages and disadvantages in terms of the information that can be retrieved. The NRM intensity of the original haematite bearing rock is very low, making the precise definition of the direction of magnetisation difficult. The conversion of haematite to magnetite close to the heated surface increases the NRM intensity by several orders of magnitude greatly improving the precision of the determination of the direction of magnetisation. The increases NRM intensity coupled with negligible deflection effect (almost no refraction or magnetic fabric) suggests that fired red sandstone is useful for dating *in situ* artefacts using the directional secular variation method. The determination of geomagnetic field intensities usually requires the magnetisation of samples to have only a thermal origin, ie a TRM, this is also necessary for the determination of the temperature of firing. The growth of magnetite during firing may preclude the formation of a simple TRM and therefore care must be exercised when using fired red sandstone for field intensity and temperature determinations.

Acknowledgements

The authors would like to thank Jennifer King, Drs T.C. Rolph and D.J. Robertson for guidance and assistance during sampling and for discussions. We would also like to thank Chester City Council for permission to take samples.

References

Abrahamsen, N. (1986). On shape anisotropy. In: *Twenty-five Years of Geology in Aarhus*, ed. J.T. Moller, pp.11-21. Geoskrifter, 24.

Aitken, M.J. & Hawley, H.N. (1971). Archaeomagnetism: Evidence for magnetic refraction in kiln structures. *Archaeometry* 13, 83-85.

Clark, A.J., Tarling, D.H. & Noel, M. (1988). Developments in archaeomagnetic dating in Britain. *Journal of Archaeological Sciences* 15, 645-667.

Hounslow, M.W., Noel, M. & Bootes, P.A. (1988). Sensitivity and sample-shape related measuring effects on the MOLSPIN Susceptibility Anisotropy Meter. *Geophysical Journal* 94, 355-363.

Hrouda, F. (1982). Magnetic anisotropy of rocks and its application in Geology and Geophysics. *Geophysical Reviews* 5, 37-82.

O'Reilly, W. (1984). *Rock and Mineral magnetism*. Blackie, Glasgow.

Stacey, F.D. & Banerjee, S.K. (1974). *The Physical Principles of Rock Magnetism*. Elsevier, Amsterdam.

Tarling, D.H. (1989). Secular variation during the last 2000 years. In: *Geomagnetism and Palaeomagnetism. NATO Advanced Study Institute on Geomagnetism and Palaeomagnetism*, Newcastle upon Tyne, 1988, ed. F.J. Lowes. Kluwer Academic, Dordrecht, 1989.

Thompson, R. & Oldfield, F. (1986). *Environmental Magnetism*. Allen & Unwin, London.

SOME COMPARISONS BETWEEN DENDROCHRONOLOGY AND OTHER DATING METHODS

M.G.L. Baillie

Palaeoecology Centre, Queen's University, Belfast, BT7 1NN.

Introduction

Dendrochronology in the British Isles has developed along certain well defined lines. Chronologies had to be constructed and replicated before precise dating could begin. Most of the emphasis, in Belfast, was on using the precisely dated samples to refine radiocarbon through calibration, ie. using a very precise method of limited applicability to refine a universal but less precise method. Direct dating of archaeological samples tended to form an attractive bonus.

This paper looks briefly at some recent results, developing out of the tree-ring work, which throw light on routine radiocarbon dating quality and on what may be ancient catastrophic events. While these topics may seem unrelated, they both point out the deficiencies in conventional chronology - historical or radiocarbon - exposed by comparisons with dendrochronology. High-precision radiocarbon dating, used in a wiggle-matching mode, currently offers the only hope of matching tree-ring or absolute historical dating quality.

Accuracy in Routine Radiocarbon Dating

Radiocarbon has always been a difficult method for archaeologists to assess realistically. Their expectations have always been high but their ability to critically assess the method has always been low. At least in part this is because of the inevitable 'blind' nature of the dating process, but it is also tied in with the relatively broad windows of acceptance involved in many archaeological dating problems.

Arguments about accuracy and precision are largely lost on an archaeologist who is asking the question 'is this a Neolithic or Bronze Age feature?'. In terms of prehistoric chronology most of the questions, until recently, have tended to be loosely couched. However, it would seem reasonable that any given date, supplied by a radiocarbon laboratory, should be broadly within the statistical error quoted viz that 95% of results should be within two standard deviations.

It became clear, following the first Interlaboratory Study (International Study Group, 1982), that the true state of affairs was rather different. The results of the study suggested that laboratories were underestimating their errors and that the quoted errors might have to be multiplied by some factor to account for the spread of results. This was of course a rather abstract argument because not all laboratories were equally 'bad'. Unfortunately, the anonymous nature of the Interlaboratory Study meant that each laboratory had to be treated as a worst case. Interestingly many archaeologists sided with 'their' radiocarbon lab (ie. the one they used) and adopted a 'best of all possible worlds' approach. That is, they viewed the whole exercise uncritically or, ostrich-like, buried their heads in the sand. By and large the recommendations coming out of the Interlaboratory Study have not been acted upon by archaeologists. Radiocarbon dates have still been treated as if written in stone.

Part of the reason may be that it is surprisingly difficult to lay hands on any objective archaeological assessment of the performance of radiocarbon laboratories. It was this fact, together with a desire not to make their dates any worse, which allowed archaeologists to swallow the bitter pill of the Interlaboratory Study. Thus it is of interest when any independent evidence throws light on the performance of radiocarbon laboratories.

Normally dates which can be tested historically tend to be single dates. An archaeologist 'knows' the earliest possible use date for a sample and sees what has to be done to the radiocarbon date in order to make it encompass the known date. An example of this was presented by Warner (1985) where a radiocarbon date of 1165 $\pm$ 55 BP (UB 2524) was produced on the wooden fabric of a house which had to be later than AD 1180 on coin evidence. Only by using multiplication factors (Scott *et al.*, 1983) and two standard deviation limits (Baillie & Pilcher, 1983) and an age-lapse factor (Warner, pers. comm.) can the radiocarbon date range be sufficiently extended to accommodate the post-AD 1180 use date. Unfortunately such single date examples prove little. Typically the other dates from this site were not well dated archaeologically and formed no test of the radiocarbon method.

It was therefore interesting to stumble upon a relatively large body of radiocarbon dates which had been applied to a group of English prehistoric tree-ring chronologies. With the precise dating of the chronologies, first by high-precision wiggle matching (for a discussion of this procedure see Pearson, 1986) and subsequently by dendrochronology, it has been possible to review the various routine radiocarbon measurements associated with the chronologies themselves or with closely related samples.

The chronologies derived from three sources and were linked to form a Neolithic complex. The constituent chronologies came from the Sweet Track, the River Trent gravels and a submerged forest at Stolford, Somerset (Morgan *et al.*, 1987). The Neolithic chronology complex was of interest because in the course of outlining an English prehistoric oak chronology (Baillie & Brown, 1988) it has been noted that around 4000 BC there was a singular scarcity of bog oaks. Oaks from Co. Durham, East Anglian and Lancashire yielded chronology sections which spanned 381 BC to 1584 BC, 1681 BC to 3807 BC and 4165 BC to 4989 BC. The 17th century BC gap was resolved using the Hasholme 1362-1687 BC bog oak chronology supplied by J. Hillam at Sheffield.

So until 1989 essentially random sampling of English bog oaks had failed to identify any material across the period 3807-4165 BC. The English Neolithic complex offered a possible solution to this problem. However, the chronology, which spanned 631 years as published, proved difficult to tie down against the Irish and German chronologies.

The details of the resolution of this dating problem are given elsewhere (Baillie, 1990; Hillam *et al.*, 1990). It is sufficient here to mention that an assemblage of some 62 routine radiocarbon dates could be identified with this Neolithic chronology complex. Once the exact dating of the chronologies was resolved it became possible to look at the quality of these routine radiocarbon determinations. The dates can be treated as a group, and this gives some feel for the general quality of radiocarbon dates in the literature. In addition some interesting sub-groups can be identified, in particular one group which was used in a previous 'wiggle matching' exercise. The result of that exercise allows comparison with the results of an actual high-precision wiggle match on an associated chronology.

It became apparent that the River Trent 576 year chronology (Salisbury *et al.*, 1984) was in fact made up of two chronologies. One, Old Loop 1, spanned 354 years and cross dated with Sweet and Stolford as published. The other, Colwick Hall 1, spanned 349 years and was placed by a tree-ring match at 2697-3045 BC against both the East Anglian chronology and Lancashire chronology. The correlation values (t = 6.1 and t = 5.0 respectively), using the Cross84 tree-ring program (Munro, 1984), are both highly significant. This new placement was confirmed by a new radiocarbon date (UB 3055) of 4167 $\pm$ 40 BP (Pearson, pers. comm.).

Our interest in this splitting of the River Trent chronology was that the original linking of the Old Loop and Colwick Hall chronologies had been at least partly conditioned by two radiocarbon determinations of 5110 $\pm$ 45 BP and 5335 $\pm$ 50 BP respectively. It was now apparent that the 5335 $\pm$ 50 BP date was in error by more than 1000 radiocarbon years.

As research continued into the dating of the reduced 438 year Neolithic chronology, there was a further revelation. It was discovered that the Stolford timber, which represented the outer 272 years of the complex, had been used in the first Glasgow inter-laboratory radiocarbon comparison exercise (International Study Group, 1982). Eight samples from this timber had each been dated by up to 20 laboratories, some at high precision. The resulting weighted mean dates (oldest to youngest) were 5168, 5196, 5173, 5094, 5070, 5031, 5025 and 5030 BP. All means had standard error values of less than 10 years and could thus be classed as high-precision dates. It therefore became possible to 'wiggle match' this component of the Neolithic chronology against the high-precision calibration curve (Pearson et al., 1986). This exercise showed that the young end of the 410 year Sweet Track chronology (which cross-dated with the Stolford chronology) was likely to fall within about 10 years of the calendar date 3800 BC (Baillie & Pilcher, 1988). With some recent extensions of the dendrochronological work, it is now known that the actual end-date of the Sweet Track chronology is 3807 BC (Hillam et al., 1990).

With the chronologies tied down and their exact calendar ages known, we can now assess a previous wiggle matching attempt which used eleven routine radiocarbon dates on wood from the Sweet Track itself (Morgan, 1988). The attempt, by Clark and Morgan, suggested an end-date for the Sweet Track in the range 3685-3415 BC, with 95% confidence (Morgan, 1988). It is now clear that this range does not include the correct date and the centre point of this range is in excess of three centuries too young. Even allowing for the original laboratory disclaimer on one replicate sample (which was approximately 1000 years too young) and the general statement about root and insect

contamination, it has to be recognised that these dates were obviously considered good enough to be presented in the literature. The case demonstrates that wiggle matching exercises only mean anything if the radiocarbon determinations are accurate.

In addition to the dates on the wood of the track itself, there are a further ten samples on peat and stakes associated with the track. With the track constructed using timbers felled in 3807 BC, reference to the calibration curve shows that all samples younger (closer to the present) than the track should give radiocarbon ages younger than 5000 BC, while all samples older (further back in time) than the track should give radiocarbon ages older than 5000 BP.

Two samples on peat from below the Sweet track are interesting in this respect. One sample 'peat below track' gave a date of 4744 $\pm$ 45 BP while one 'well below track' gave 4848 $\pm$ 45 BP. These dates are significantly too young and suggest either poor pretreatment or a substantial laboratory bias. Suggestions that 'peat is a notoriously difficult medium for dating' (Harkness, pers. comm.) hardly encourage the archaeologist as there appears to have been no prior warning that the samples might give erroneous dates. It hardly befits the radiocarbon laboratories to discount dates which were regarded as perfectly adequate in a 'blind dating' exercise, ie. where the customer did not know the correct answer!

When the total sample of 62 dates associated with the River Trent, Stolford and Sweet chronologies are grouped together, they form a perfectly good guide to the general performance of typical routine radiocarbon laboratories up to the early 1980s. The simplest analysis of the results is presented in Table 1 where the distance of the radiocarbon date from its true radiocarbon age is accumulated for the 62 samples.

Table 1

In a sample of 62 routine dates:

42% fall within 100 radiocarbon years
58% lie outside 100 radiocarbon years
34% of the total lie outside 200 radiocarbon years
19% of the total lie outside 300 radiocarbon years
13% of the total lie outside 400 radiocarbon years
 5% of the total lie outside 500 radiocarbon years.

Of these figures the most damning, for routine radiocarbon analysis, are the 34% outside 200 radiocarbon years. On this basis it would seem that a full one-third of all radiocarbon dates are effectively useless from the point of view of tight chronological research. We need to give some thought to the implications of these results and there are a series of points which can be made; some of which are listed below.

1. The quoted errors are highly unreliable in about one third of all samples.

2. Dates on dubious samples, particularly from the point of view of possible contamination, (be it mobile humic acid or rootlets or both) were allowed to go forward into the literature with little comment.

3. The blind nature of most dating exercises - where the customer has no idea of the true age - means that bad dates are neither suspected nor easily detected.

4. There is no reason to assume that there is anything atypical about these 62 dates. They should be a perfectly typical sample of all the routine dates in the literature.

5. It should be noted that 95% confidence, as applied in the case of Clark and Morgan's wiggle match of the Sweet dates, means nothing if the dates are wrong or systematically biased.

6. On the positive side it should be noted that only 5% of the dates were in error by more than 500 years.

There is no intention here to embarrass the labs concerned. There is no doubt that the intention to improve is there and indeed things may have improved in the meantime. However, all things considered, I see no reason to suppose that this whole assemblage of dates is unrepresentative of the vast bulk of archaeological dates in the literature. If this is the case, then the figures in Table 1 represent a basis for assessing the past ability of a wide range of radiocarbon laboratories and should be taken into account when archaeologists attempt to infer absolute dates from radiocarbon dates - particularly old, published radiocarbon dates.

The Making of a Catastrophe

After many years of uniformitarianism in archaeology we may now be seeing a partial resurgence in catastrophism. What is perhaps surprising is that, to some extent, this resurrection of catastrophe comes not from any better understanding of archaeological processes but from an excursion into scientific dating, ie. an offshoot of dendrochronology.

A recent paper in *Current Archaeology* entitled 'Do Irish bog oaks date the Shang dynasty?' (Baillie, 1989) was couched as a low level attempt to draw attention to the remarkable possibility that previously unregarded volcanoes may have given rise to highly significant dust veil 'events' with grave consequences for human populations on a hemispherical scale. Of course the case is not proven, but it is fair to say that it requires further investigation.

These results derived from a study which identified 'narrowest ring' events in the Irish bog oak collections at 4370 BC, 3195 BC, 2345 BC, 1628 BC, 1159 BC, 207 BC and AD 540 (Baillie & Munro, 1988). The archaeological case for ascertaining just what did happen at these dates is compelling. If these events were substantiated it would require the re-examination of numerous theories relating to folk movement and population collapse. Indeed, if they were substantiated and, in particular, if the dates 1628 BC and 1159 BC did delineate the Shang dynasty in China, then we would have an integrated world chronology back to around 2000 BC. In this paper I will give a brief review of the way some of these catastrophic events have developed from effective invisibility in 1980.

In 1980 Hammer et al. published their dates for ice-core acidity layers in Greenland and drew attention to, among others, events at 210 ± 30 BP, 1120 ± 50 BC and 1390 ± 50 BC. At that time they suggested that the latter acidity layer might be the Santorini eruption in the Aegean. In 1984 Lamarche and Hirschboeck produced a paper on the partial coincidence between bristlecone pine frost-damage rings and volcanoes pointing out the coincidence between their only 2nd millennium BC frost 'event', at 1627 BC, and the radiocarbon dates for the Santorini eruption. In 1985 Pang and Chou pointed out the interesting possibility that volcanic dust veils might explain environmental references in Chinese histories associated with the beginning and end of the Shang dynasty (unfortunately not precisely dated).

As a result of the precise date produced by Lamarche and Hirschboeck it was possible to search the Irish bog oak record at 1627 BC. It was discovered that there was a fairly clear 'narrowest ring' event in the Irish oaks beginning in 1628 BC. This was one of only a handful of dates where significant numbers of oaks, on different Irish bogs, simultaneously showed their narrowest rings (Baillie & Munro,

1988). These Irish events included 1628 BC (Baillie, 1989a), 1159 BC (Baillie, 1989b) and 207 BC. In 1987 Hammer et al. withdrew their former 1390 ± 50 BC date and, on the basis of the analysis of a new Greenland core, suggested a new 'Santorini' date of 1645 ± 20 BC (the highest peak being in 1644 BC). It was clear that a package of evidence was emerging which tended to reinforce three basic episodes in the 17th, 12th and 3rd centuries BC. Additional evidence has tended to add to the story rather than subtract from it. For example, an additional bristlecone frost-ring event was uncovered at 206 BC (Lamarche, pers. comm.); Pang and Chou have refined the best estimate for the start of the Shang dynasty to 1600 ± 30 BC purely on the grounds of astronomical references (Pang, pers. comm.). As a result it is possible to summarise the evidence as follows:

For the 17th century BC

1628 BC Irish narrowest ring event starts
1627 BC Bristlecone frost-ring event
1644 ± 20 BC Greenland ice core acidity peak
1600 ± 30 BC End Xia dynasty, start Shang.

For the 12th century

1159 BC Irish narrowest ring event starts
1100 ± 50 or 1120 ± 50 BC Greenland ice core acidity peak
circa 1100 BC End Shang dynasty.

For the 3rd century BC

207 BC Irish narrowest ring event
206 BC Bristlecone frost-ring event
210 ± 30 BC Greenland ice core acidity peak
208 BC Chinese record 'stars lost from view for 3 months'
207-205 BC Chinese famines
202 BC Chinese dynastic change, start Han dynasty.

The significance of these dates is as follows:

-    The Irish oak dates, 1628 BC, 1159 BC and 207 BC are the only significant 'narrowest ring' events in the Irish tree-ring record between 1800 BC and 100 BC.

- The bristlecone dates at 1627 BC and 206 BC are the only 'frost ring' events between 1800 BC and 100 BC.

- The ice core volcanic events at 1644 $\pm$ 20 BC, 1100 $\pm$ 50 BC and 210 $\pm$ 30 are three of only four significant events between 1800 BC and 100 BC (the other is at 260 $\pm$ 30 BC).

- The Chinese dynastic changes at 1600 $\pm$ 20 BC, circa 1100 BC and 202 BC are three of only four such changes between 1800 BC and 100 BC (the other is at 221 BC). Pang and Chou have been instrumental in providing the link between volcanoes and Chinese dynastic change through their suggestion that the effects of major dust veils might provide an explanation for the concept of 'Mandate of Heaven' which appears to be associated with such change. What better mandate than dark skys, failed crops, famine and resulting death on a massive scale.

However, if we are going to play these games, we have to ask why there are so few such events in the first two millennia BC when we are told that volcanoes are common - 3 per decade on average with a 'Volcanic Explosivity Index' greater than magnitude 4 (Pyle, 1989). This does raise the question, hinted at by other workers, whether there is something special about these three eruptions or whether these events are the results of multiple 'simultaneous' eruptions. Presumably several large eruptions happening within a few years could produce a cumulative effect on climate.

To give an example, the main criticism of the 17th century BC event has not been of the event itself but the relationship of the important Santorini eruption to the event. Obviously the tree-ring and ice-core workers have tended to suggest that the events may indicate a causal relationship between the

17th century Santorini eruption and the 17th century environmental effects - Santorini was a large eruption and at about the right time. However others have rightly pointed out that there were other eruptions broadly at the same time which, because of their 'better' sulphur signatures, might be better candidates to have produced the environmental effects (Nelson et al., 1989). It seems to be widely accepted that the environmental effects are due not to the dust in a dust veil but to the sulphuric acid droplets which result from sulphur dioxide injected into the upper atmosphere. However, since this criticism is only about which eruption, and indeed suggests more eruptions, it does not weaken - but in fact reinforces - the argument for something in the 17th century BC. The arguments over the dating of Santorini look set to continue for some time. The radiocarbon evidence, despite the large number of dates from different laboratories, does not specify the date of the Santorini eruption exactly. It could be argued that, with the shape of the calibration curve around 1500-1700 BC, It may not be technically feasible to specify the calendar date exactly no matter how many routine dates are applied to the problem.

Irrespective of the arguments about Santorini, we are, without doubt, seeing clustering of evidence in the 17th 12th and 3rd centuries BC with no other equivalent clustering in the overall period. Given this tight clustering of lines of evidence all of which appear to have some association with major volcanic dust veil events, it seems reasonable to assume that they may be causally related.

Conclusion

We have looked at two areas where precise dating by dendrochronology throws light on chronological problems. There are three things that can be said about radiocarbon dating in comparison. First, radiocarbon is a very useful method for putting most samples into the right chronological ballpark. Used as a fairly blunt tool it works most of the time - hence its success.

Second, as exemplified by the body of routine dates associated with the English Neolithic chronologies, it is clear that routine radiocarbon dates are often inherently inaccurate. Great care must be exercised when attempting to pinpoint any event in real time using routine radiocarbon dates - even where large numbers of dates are involved. The various attempts to date the Santorini eruption by radiocarbon represent another example where the outcome is simply not calendrical in a tree-ring sense.

Third, as has been shown with the wiggle matching of the English Neolithic complex, the date of the Sweet Track could be successfully specified within a decade or so using high-precision dates, 'this puts the date of construction of the Sweet Track close to 3800 BC' (Baillie and Pilcher, 1988) it is now known that the last ring of the Sweet Track chronology dates to 3807 BC purely on the grounds of tree-ring comparisons with new chronologies (Hillam *et al.*, 1990). In this case the high-precision placement of the tree-ring chronology was extremely accurate. However that has to be compared with the previous attempt at wiggle matching using routine radiocarbon dates which suggested an end date for Sweet between 3685-3415 BC (Morgan, 1988). That result was in error because the radiocarbon dates which underpinned it were inaccurate. So high-precision wiggle matching was to be recognised as the one technique which produces dates which are closest to being compatible with calendar dates.

Simplified, these findings suggest that, with the exception of high-precision dates, radiocarbon errors are not adequately reflected in the quoted standard deviations. Unfortunately radiocarbon workers have allowed users to fall into the belief that their results are within the statistical limits quoted. This is most clearly seen in the fact that most laboratories calibrate their dates using the quoted errors - they do not habitually use multiplication factors to allow for additional errors.

The spread of routine dates associated with the neolithic chronologies suggests strong support for the concept of multiplication factors - a concept which has so far found little favour with archaeologists but which would at least inject some realism into the study of routine dating. Put another way, multiplication factors are strongly recommended when attempting to relate routine radiocarbon dates to absolute chronology.

This study, as with the Interlaboratory Study, leads to the conclusion that laboratories should not make extravagant claims about the accuracy and precision of their dates unless they have demonstrated proof for such claims.

In the case of the 'events' at 1628 BC and 1159 BC, with their implied circumstantial dating of the Shang dynasty, it is clear that neither the historical records nor the ice-core dates, currently available, conform to the same standards of absolute dating as dendrochronology. Thus we can isolate one major chronological hurdle. While the tree-ring dates frequently specify the problem exactly - 'did 1159 BC mark the end of the Shang dynasty?' - most other dating evidence, relevant to such questions, is based on radiocarbon or inadequately dated historical evidence. While by the 3rd century BC the historical evidence is precisely dated, and hence is fully compatible with the tree-ring evidence, the earlier histories leave a lot to be desired. The only way to test the hypothesis - that the events in the Irish bog oaks are associated with the start and end of the Shang dynasty - is to pin down the Shang by other techniques. In high- precision wiggle matching we have one technique which could be used on suitable timbers from China, or indeed from any area where comparisons are being made with a precise calendar. Pang and Chou are also on the right track in their use of astronomical configurations to date ancient Chinese texts. Beyond these we presumably have to await new discoveries.

References

Baillie, M.G.L. (1989). Do Irish bog oaks date the Shang dynasty? *Current Archaeology* 117, 310-313.

Baillie, M.G.L. (1989a). Irish tree-rings and an event in 1628 BC. *Thera Conference*, (September 1989, pre-publication papers). The Thera Foundation.

Baillie, M.G.L. (1989b). Hekla 3 - just how big was it? *Endeavour* 13(2), 78-81.

Baillie, M.G.L. (1990). Checking back on an assemblage of published radiocarbon dates. *Radiocarbon* (Proceedings of the Glasgow Conference, September 1989, forthcoming).

Baillie, M.G.L. & Brown, D.M. (1988). An overview of oak chronologies. *British Archaeological Reports*, British Series 196, 543-548.

Baillie, M.G.L. & Munro, M.A.R. (1988). Irish tree-rings, Santorini and volcanic dust veils. *Nature* 332, 344-346.

Baillie, M.G.L. & Pilcher, J.R. (1983). Some observations on the high- precision calibration of routine dates. In: *Archaeology, Dendrochronology and the Radiocarbon Calibration Curve*, Occasional Paper no. 9, ed. B.S. Ottaway, pp.51-63. University of Edinburgh.

Baillie, M.G.L. & Pilcher, J.R. (1988). Make a date with a tree. *New Scientist* 117, No. 1604, 48-51.

Hammer, C.U., Clausen, H.B. & Dansgaard, W. 91980). Greenland ice sheet evidence of post-glacial volcanism and its climatic impact. *Nature* 288, 230-235.

Hammer, C.U., Clausen, H.B., Friedrich, W.L. & Tauber, H. (1987). The Minoan eruption of Santorini· in Greece dated to 1645 BC? *Nature* 328, 517-519.

Hillam, J., Groves, C.M., Brown, D.M. & Baillie, M.G.L. (1990). Dendrochronology of the English Neolithic. *Antiquity* 64, 210-220.

International Study Group (1982). An Inter-laboratory comparison of radiocarbon measurements in tree-rings. *Nature* 298, 619-623.

Lamarche, V.C. Jr. & Hirschboeck, K.K. (1984). Frost rings in trees as records of major volcanic eruptions. *Nature* 307, 121-126.

Morgan, R.A. (1988). Tree-ring studies of wood used in Neolithic and Bronze Age trackways from the Somerset Levels. *British Archaeological Reports* Series 184, 1988.

Morgan, R.A., Litton, C.D. & Salisbury, C.R. (1987). Trackways and tree trunks - dating Neolithic oaks in the British Isles. *Tree-Ring Bulletin* 47, 61-69.

Munro, M.A.R. (1984). An improved algorithm for crossdating tree-ring series. *Tree-Ring Bulletin* 44, 17-27.

Nelson, D.E., Vogel, J.S. & Southon, J.R. (1989). Another suite of confusing radiocarbon dates for the destruction of Akrotiri. *Thera Conference* (September 1989, pre-publication papers). The Thera Foundation.

Pang, K.D. & Chou, H.-h. (1985). Three very large volcanic eruptions in antiquity and their effects on the climate of the ancient world. *Eos* 66, 816.

Pearson, G.W. (1986). Precise calendrical dating of known growth-period samples using a 'curve fitting' technique. *Radiocarbon* 28, 292-299.

Pearson, G.W., Pilcher, J.R., Baillie, M.G.L., Corbett, D.M. & Qua, F. (1986). High-precision 14-C measurement of Irish oaks to show the natural 14-C variations from AD 1840 to 5210 BC. *Radiocarbon* 28, 911-934.

Pyle, D.M. (1989). Ice core acidity peaks, retarded tree growth and putative eruptions. *Archaeometry* 31, 88-91.

Salisbury, C.R., Whitley, P.J., Litton, C.D. & Fox, J.L. (1984). Flandrian courses of the River Trent at Colwick, Nottingham. The Mercian Geologist 9(4), 189-207.

Scott, E.M., Baxter, M.S. & Aitchison, T.C. (1983). 14-C Dating reproducibility: evidence from a combined experimental and statistical programme. *PACT* 8 - II.7, 133-145.

Warner, R.B. (1985). Observations on the radiocarbon dates from Rathmullan. *Ulster Journal of Archaeology* 48, 142-143.

MAGNETIC STUDIES OF ARCHAEOLOGICAL SEDIMENTS

C.M. Batt & M. Noel.

Department of Archaeology, University of Durham, 46 Saddler Street, Durham, DH1 3NU.

Introduction

The dating of fired structures using archaeomagnetism is a well established technique (Aitken, 1970; Clark et al., 1988). It has also been shown that, under certain conditions, sediments can retain a record of the Earth's magnetic field (Nagata, 1962). This paper describes a research programme which aims to extend the range of material studied using archaeomagnetism to include water lain sediments. This would enhance conventional archaeomagnetism by increasing the number and type of archaeological features which can be dated and by providing new information on the environment of deposition.

Most sediments contain magnetic minerals which can become aligned with the geomagnetic field if they are mobilised in fluid suspension. If the grains are then fixed in position by compaction or cementation, they will provide a permanent record of the field at the time of deposition. The sediments being investigated come from a number of contexts where an archaeological event has caused grain mobilisation, possibly over a very short period of time, with subsequent immobilisation; for example well, ditch and flood deposits as illustrated in Figure 1.

Some contexts dateable by archaeomagnetism

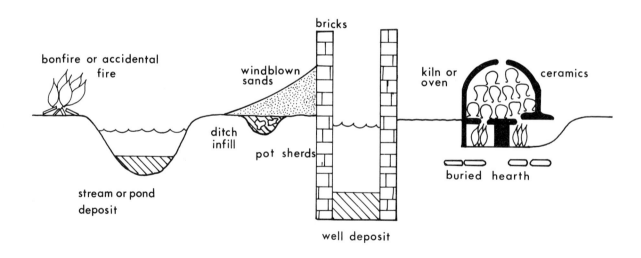

Figure 1. Examples of some of the archaeological contexts which may be dated by a study of their remanent magnetisation.

234

Field and Laboratory Methods

The object of sampling is to remove undisturbed sediment, oriented within a geographical reference frame. To achieve this a number of techniques have been adapted from palaeomagnetism. If a sediment is lithified it can be sampled by the button method (Clark *et al.*, 1988). Softer sediments are sampled by pushing 2.5 cm diameter plastic cylinders horizontally or vertically into a cleaned sediment face. At least 9 samples are required from each context to define a magnetic direction. It is necessary to orient the samples to within 2° and a number of methods are available. A magnetic compass or sun compass are the simplest, but steel shoring, local magnetic anomalies or weather conditions may necessitate the use of a gyrotheodolite, which locates true North using the Earth's rotation. This provides very accurate orientation but is an expensive piece of equipment.

Samples are then cut away from the section. In the laboratory the open face is cleaned and sealed to prevent drying out which would cause reorientation of the grains (Noel, 1983; Otofuji *et al.*, 1982).

Most sediment samples have an intensity of at least 1×10^{-6} Am^2/kg, hence their natural remanent magnetisation (N.R.M.) can be measured using a fluxgate magnetometer (Molyneux, 1971). Particularly weak samples can be measured in a sensitive cryogenic (S.Q.U.I.D.) magnetometer. Samples from the same context are plotted together on a stereographic projection, where declination is shown as an angle measured clockwise from North and inclination as a distance towards the centre. Figure 2a shows a stereo plot for a group of samples taken from a suspected flood deposit, encountered during excavations by York Archaeological Trust (Finlayson & Pearson, 1988). Samples taken from a vertical section may be plotted as declination and inclination versus depth to reveal changes in the magnetic parameters through the section.

However, the N.R.M. measured initially may not be that of the original field as this may have been overprinted by a later magnetisation caused by partial grain remobilisation, iron related chemical changes or even lightning strikes. This secondary magnetisation can be preferentially removed if it is of lower stability, either by heating in zero field or tumbling in an alternating field

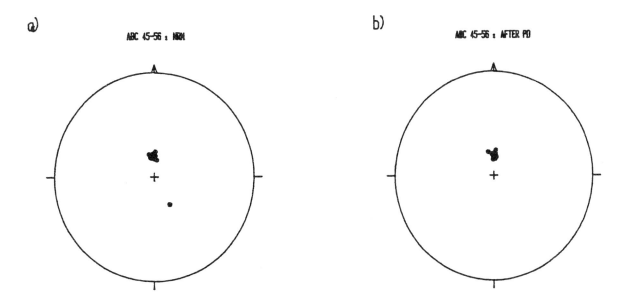

Figure 2. Stereographic projection of directions of samples 45-56 from the A.B.C. Cinema site (equal angle projection). a) The N.R.M. values which are well grouped with one outlier. b) After partial demagnetisation, showing improved grouping particularly of the outlier.

(Thellier, 1966). Alternating field demagnetisation is more appropriate to sediments and some examples can be shown in Figure 3a and 3b. After demagnetising a number of pilot samples in this fashion, a suitable field is chosen to isolate the original component and all the samples are demagnetised at this field (Figure 2b).

Many magnetic properties of a sediment contain further important environmental information, for example concerning the fabric structure. This gives an indication of the relative ease with which the sample can be magnetised in different directions, a factor which depends upon the alignment of non-spherical magnetic grains. These grains are not only oriented by the magnetic field but are also affected by gravitational and hydrodynamic forces. If there is a water current at the time of deposition it will cause

a hydrodynamic torque which will rotate the grains away from the geomagnetic vector; for example long thin grains will tend to be aligned parallel to the direction of flow (Rees, 1961 & 1968). Magnetic fabric measurements can be used to determine this rotation. The fabric can be measured in a number of ways, for example by determining the direction of magnetic susceptibility in a modified spinner magnetometer (Collinson, 1983) using the same samples as for N.R.M. measurements.

Archaeological applications

There are a number of ways in which magnetic measurements can augment sedimentological and archaeologial information, enabling better interpretation of a deposit.

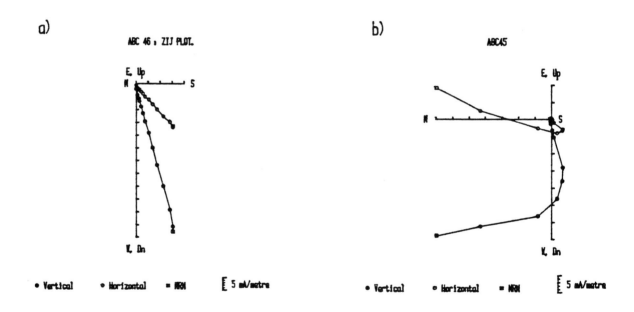

Figure 3. Zijderveld projection (Collinson, 1983) of the demagnetisation of two samples from the group 45-56. The N.R.M. components have been isolated according to their coercivity.

a) A typical sample shows little change of vector end point except for a small component, probably of viscous origin. b) The outlier exhibits behaviour characteristic of at least two components, with the original stable component consistent with other samples in the group.

A reference curve has been compiled, recording the changes in the Earth's magnetic field over archaeological time. For Britain it is built up from observatory records back to 1600 AD (Malin & Bullard, 1981), lake sediment data (Turner & Thompson, 1982) and by archaeomagnetic measurements on earlier dated features, currently extending back to 1200 BC (Clark et al., 1988). The most familiar use of archaeomagnetic information is in dating the sediments by comparing an average of their stable magnetic vectors, calculated using Fisher statistics (Fisher et al., 1987), with the archaeomagnetic reference curve.

This has proved successful for a number of deposits from York, as shown in Figure 4. The archaeomagnetic dates of 130 ± 70 AD, 100 ± 50 AD, 50 ± 40 AD and 400 ± 100 BC were consistent with pottery dates (Finlayson & Pearson, 1988). The technique was also

applied to sediments from the fill of a moat surrounding an extensive neolithic (5000-4000 BC) agricultural settlement at Banpo in Shaanxi Province, China. Three silts were sampled from layers identified as cultural horizons. These silts could not be dated as the calibration curve for that region of China as not yet been established, but they can be used in building up that curve when the radiocarbon dates are available.

However, the aspect of investigations which may be of most interest to the archaeologist lies in the determination of depositional conditions. Magnetic fabric measurements are diagnostic of depositional environment (Figure 5). Numerical criteria have been established in theoretical, laboratory and field studies which enable a sediment to be classified as being of a primary sedimentary fabric, if q (the ratio of lineation to foliation) falls in the range of 0.06-0.67. A value

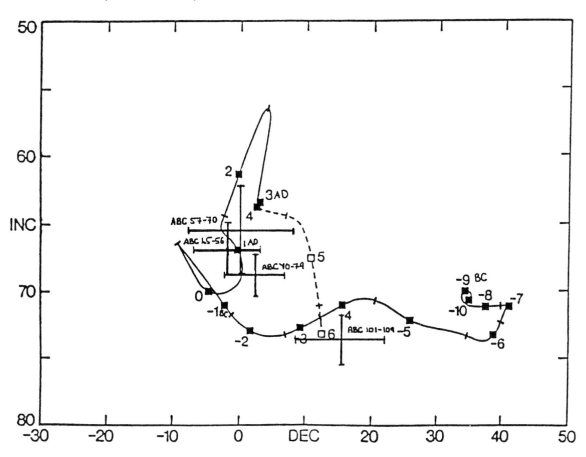

Figure 4. The British archaeomagnetic reference curve (cal. BC 1000 - cal. AD 600, normalised to Meriden) overlain by the magnetic directions from four suspected flood deposits with their associated errors.

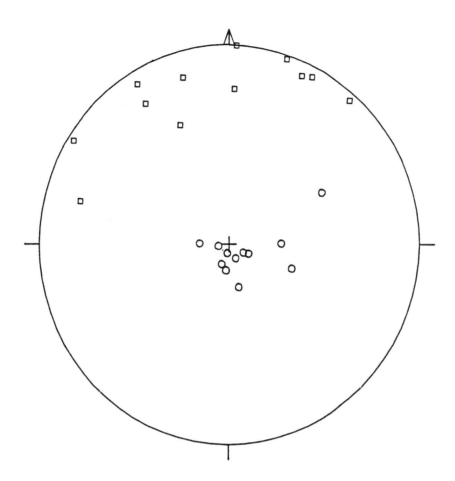

Figure 5. Directions of maximum (square) and minimum (circle) axes of susceptibility of anisotropy of samples 45-56. Minimum axes are well grouped indicating a near horizontal magnetic foliation plane and grouping of maximum directions is typical of a primary depositional style fabric, where foliation is due to gravitational forces and lineation is due to magnetic and hydrodynamic forces. (Stereographic projection, open symbols indicate negative inclination.)

outside this range is generally considered to be due to coring disturbance or bioturbation (Tarling, 1983). As discussed previously, measurements of susceptibility anisotropy and the stable remanence direction can be used to determine the rotation of grains caused by current flow (Rees, 1961; Noel & Rudnicki, 1988). An average of rotations for a group of samples gives an estimate of the direction of current on deposition. Figure 6 shows the application of these ideas to a group of samples from the A.B.C. suite. It must be noted that this method is only valid for material in which the susceptibility is mainly associated with ferromagnetic minerals of high susceptibility, such as magnetite, which is usually the case in the sediments investigated here.

Discussion of the accuracy

A number of sources of uncertainty must be taken into consideration when using archaeomagnetic techniques. The remanence may not be representative of the field of deposition due to overprinting or the absence of sufficient magnetic grains capable of retaining a record. There may be physical disturbances of the original material: it may

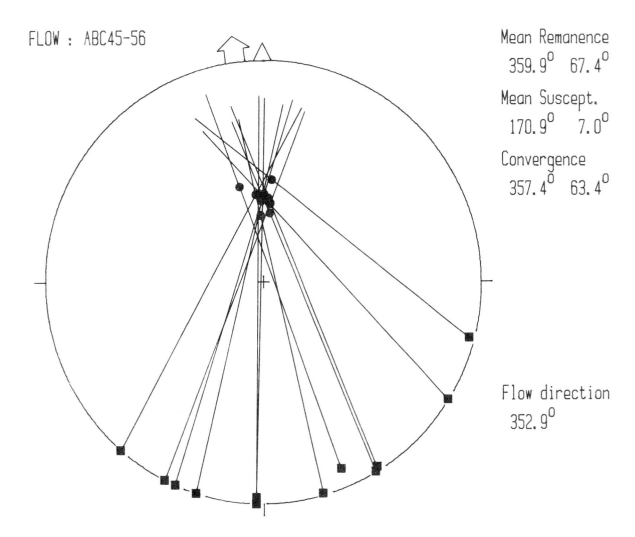

FLOW : ABC45-56

Mean Remanence
359. 9° 67. 4°

Mean Suscept.
170. 9° 7. 0°

Convergence
357. 4° 63. 4°

Flow direction
352. 9°

Figure 6. Directions of remanence (circles) and maximum susceptibility (squares) for samples 45-56, joined by small circles giving flow direction during deposition. (Orthographic projection, hence small circles transform to straight lines.)

have been deposited on a slope, have subsequently slumped or been subjected to post-depositional compaction or bioturbation. Sampling will inevitably cause some disturbance (Gravenor, 1984) and errors will occur in sample orientation. The accuracy with which the magnetic vector can be determined depends on the sample orientation and the instrumentation used. The latter is only likely to be a problem if the samples are very weak.

The calibration curve itself has uncertainties. The determinations of magnetic moment used to construct it have associated errors and the position of the date markers is open to discussion, as they rely on other dating techniques. As more archaeomagnetic measurements are made the curve will begin to 'draw itself' and these errors will be reduced. The geomagnetic field varies spatially as well as temporally and hence some correction must be made when comparing magnetic directions from different localities (Shuey et al., 1970; Noel & Batt, 1990). A region must be defined within which errors arising from spatial variation are considered acceptable. This will depend on the uncertainty in other measurements; an area of radius 900 km would cover Great Britain and introduces an error of 1.2° at most (Noel & Batt, 1990; Tarling, 1989). Even if the current reference curve (Clark et al., 1988) is accepted as completely accurate, mathematical difficulties arise in the objective comparison of a single magnetic vector with the curve and the determination of an error bound. With current techniques, the total uncertainty in a direction can be less than 2° (Tarling, 1983). This gives an error of at least ±20 years, although the magnitude depends on the region of the curve under consideration as its velocity and definition vary.

Conclusions

Initial investigations into the archaeomagnetism of sediments have suggested that this may prove a powerful tool, both for dating and for determining the environment of deposition. However, much scope remains for further work. Of particular interest will be the investigation of modern sedimentary analogues and experiments to determine the magnetic carriers, which will enable predictions to be made as to what type of archaeological contexts it is most valuable to examine in this way.

Acknowledgements

C.B. acknowledges funding from the Science and Engineering Research Council. The authors would like to thank York Archaeological Trust for their cooperation with, and interest in, this research.

References

Aitken, M.J. (1970). Dating by archaeomagnetic and thermoluminescent methods. Phil. Trans. R. Soc. A269, 77-88.

Clark, A.J., Tarling, D.H. & Noel, M. (1988). Developments in archaeomagnetic dating in Britain. Journal of Archaeological Science 15, 645-667.

Collinson, D.W. (1983). Methods in rock magnetism and palaeomagnetism. Chapman and Hall, London.

Finlayson, R. & Pearson, N. (1988). The A.B.C. Cinema excavation. Level 3 Archive Report, York Archaeological Trust.

Fisher, N.I., Lewis, T. & Embleton, B.J.J. (1987). Statistical Analysis of Spherical Data. C.U.P.

Gravenor, C.P., Symons, D.T.A. & Coyle, D.A. (1984). Errors in the anisotropy of magnetic susceptibility and magnetic remanence of unconsolidated sediments produced by sampling methods. Geophysical Research Letters 1(9), 836-839.

Malin, S.R.C. & Bullard, E.C. (1981). Direction of the Earth's magnetic field at London, 1570-1975. Phil. Trans. R. Soc., Lond. A299, 357-423.

Molyneux, L. (1971). A complete result magnetometer for measuring the remanent magnetisation of rocks. Geophys. J. R. Astr. Soc., 24, 429-433.

Nagata, T. (1962). Notes on detrial remanent magnetisation of sediments. J. Geomag. Geoelec. 14, 99-106.

Noel, M. (1983). The magnetic remanence and anisotropy of susceptibility of cave sediments from Agen Allwedd, South Wales. Geophys. J. R. Astr. Soc. 72, 557-570.

Noel, M. (1986). The palaeomagnetism and magnetic fabric of sediments from Peak Cavern, Derbyshire. Geophys. J. R. Astr. Soc. 84, 445-454.

Noel, M. & Batt, C.M. (1990). A method for correcting geographical separated remanence directions. Geophys. J.R. Astr. Soc. 102, 753-756.

Noel. M. & Rudnicki, M.D. (1988). A computer program for determining current directions from rock magnetic data. *Computers and Geosciences* 14(3), 321-338.

Otofuji, Y., Katsura, I. & Sasajima, S. (1982). Decay of a post- depositional remanent magnetisation in wet sediments due to the effect of drying. *Geophys. J. R. Astr. Soc.* 70, 191-203.

Rees, A.I. (1961). The effect of water currents on the magnetic remanence and anisotropy of susceptibility of some sediments. *Geophys. J.R. Astr. Soc.* 5, 235-251.

Rees, A.I. (1968). The production of preferred orientations in a concentrated dispersion of elongated and flattened grains. *J. Geol.* 76, 457-465.

Shuey, R.T., Cole, E.R. & Mikulich, M.J. (1970). Geographic correction of archaeomagnetic data. *J. Geomag. Geoelec.* 22(4), 485-489.

Tarling, D.H. (1983). *Palaeomagnetism.* Chapman & Hall.

Tarling, D.H. (1989). Geomagnetic secular variation in Britain during the last 2000 years. In: *Geomagnetism and Palaeomagnetism*, ed. F.J. Lowes, NATO ASI series.

Thellier, E. (1966). Methods of alternating current and thermal demagnetisation. In: *Methods and Techniques in Geophysics*, Vol. 2, ed. S.K. Runcorn. Interscience, London.

Turner, G.M. & Thompson, R. (1982). Detransformation of the British geomagnetic secular variation record for Holocene times. *Geophys. J.R. Astr. Soc.* 70, 789-792.

241

TEPHROCHRONOLOGY AND UK ARCHAEOLOGY

A.J. Dugmore

Department of Geography, University of Edinburgh, Drummond Street, Edinburgh EH8 9XP.

Abstract

Tephrochronology is a well established and powerful dating technique that has been used to great effect in areas covered by visible layers of volcanic fallout. Tephra-falls create time-parallel or isochronous marker horizons that enable associated artifacts, monuments or sediments to be dated with unusual precision. This provides effective correlations and an efficient method of testing apparently synchronous archaeological or environmental changes.

Fine grained (smaller than 200 microns) air-fall tephras of Icelandic origin have been found through Scotland at least as far south as Aberdeen and Fort William. The identification and correlation of these discrete layers requires a refinement of tephrochronological techniques and highlights the need for detailed knowledge of the geochemical characteristics of major Icelandic tephras. Given this knowledge, accurate correlations may be made not only between UK sites and Iceland, but also between the UK and other areas affected by Icelandic ashfalls in the NE Atlantic region, mainland Scandinavia and North Germany.

In Sweden, it is possible that tephras of both historic and pre-historic age may be traced into varve sequences, enabling the correction of the Swedish time scale to present to be tested, and the age of prehistoric tephras to be determined independent of the radiocarbon method. This work is in progress in collaboration with Swedish scientists.

Although holocene tephras identified in Scottish peat bogs are invisible to the naked eye, they can be identified on X-ray pictures of the peat. Pollen sampling may therefore be undertaken along clear, millimetre scale isochrones, the dispersal of pollen within a peat profile may be assessed and possible ecological impacts associated with volcanic eruptions may be considered.

Additional tephrochronological data in the UK may be obtained from the ocean-rafted, cobble-grade pumices that occur in coastal archaeological sites in the north-west British Isles and throughout the NE Atlantic region.

Introduction

Tephrochronology is a dating technique based on the identification, correlation and age determination of layers of volcanic ash, or tephra, and the use of these layers to date associated materials (Thorarinsson, 1974). The dispersal of tephra is usually very rapid, and it can be very extensive resulting in the creation of a widespread marker horizon effectively formed at a moment in time. The great power of tephrochronology stems not from the frequency of eruptions, although in places such as Iceland this may be considerable, but from the great extent of comparatively few layers. There are probably about twenty identifiable tephra layers in soils of NW Europe (Dugmore, 1989; Mangerud et al., 1984, 1986; Persson, 1971), but these layers, forming discrete identifiable horizons 1 000 - 2 000 km from their Icelandic sources, cover very large areas, defining planes of equal, and precise ages (Figure 1).

These extensive isochrones have a significance much greater than their comparatively limited number may imply, because they provide reference points in the stratigraphy that may be precisely correlated not only within a small area of archaeological interest, but also between sites and their surroundings, between regions or even between the UK, Scandinavia and North Germany.

242

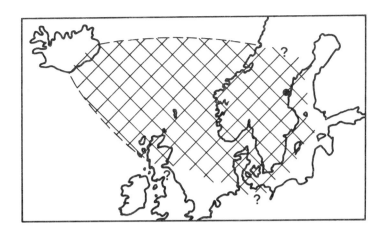

Figure 1. The minimum area covered by the combined distributions of Holocene tephras from Iceland. The circle in W. Sweden locates the Angermanalven valley where varved sediments have been accumulating continuously since before 9 000 BP. (Cato, 1984).

Distribution data from: Dugmore, 1989; Mangerud *et al.*, 1984, 1986; Merkt, written comm., 1989; Persson, 1971 & Thorarinsson, 1980, 1981.

The development of tephrochronology for UK archaeology faces a number of challenges. The first is the refinement of existing techniques in order to make use of tephra layers in areas where they are very thin.

Tephrochronology based on very thin layers

Individual microscopic grains of tephras may be identified by optical examination, because of the isotrophic character of volcanic glass and the distinctive morphologies of separate shards. An example of the highly vesicular nature of rhyolitic grains from mid to late-Holocene Scottish deposits is illustrated in the scanning electron micrograph presented in Figure 2.

At present in Scotland two main methods are currently used to extract fine grained tephras from the material into which they have been deposited, in order to prepare them for microprobe analysis. The most straightforward materials to work with are peats and organic gyttja. The destruction of the whole organic fraction leaves a small inorganic residue, and at levels where volcanic ash falls have occurred 80-90% of the inorganic content can be tephra. Organic material may be removed with hydrogen peroxide (H_2O_2) but this treatment can alter the chemical composition of the tephra (Steen-McIntyre, 1977). Most Scottish examples are currently being treated with a modified diatom extraction technique, first used on Icelandic tephra by Persson (1966). A combination of sulphuric acid (H_2SO_4) and potassium nitrate (KNO_3) is used to oxidise the organic content, and produce a clean sample of volcanic glass (Dugmore, 1989). Tests have been undertaken to determine the extent to which the extraction process affects the composition of the tephra. In one such test samples of separate phases of the Hekla-4 eruption were collected from a site in southern Iceland; each sample was split, one half was immediately prepared for microprobe analysis, whereas the other half was treated with H_2SO_4/KNO_3. Microprobe analyses of thin sections of treated and untreated shards produced comparable results, so it is possible to be confident that the extraction technique

243

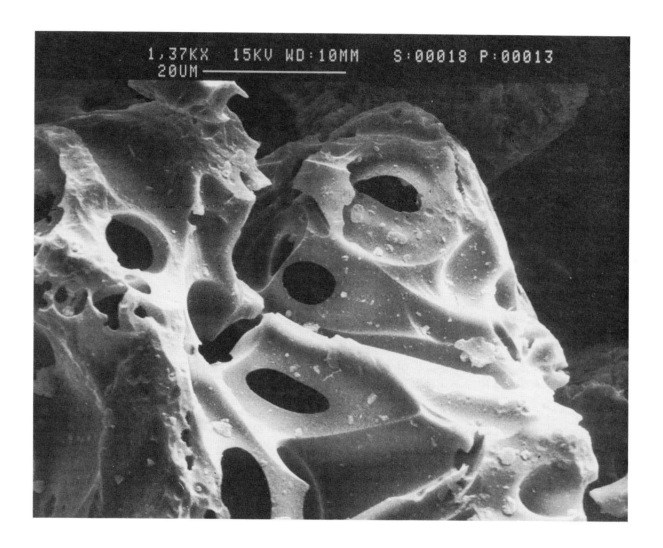

Figure 2. Scanning electron micrograph of an Icelandic tephra grain, part of the layer shown in Figure 4.

is not significantly biasing subsequent analyses.

· Samples that contain a significantly sized inorganic fraction need additional treatment in order to increase the concentration of tephra. If the modal particle size or the tephra is known sieving may reduce the non-tephra fraction of the sample; alternatively, or in addition heavy liquids may be used to separate grains with different densities to tephra.

244

Once tephra grains have been identified in a particular horizon and separated from the bulk of the enclosing material, geochemical analysis can be undertaken in order to determine the likely volcanic source area. The volcanically active areas of Iceland, Jan Mayen and the Eiffel district of Germany are possible sources of tephra found in the UK. The geochemical characteristics of these regions are quite distinct and their pyroclastic products may be discriminated with ease (Mangerud et al., 1984). To date, only tephra from Iceland has been found in the British post-glacial (Dugmore, 1989). In Iceland active central volcanoes and fissure swarms may be grouped into volcanic systems, or spatially distinct clusters of eruption sites that have particular petrological and geochemical characteristics (Jakobsson, 1979). The characteristics of these systems do change, but on a timescale considerably longer than that of relevance to archaeological studies. Given a reasonable geochemical knowledge of the pyroclastic products of Icelandic volcanic systems, it is therefore possible to use major and minor element abundances to determine likely source areas for the Icelandic tephra found in the UK. Furthermore individual Icelandic eruptions can have discrete characteristics that enable their products to be correlated on the basis of geochemical data alone (Annertz et al., 1985; Larsen, 1981, 1982).

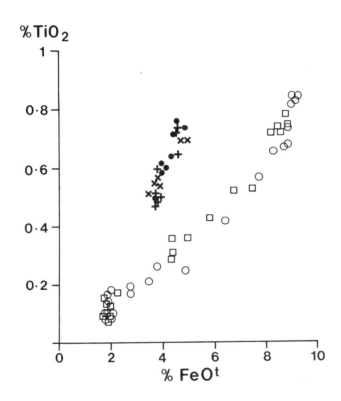

Figure 3. TiO2/FeOt for five tephra deposits determined by microprobe analysis performed on a Cambridge Instruments Microscan V. A standard WDS technique was employed using an accelerating voltage of 20 kV and a probe current of 30 nA.

□   Tephra from Altnabreac, Caithness, UK.
o   Tephra from Kalfafell, Vestur-Skaftafellssysla, Iceland.
+   Tephra from Glenne'mbeist, Caithness, UK.
●   Tephra from New Pitsligo, Grampian, UK.
x   Tephra from Glen Garry, S. Highlands, UK.

Figure 4a

Figure 4b

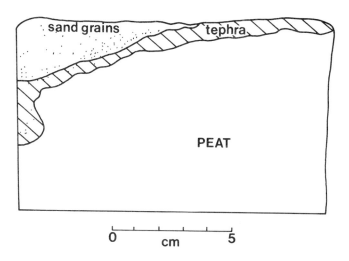

Figure 4. X-radiograph of peat from Altnabreac, Caithness, showing tephra layer in cross-section. One grain from this layer is shown in Figure 2 and some results of microprobe analyses are shown in Figure 3.

Knowledge of the geochemical characteristics of each tephra layer in the UK is essential in order to establish a tephrochronological framework; the need to correlate to source in Iceland on major and minor element characteristics alone has prompted further study of the major Icelandic tephras. Gudrun Larsen is currently establishing thoroughly documented reference sections in Iceland, and joint work with the author is focused on the detailed analysis of layers known to have spread into the UK and mainland Europe. Comparability of results is a major concern when attempting international correlations; and as part of the collaboration between the Universities of Iceland and Edinburgh analytical methods are being standardised as much as possible in order to minimise potential sources of inaccuracy and inprecision. Original results presented in this paper (Figure 3) have all been produced using the same methods by the author working on one machine. Some inaccuracy is inevitable, but the results are reproducible, and since the analytical conditions have been standardised the comparability of separate analyses on the same tephra layer should be good. Furthermore, analyses performed at Edinburgh on samples of Hekla 3 and Hekla 4 tephra provided by Gudrun Larsen reproduced results obtained by Annertz et al. (1985). Data in Figure 3 illustrates potential correlations between five separate tephra deposits. The lower cluster of points shows an example of a complete overlap of two data sets; one consisting of analyses of a tephra layer at Altnabreac in Caithness, and the other from analyses of tephra at Kalfafell in Iceland both correlated to the Hekla-4 eruption of ca. 4 000 bp by comparison with data from Annertz et al. (1985). (Dugmore, 1989; Larsen and Thorarinsson, 1977). The sites are about 1 100 km apart and are here linked on the basis of geochemistry. Since the initial phase of the eruption is present at both sites, it is possible to sample peat from the layers immediately below the ash-fall in Iceland and Scotland and be confident that the samples will be very nearly, if not precisely the same age.

The second cluster of points in Figure 3 is formed by data from three sites in Scotland.

Major and minor element abundances of a thin layer of tephra younger than Hekla 4 at New Pitsligo in Grampian, Glen Garry in the Southern Highlands and the Flow Country in Caithness form a mutually overlapping but discrete data set, implying that these sites all contain fallout from the same Icelandic source. These three deposits are likely to be part of another extensive isochrone covering much of northern Scotland. The precise Icelandic source of this tephra is not yet apparent. This uncertainty reinforces the need for further research in Iceland, because it shows how a tephra that is perhaps marginal to Icelandic tephrochronology is of considerable importance in the UK. An eruption capable of spreading ash from Iceland through northern Scotland is clearly an important event, but at present there is insufficient data available in Iceland to make a definitive correlation between distal and proximal parts of this tephra. It is this type of difficulty that the joint work between the author and Gudrun Larsen is seeking to eliminate.

The most commonly applied methods for determining the age of tephras in archaeological contexts are either historical research or radiocarbon assays. Numerous historical records exist referring to ash-fall in the Northern Isles and mainland Scandinavia and for certain periods since the Norse Settlement Icelandic records are outstanding; there are, however, periods such as early Icelandic history and the fifteenth century from which few records survive, and as a result precise historical dating is difficult, if not impossible. Holocene, pre-historic tephras are usually dated through radiocarbon assays on associated organic remains; although relative dates defined by a pre-historic tephra layer may be accurate to within a decade, 'absolute' dates can be no more accurate than the limits of the radiocarbon method. The tephra can, however, be traced to the best possible place for dating - which may, for example, be a piece of buried woodland close to source.

An exciting possibility exists that definite links may be made between volcanic eruptions in Iceland and periods of reduced growth in oak trees recovered from Irish bogs (Baillie & Munro, 1988). Direct, incontrovertible links between UK tephrochronology and dendrochronology will be difficult to establish because of the nature of the two dating techniques, but definite links may be demonstrated rapidly between tephrochronology and the Swedish varve sequences. Reference to Persson's (1966, 1967) work shows that a number of later Holocene tephras have fallen in central eastern Sweden. This area contains varved sediments that have been accumulating throughout the last 9 000 years (Figure 1) (Nilsson, 1968) and it is here that the Swedish geochronological time scale has been correlated to present (Cato, 1984). The identification of historically dated tephra in the recent varve sequences could be used to critically test the 'zeroing' of the Swedish time scale and, once tested, this time scale could be used to date any enclosed traces of prehistoric tephra. Joint work being undertaken by the author and Ingmar Cato could lead to the exciting possibility of determining absolute dates for the tephra layers prominent in the British pre-historic record and identifiable through extensive parts of the NE Atlantic region.

Ocean-rafted pumice

Additional tephrochronological data in the UK may be obtained from the ocean-rafted, cobble-grade pumices that occur in coastal archaeological sites in the north-west British Isles and throughout the NE Atlantic region (Binns 1972A, 1972B). Much of the pumice recovered from British sites shows signs of having been used as an abrasive. It has been collected from natural deposits, and kept for some unknown length of time before being discarded. The pumice may have been gathered from the high-water mark soon after it washed ashore, or from older beach deposit; large pieces can float for decades although smaller fragments are likely to waterlog and sink after a few weeks. Geochemical analysis of major, minor and trace elements can be used to group the pumices according to likely source area, and even distinct eruptions, but the time delay

between formation and final deposition may be decades, centuries or even millennia. Work in progress around likely source areas could soon match the various pumice drifts with airfall deposits created by the same eruptions. Since the scale of volcanic activity required to deposit large quantities of pumice along the shores of NW Europe is considerable, there is a possibility that airborne ash generated by the same events may have also reached mainland scandinavia and the UK.

Archaeological applications

Archaeological applications of tephrochronology in the UK are being developed at Edinburgh University in a number of collaborative projects with Sheffield University and the Scottish Development Department's Archaeological Operations and Conservation unit. Although the thin mid- late Holocene layers of the tephra present in the UK are invisible to the naked eye in favourable circumstances they can be clearly revealed through X-radiography (Dugmore & Newton in prep.). Figure 4a shows an X-radiograph of a slice of peat 2 cm thick taken from a section at Altnabreac. The dark band is tephra (Figure 4b); one of the grains is shown in the micrograph in Figure 2 and chemical analysis of 25 grains from this layer provided the data shown in Figure 3. Using X-radiographs the location of thin tephra layers in peat may be accurately determined, providing unusually precise dating control of considerable use in studies of both local and regional vegetation change.

Tephrochronology has many potential applications in UK archaeology, especially since major eruptions in Iceland occur at time of particular interest such as the twelfth century BC towards the end of the British Bronze Age. It has been suggested that volcanic eruptions not only provide useful marker horizons, but may also have significant impacts on the environment and human society (Baillie, 1989). Colin Burgess (1989) has suggested that there may be a causal link between major changes in the late British Bronze Age and volcanic activity in Iceland.

The idea that settlement change more than 1000 km from Iceland was triggered by the Hekla-3 eruption is thought-provoking. There are hints that some environmental changes in Britain may have coincided with the largest Icelandic eruptions; the X-radiograph in Figure 4a shows many black specks above the dark band that is tephra. Detailed examination showed that these grains are part of a fine influx of sand that occurred in the peat profile immediately after the ash-fall. The influx was short-lived and could have been coincidental. Alternatively it might have been triggered by direct or indirect effects of the eruption that produced the tephra layer. Using tephra layers as dating control the possibility that particular eruptions have produced widespread ecological impact is being critically tested in a joint project between the author, Paul Buckland and Kevin Edwards.

Conclusion

Preliminary work has established that Scotland has been repeatedly covered with fine grained volcanic fallout, which has left behind discrete, geochemically distinct layers of tephra. These layers may be used to create a tephrochronological framework for northern Britain of particular use in archaeology.

Acknowledgements

The initial work on tephrochronology in the UK was made possible by the Principal's Fund at the University of Edinburgh, and the Carnegie Trust for the Universities of Scotland and the Shetland Amenity Trust. I very much appreciate the support of Dr P.C. Buckland, Sir John Burnett, Professor G.S. Boulton and Professor D.E. Sugden that made the initial studies possible. The tephrochronological work is now being continued thanks to an award from the Science Based Archaeology Committee of SERC, which is very gratefully acknowledged. Many thanks to Camilla Erskine for typing up the manuscript.

References

Annertz, K., Nilsson, M. & Sigvaldason, G.E. (1985). The post-glacial history of Dyngjufjoll. *Nordic Volcanological Institute* 8503, 1-22.

Baillie, M.G.L. (1989). Do Irish bog oaks date the Shang Dynasty? *Current Archaeology* 117, 310-13.

Baillie, M.G.L. & Munro, M.A.R. (1988). Irish tree rings: Santorini and volcanic dust veils. *Nature* 332(6162), 344-346.

Binns, R.E. (1972A). Flandrian strandline chronology for the British Isles and the correlation of some European post-glacial strandlines. *Nature* 235(5335), 206-210.

Binns, R.E. (1972B). Composition and derivation of pumice on post-glacial strandlines in northern Europe and western Arctic. *Geol. Soc. Am. Bull.* 83(8), 2303-2324.

Burgess, C. (1989). Volcanoes, catastrophe and the global crisis of the Late Second Millennium BC, *Current Archaeology* 117, 325-329.

Cato, I. (1984). The definitive connection of the Swedish geochronological timescale with the present, and the new date of the zero year in Doviken, northern Sweden. *Boreas* 14, 117-122.

Dugmore, A.J. (1989). Icelandic volcanic ash in Scotland. *Scottish Geographical Magazine* 105, 168-72.

Dugmore, A.J. & Newton, A.J. (in prep.) Thin tephra layers in peat revealed by X-radiography.

Jakobsson, S.P. (1979). Petrology of recent basalts of the Eastern Volcanic Zone, Iceland. *Acta Naturalia Islandica* 26, 1-103.

Larsen, G. (1981). Tephrochonology by microprobe glass analysis. In: *Tephra Studies*, eds. S. Self & R.S.J. Sparks, pp. 95-102. Reidel, Dordrecht.

Larsen, G. (1982). Gjoskutimatal Jokuldals og nagrennis. In: *Eldurier i Nordri*, eds. H. Thorarinsdottir, O.H. Oskarsson, S. Steinthorsson & Th. Einarsson, pp. 51-65. Sogufelag, Reykjavik.

Larsen, G. & Thorarinsson, S. (1977). H-4 and other acid Hekla tephra layers. *Jokull* 27, 28-46.

Mangerud, J., Lie, S.E., Furnes, H., Kristiansen, I.L. & Lomo, L. (1984). A Younger Dryas ash bed in western Norway, and its possible correlations with tephra in cores from the Norwegian sea and the North Atlantic. *Quaternary Research* 85-104.

Mangerud, J., Furnes, H. & Johansen, J. (1986). A 9000 year-old ash bed on the Faroe Islands. *Quaternary Research* 26, 262-265.

Nilsson, E. (1968). Sodra Sveriges senkvartara historia geokronologi, issjoar och landhojning. *K. Sven. Vetenskapsakad. handl.* 4, 1-117.

Persson, C. (1966). Forsok till tefrokronologisk datering av nagra Svenska torvmossar. *Geologiska Foreningens i Stockholm Forhandlingar* 88, 361-394.

Persson, C. (1967). Forsok till tefrokronologisk datering i tre Norska myrar. *Geol. Foren. Stock. Forhand.* 89, 181-197.

Persson, C. (1971). Tephrochronological investigations of peat deposits in Scandinavia and on the Faroe Islands. *Sveriges Geologiska Undersokning* 65, 3-34.

Steen-McIntyre, V. (1977). *A manual for tephrochronology*, Colorado School of Mines Press, Colorado.

Thorarinsson, S. (1974). The terms tephra and tephrochronology. In: *World Bibliography and Index of Quaternary Tephrochronology*, ed. J.A. Westgate & C.M. Gold, pp. XVII-XVIII. University of Alberta, Alberta.

Thorarinsson, S. (1980). Langleidir gjosku ur thremur Kotlugsum. *Jokull* 30, 65-72.

Thorarinsson, S. (1981). Greetings from Iceland; ashfalls and volcanic aerosols in Scandinavia. *Geografiska Annaler* 63a, 109-118.

URANIUM SERIES DATING OF CONTAMINATED CALCITE USING MULTIPLE LEACHATES AND ITS USE IN ARCHAEOLOGY

A.G. Latham[1] and H.P. Schwarcz[2]

[1]Institute of Prehistoric Sciences and Archaeology, PO Box 147, Liverpool University, L69 3BX, UK.

[2]Department of Geology, McMaster University, Hamilton, Ontario, L8S 4M1. Canada.

Abstract

Uranium-Series dating has come to play an increasingly important role in helping to establish a hominid chronology in the range 350 Ka to present. Many secondary carbonates associated with early-man sites are, however, contaminated with detritus which has, until recently, prevented the production of reliable dates. We present here the theory to the 'leachates' method which enable the production of much more accurate U-Th dates, together with an application to the Vertesszöllös (Hungary) hominid site. This development and the development of U-Th mass-spectrometry represents significant extensions to the U-Th method and, hence, of its use for early-man chronology back to 500 Ka.

Introduction.

The method of determining the age of young (< 350 Ka) precipitated calcite by the uranium disequilibrium method is well established (eg. Ivanovich & Harmon, 1982). The technique takes advantage of the observation that calcite, when freshly precipitated, contains traces of uranium, but is free of any daughter thorium because of the latter's insolubility in groundwater. The age is determined from the ^{230}Th/^{234}U and ^{234}U/^{238}U activity ratios using the equation:

$$\frac{^{230}Th}{^{234}Th} = \frac{^{230}U}{^{234}U}(1-e^{-\lambda_o t}) + \frac{\lambda_o}{\lambda_o - \lambda_4}\left(\frac{1-^{238}U}{^{234}U}\right)(1-e^{(\lambda_4 - \lambda_o)t}) \quad (1)$$

where λ_0 and λ_4 are the decay constants for ^{230}Th and ^{234}U, respectively, and t is the age.

Many geologically important calcite deposits are, however, contaminated with detrital material that contributes some ^{230}Th, ^{234}U and ^{238}U to the solution (leachate) prepared during isotopic analysis. This results in a deviation of the observed ^{230}Th/^{234}U ratio from that of the chemically precipitated calcite component alone and, if not corrected for, produces an apparently greater age. Such contamination is easily recognised by the presence of a ^{232}Th peak in the alpha-particle spectrum; this long-lived isotope is present only in the detritus.

Various leaching methods have been proposed to "correct" for this detrital contaminant, many of which make use of various assumptions about the relative solubility of the respective radioisotopes from the detritus. (Kaufman, 1971; Schwarcz, 1980, 1982; Ku et al., 1979; Ku & Liang, 1984)

The 'leachates' method presented here has the advantage that it gets around the two main problems encountered in previous approaches: 1) There is no necessity to know whether the isotopic ratios of the detritus are in equilibrium or not, and 2) It does not matter if one isotope is differentially leached during the laboratory experiment. It is only necessary that differential isotopic fractionation (DIF) be constant - and this is testable.

251

In this paper we shall show that we can obtain the age of such deposits from the leachates alone, whether differential isotopic fractionation is present or not. The prior reason for our advocacy of this approach is the recognition that some sediments, in the calcite matrix, may consist of phases with easily removable (labile) isotopes together with a more resistate component such that laboratory leaching may thereby produce isotopic fractionation of these different isotopes. Our approach has the advantage that fractionation of the isotopes from the detritus to the leachate, if it should occur, does not affect the gradients of the isochron plots from which the calcite age is obtained.

Derivation of Isotopic Relationships

We shall assume that our sample consists of a mixture of two components: C; a pure chemically precipitated calcite, aragonite or some other secondary carbonate which incorporated, at the time of deposition, a certain amount of uranium but no thorium: D; a mixture of one or more non-carbonate, detrital minerals, physically entrapped by component C at the time of the latter's deposition. They may contain some U and Th either in the crystal lattices, adsorbed on the surfaces of these minerals or chelated by related organic materials (eg. as detected by microscopy).

It is important that neither component should contain detrital carbonate material since then it will be impossible to separate its isotopic ratios from those of the authogenic carbonate. This would preclude the use of any of the leaching methods in trying to date the calcite cement of sands or calcites of cave mouths if these have a windblown carbonate component.

Generally, a dilute acid, eg. $2N$ HNO_3, is used to dissolve the sample prior to separation of Th from U in the isotopic analysis. In dissolving just the carbonate component, whether the acid is used in excess or not, it is expected that some detritrus will also be leached.

After chemical separation of Th and U, their alpha activities are measured and can be represented as follows:

Let the activities in dpm/gm of the calcite be,

A_{C0} = activity of ^{230}Th
A_{C4} = activity of ^{234}U
A_{C8} = activity of ^{238}U
A_{C2} = activity of ^{232}Th (= 0)

For the detritus, the analogous activities are A_{D0}, A_{D4}, A_{D8} and A_{D2}. The activity ratios may then be given by,

$R_{C02} = A_{C0}/A_{C2}$, etc.

Also; let x = the weight fraction of carbonate in a sample.

Therefore, 1 - x is the weight fraction of detritus in the same sample.

During dissolution, let different fractions r_0, r_4, r_2 and r_8 of the isotopes be taken into the leachate from the detritus. Then the solution which is produced will have isotopic activities per gram of sample dissolved, as follows (L = leachate):

$$A_{LO} = A_{CO} + r_0(1-x)A_{DO} \qquad (2)$$

$$A_{L4} = x A_{C4} + r_4(1-x)A_{D4} \qquad (3)$$

$$A_{L2} = r_2(1-x)A_{D2} \qquad (4)$$

$$A_{L8} = x\,A_{C8} + r_8(1-x)A_{D8} \qquad (5)$$

Thus, we shall simultaneously analyse a number of sub-samples of the same geological unit, each of which contains a different proportion, x, of carbonate. Although a given leachate (representing each sub-sample) has a different chemical fraction, r_0, etc, of each isotope, from the $1 - x$ portion of the detritus, we shall assume that these different fractions are repeated from one leachate to the next. That is, $r_0{:}r_4{:}r_8{:}r_2$ are in a fixed ratio to each other from one leachate to the next.

Derivation of Expressions for the Isotope Ratios

The isotope ratios of the leachates are given by;

$$R_{L02} = \frac{A_{L0}}{A_{L2}} = \frac{x\,A_{C0} + r_0(1-x)A_{D0}}{r_2(1-x)A_{D2}} = \frac{x\,A_{C0}}{r_2(1-x)A_{D2}} + \frac{r_0 R_{D02}}{r_2} \qquad (6)$$

and similarly,

$$R_{L42} = \frac{A_{L4}}{A_{L2}} = \frac{x \cdot A_{C4}}{r_2(1-x)\,A_{D2}} + \frac{r_4 R_{D42}}{r_2} \qquad (7)$$

$$R_{L82} = \frac{A_{L8}}{A_{L2}} = \frac{x\,A_{C8}}{r_2(1-x)\,A_{D2}} + \frac{r_8 R_{D82}}{r_2} \qquad (8)$$

Note that equations (6) and (7) have a common term;

$$B = \frac{x}{r_2(1-x)\,A_{D2}}$$

so we can rewrite (6) and (7) to get,

$$R_{L02} \;=\; BA_{CO} \;+\; r_0/r_2 . R_{D02} \tag{6b}$$

and

$$R_{L42} \;=\; BA_{C4} \;+\; r_4/r_2 . R_{D42} \tag{7b}$$

Solving (7b) for B, substituting into (6b) and rearranging, we get:

$$R_{L02} \;=\; R_{C04} . \; R_{L42} \;+\; r_0 r_2 . R_{D02} \;-\; r_4/r_2 . R_{C04} . R_{D42} \tag{9}$$

which, in conventional isotope notation, may be written,

$$(^{234}U/^{232}Th)_{\,L} \;=\; (^{230}Th/^{234}U)_{\,c} . \; (^{234}U/^{232}Th)_{\,L} \;+\; E \tag{10}$$

where

$$E \;=\; r_0/r_2 . \; (^{230}Th/^{232}Th)_{\,D} - r_4/r_2 . \; (^{230}Th/^{234}Th)_{\,c} . \; (^{234}U/^{232}Th)_{\,D} \tag{11}$$

Linear equation (10) gives the slope of the best-fit line ("isochron") to a plot of $(^{230}Th/^{232}Th)_L$ v $(^{234}U/^{232}Th)_L$ for sub-sample leachates of a single dirty calcite deposit.

Hence the slope is $(^{230}Th/^{234}U)_c$, which may be input into equation (1).

Equations for the uranium isotope ratios can be similarly obtained from equations (3) and (5), yielding

$$R_{L42} \;=\; R_{C48} . R_{L82} \;+\; F, \tag{12}$$

where the intercept F is given by

$$F \;=\; r_4/r_2 . R_{D42} \;-\; r_8/r_2 . R_{C48} . R_{D82} \tag{13}$$

254

Linear equation (12) gives the slope of the best-fit line from several leachates from one sample for a plot of $(^{234}U/^{232}Th)_L$ v $^{238}U/^{232}Th)_L$. Thus the slope is $(^{234}U/^{238}U)_C$, which may be input into equation (1). Now the age, t, can be computed.

Equations (10) to (13) show that the DIF is located in the intercept terms, E and F, and not the gradients, and this is the advantage of this method. The value of $^{230}Th_C/^{234}U_C$ is invariant under DIF. (It is therefore not possible to detect DIF by examining several leachates without also examining the isotopic character of the residue, if this is required to be known.)

We have tested the leachates method on a series of artificial calcite-detritus mixtures in which the age of the pure calcite was known a priori (Pryzybilowicz et al., 1990). In each case, the prior known age of the calcite was recovered even from heavily contaminated mixtures. From the experimental evidence we believe that it should always be possible to retrieve $^{230}Th/^{234}U_C$ and $^{234}U/^{238}U_C$ from straight line sections of the gradients, in particular, from the weak-leaching end of the isochrons, even when the detritus is multi-component.

(For comparison of this method with the Ku and Liang 'Leachate-Residue' method, see Schwarcz & Latham (1989).)

Applications of the Method

The calvaria from the quarry at Vertesszöllös (Hungary) has apparent affinities with homo erectus and Neanderthal forms, and first attempts to date the enclosing loessic/hot spring deposits were made by Cherdyntsev et al., (1965) who obtained ages mainly in excess of 350 Ka by the U-Th method. Detrital contamination was a severe problem. Single and double leachates were obtained from three hot-spring travertines from Vertesszöllös (Schwarcz & Latham, 1984) and these were all combined in a single pair of isochron plots. Samples were contaminated to varying degrees with windblown silt (loess). This was evident from the presence of a significant fraction (1 to 5%) of insoluble residue in the travertines, and from low sample $^{230}Th/^{232}Th$ ratios. The travertines were interstratified between layers of loess or loessic marl. The samples do not, in fact, consist of various aliquots of the same travertine, as required by strict application of the method, but are taken from three deposits believed to have been formed over a short time period and having the same detrital component. Variation in U concentrations between the pairs of carbonate and detritus is mathematically equivalent to different fractions of carbonate and detritus, as set out in the original equations (2 to 5). The data are shown in figures 1a and 1b. The slope of the least squares line for the plot of $^{230}Th/^{232}Th$ v $^{234}U/^{232}Th$ is 0.866 ± 0.030 while the $^{234}U/^{238}U$ ratio of the calcite component is 1.122 ± 0.040. Inserting these values into equation (1) yields an age of 203 ± 24 Ka.

Discussion

It is now clear from the theoretical and experimental work of Ku & Liang (1984) and others and from our own work, that reliable ages are recoverable from contaminated carbonates. The leachates method also allows ages to be recovered when the isotopes are fractionated differentially from the detritus into the leachates.

For each different sample, good precision on the derived age depends mainly upon data points showing a good spread on the leaching plots, and this in turn depends on the relative concentrations of isotopes in the carbonate and detritus.

The morphology of some cave and hot-spring travertines allows choice of samples having varying degrees of contamination. Compact samples are preferred over porous ones, since the latter are likely to have been leached or isotopically overprinted at some later date. The problem of porosity and isotopic variation is frequently caused by plant roots which later rot away leaving behind decayed organic matter and hydrologic pathways across different strata.

Although not strict examples of the

leachates-alone methodology, the application presented here does go some way toward justifying its adoption. For a much fuller justification we present the experimental results in the paper by Pryzybilowicz *et al.*, (1990).

Until fairly recently most U-Th analyses were carried out using alpha- spectrometry whose dating limit was about 350 Ka, depending on the concentration of available U. U-Th mass-spectrometry has now been developed to a point where a tenfold dating increase in precision is now possible on even quite high ages and this has had the effect of pushing the dating limit back to at least 500 Ka. Therefore the new leaching techniques together with the use of mass-spectrometry promises a bright future for the chronology of those early-man sites associated with secondary carbonates.

Acknowledgements

We thank A. Bietti, for stimulating our thoughts on this problem. We also thank R. Grun, M. Ivanovich, A. Kaufman, and T-L Ku for critical comments.

References

Cherdyntsev, V.V., Kazachevskii, I. & Kuzmina, E.A. (1965). Dating of Pleistocene carbonate formations by the uranium and thorium isotopes. *Geochemistry International.* 2, 794-801.

Ivanovich & Harmon (eds) (1982). *Uranium Series Disequilibrium: Applications to Environmental Problems.*, Oxford U. P.

Kaufman, A. (1971) U-series dating of Dead Sea Basin carbonates. *Geochimica et Cosmochimica Acta*, 35, 1269-1281.

Ku, T.-L., Bull, W.B., Freeman, S.T. & Knauss, K.G. (1979). Th230/U234 dating of pedogenic carbonates in gravelly desert soils of Vidal Valley, southeastern California. *Geological Society of America Bulletin* 90, 1063-1073.

Ku T.-L. & Liang, Z.-C. (1984). The dating of impure carbonates with decay- series isotopes. *Nuclear Instruments and Methods* 233, 563-571.

Przybylowicz, W., Schwarcz, H.P & Latham, A.G. (1990). Dirty Calcites II: U-series dating of artificial calcite-detritus mixtures (submitted to Isotope Geoscience).

Schwarcz, H.P. (1980). Absolute age determination of archaeological sites by uranium series dating of travertines. *Archaeometry* 22, 3-24.

Schwarcz, H.P. (1982). Archaeometry. in. Ivanovich M. & Harmon, R.S. (eds). *Uranium series Disequilibrium: Application to Environmental problems in the Earth Sciences* 302-325. (Oxford U Press).

Schwarcz, H.P. & Latham, A.G. (1984). Uranium serles age determinations of travertines from the site of Vertesszollos, Hungary. *Journal of Archaeological Science* 11 327-336.

Schwarcz, H.P. & Latham, A.G. (1989). Dirty Calcites I: Uranium Series Dating of Contaminated Calcite using Leachates Alone. *Chemical Geology* 80, 35-43.

Figure 1. Leachates isochrons from Vertesszöllös travertines.

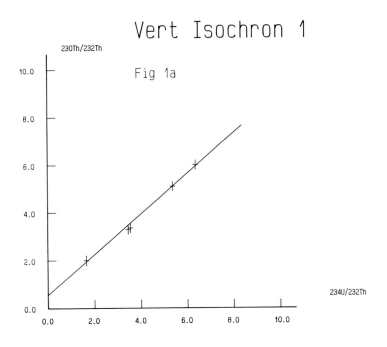

Figure 1a. The slope gives the $(^{230}Th/^{234}U)_c$ ratio to input into the age equation.

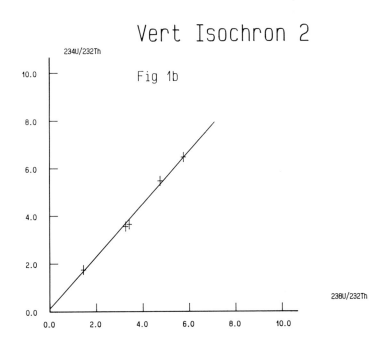

Figure 1b. The slope gives the $(^{234}U/^{238}U)_c$ ratio to input into the age equation.

AN ESR INVESTIGATION OF THE INHERENT DEFECT CENTRES IN MOLLUSC SHELLS

R. Williams[1], C.C. Rowlands, J.C. Evans[2] and A.M. Pollard[3]

School of Chemistry and Applied Chemistry, University of Wales, College of Cardiff, PO Box 912, Cardiff, CF1 3TB.

[1] Present address: Biochemistry Section, King's College London, Campden Hill Road, London W8 7AH.

[2] Deceased.

[3] Present address: Department of Archaeological Sciences, University of Bradford, Bradford BD7 1DP.

Introduction

As explained in our previous paper (Williams et al., 1988), Electron Spin Resonance (ESR) is a method of detecting species with unpaired electrons, which is increasingly being used in archaeology and geology for age determination. As in Thermoluminescence (TL) dating, the quantity of defects, caused by natural irradiation over time, is measured and compared with the rate of production of defects by artificial irradiation. This gives the Equivalent Dose (ED) that the sample has received since formation or zeroing of the signals. Hence, if the annual dose rate for the sample is known, an age can be calculated.

ESR has several advantages over TL. ESR can be used on bio-inorganic materials, where TL is not feasible because of chemiluminescence from the residual organic content (Driver, 1979). ESR is also non-destructive; both of the sample, which may only need to be mechanically broken so as to fit inside the ESR cavity (typically less than 1 cm diameter, but several centimetres long), and of the signal, which can be repeatedly remeasured, so providing higher precision. Unfortunately, however, ESR is a less sensitive technique than TL, and consequently the minimum measurable ED is larger.

This paper discusses typical ESR spectra of irradiated aragonite, bone and calcite, and some of the problems involved in evaluating EDs.

Experimental

Samples were irradiated using a Nuclear Chemical Plant Ltd. Super Hotspot 800 ^{60}Co gamma source. Samples were lowered into the centre of the radial source, and different doses were obtained by leaving the samples within the source for varying times. The source had a dose rate of approximately 500 Gy per hour.

Electron Spin Resonance spectra were recorded on a Jeol JES-RE2X spectrometer, operating at room temperature.

Data line fitting was performed using Technicurve software on an Opus V computer, and the resulting graphs plotted on a Roland DG DXY-980A X-Y plotter, connected via a serial port.

Aragonite

For ESR dating to be applicable to a material, it must have a radiation induced defect which increases linearly with the received natural radiation dose. One way of demonstrating this is to take samples of modern material and measure the signal response to artificial radiation. The situation in aragonite is complicated by there being five superimposed ESR peaks (Hütt et al., 1985; Radtke et al., 1985), with varying sensitivities to radiation.

Typical defect signals for aragonitic samples - in this case produced by six hours ^{60}Co

258

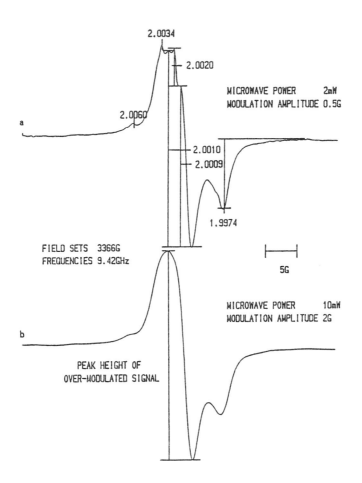

Figure 1. The g values of the signals observed in irradiated aragonite, and the methods of measuring peak heights.

gamma irradiation of modern *Helix aspersa* shells - are shown in Figure 1. ESR peaks are usually referred to by their g values, a measure of the ratio of frequency to magnetic field at resonance. There is a small peak at g = 2.0060, which is described in the literature as non-radiation sensitive, and another small peak at g = 2.0034 which is difficult to measure as it is superimposed on the main signal. The peak with an apparent g value of 2.0009 is actually the bottom half of the major perpendicular component at g = 2.0010, the top half of which is obscured by the minor signal at g = 2.0020. The downswing at g = 1.9974 contains the parallel components corresponding to g = 2.0010 and at least one of the minor peaks. It can be seen that over-modulation (Figure 1b) prevents all available information being observed, as it gives one broad, unresolved signal compared to the several different peaks visible in Figure

1a. This emphasizes the importance of the correct choice of spectrometer operating conditions to obtain the maximum information. In order to investigate how the 'signal size' depends on dose rate, a standard addition technique was employed.

Figures 2a and 2b show Additional Dose Plots for the peak heights of the over-modulated signal and one of the component peaks, g = 2.0010, measured as shown in Figure 1. Good straight lines can be fitted to the data up to 150 minutes irradiation (ie. 1.25kGy), but after this time signal saturation occurs. Second order polynomial curves can be fitted to the total data. In either case, the intercept with the x axis gives the ED. In this case, of course, the ED should be zero as these are modern samples; however the spectrum of unirradiated shell is not totally flat because of

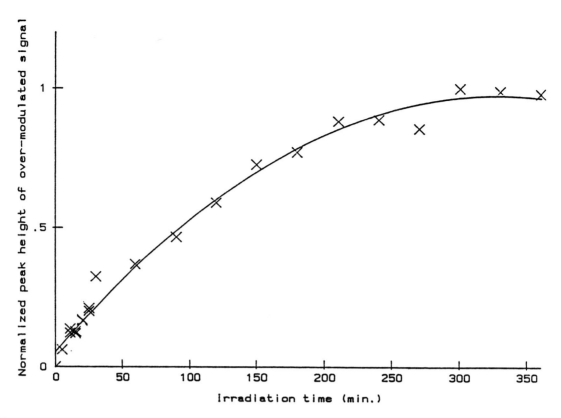

Figure 2a. The effect of irradiation time on the normalised peak height of the over-modulated signal of aragonite.

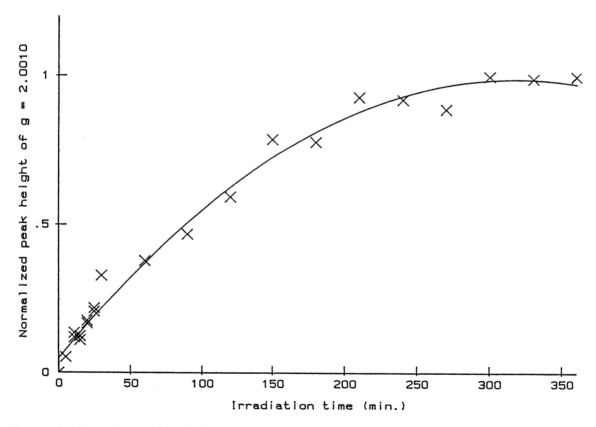

Figure 2b. The effect of irradiation time on the normalized peak height of the g=2.0010 peak of aragonite.

the wide signals due to paramagnetic impurities such as iron and manganese.

Table 1 gives EDs for the data calculated from both linear and second order line fitting. For second order fitting, including the origin as a data point does not affect the answer, but it has a considerable effect on the linear fits. The latter gave larger EDs with a wider spread. Obviously, the systematic error involved will be of less importance when examining older samples. The over-modulated signal is seen to give answers compatible to those from the individual peaks. (Grün et al. (1987) found a similar result for the signal in tooth enamel.) It is not surprising that the results for the over-modulated signal are similar to those for g=2.0010, g=2.0009 and g=1.9974, as the latter are three different ways of measuring the major signal in the spectrum. Therefore, since the errors introduced are not large, the use of over-modulation is possible for aragonite samples. This not only makes it easier to measure the peak height accurately, but also means that lower gains can be used, as the signal size is proportional to both modulation amplitude and to the square root of the microwave power.

Table 1. Apparent Equivalent doses (in minutes of gamma irradiation) for powdered samples of modern *Helix aspersa* shells

| | First Order (Straight Line) up to 150 min. | | Second Order (Curve) | |
	Correlation Coefficient	E.D.	Correlation Coefficient	E.D.
Over-Modulated	0.9838	19.6	0.9945	9.4
g = 2.0020	0.9451	17.1	0.9650	10.2
g = 2.0010	0.9822	17.4	0.9940	8.1
g = 2.0009	0.9612	24.4	0.9928	11.9
g = 1.9974	0.9227	40.4	0.9133	8.6

Bone

Defect signals can also be seen in the mineral fraction of bone and EDs can be derived. However it is not recommended that age determinations are attempted, (see Grün and Schwarcz, 1987), because of the difficulties of obtaining accurate dose rates, due to the ease of uranium migration in and out of the bone. Some workers are still attempting relative dating; either by assuming a constant dose rate for a single site which has had one or more contexts dated by another method, or by using an estimated annual dose rate of 1-2 rad (Sales et al., 1989).

Figure 3 shows the defect signal in a small piece of bone from Pont Newydd Cave, with g values of 2.0021 and 1.9968. An ED determination was attempted. A linear extrapolation gave an ED of 245 minutes of gamma irradiation (ie. 2.04kGy), with a correlation coefficient of 0.4446. A second order polynomial was also fitted to the data, (see Figure 4), this gave an ED of 27.0 minutes of gamma irradiation (ie. 225Gy), with a correlation coefficient of 0.6573. Such a large difference in the answers obtained by the two methods suggest that the signal in this particular sample is near, or may even have reached, saturation, and that an ED (and therefore an age) cannot be obtained.

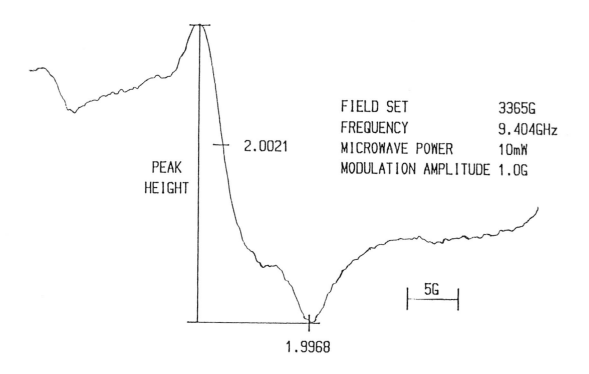

PEAK
HEIGHT

2.0021

1.9968

FIELD SET   3365G
FREQUENCY   9.404GHz
MICROWAVE POWER   10mW
MODULATION AMPLITUDE   1.0G

5G

Figure 3. The defect signal in bone sample NMW PN88 H352.

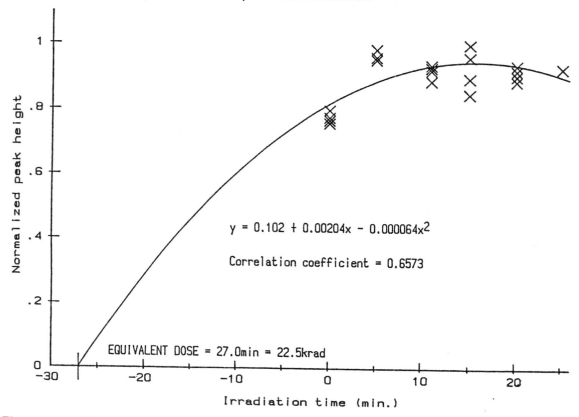

$y = 0.102 + 0.00204x - 0.000064x^2$

Correlation coefficient = 0.6573

EQUIVALENT DOSE = 27.0min = 22.5krad

Figure 4. Additional dose plot for bone sample NMW PN88 H352.

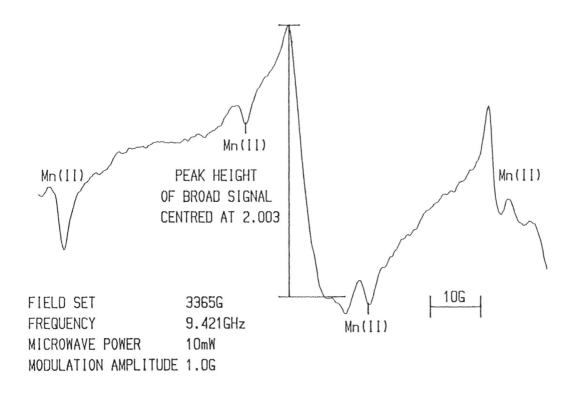

Figure 5. The defect signal in calcite sample NMW PN88 F773.

Calcite

There are at least a dozen ESR signals reported in dating studies of natural calcite; the paper by Smith et al. (1985) discusses nine of them. Figure 5 shows the defect signal in a piece of calcitic flowstone, also from Pont Newydd Cave. The only defect signal visible in this sample is the broad peak centred at $g = 2.0036$; the spectrum appears to be similar to those shown for samples PRI-8 and PARG in Hennig et al. (1985). After artificial irradiation, a very sharp peak appears at $g = 1.9998$, this is due to the occupation of defects produced by grinding. The two lines at the edges of the displayed spectrum are the third and fourth allowed transitions of Mn(II); the two dips on either side of the central defect are due to 'forbidden' transitions of Mn(II). This sample has an unusually low manganese concentration; it is frequently impossible to see the defect signals in calcite because they are hidden under a large Mn(II) signal. Irradiation has no effect on Mn(II) signals, so there should be no possibility of mistaking them for defect signals, however the overlap of defect and Mn(II) signals might produce apparent size changes in the latter.

Conclusions

ESR is a powerful technique which does not destroy the signal being measured, as does TL, but it is apparent that much care must be taken in the selection of experimental conditions. Sample preparation, instrumental parameters, peak selection and data extrapolation can all contribute to erroneous answers. However, bearing these points in mind, its use as a dating technique in archaeology and geology could be invaluable.

Acknowledgements

One of the authors (R.W.) was in receipt of a SERC Science-Based Archaeology Committee Quota Award. The samples from Pont Newydd Cave were supplied by the excavator, Dr H.S. Green, of the Department of Archaeology and Numismatics, National Museum of Wales.

References

Driver, H.S.T. (1979). The preparation of thin slices of bone and shell for thermoluminescence. *PACT* 3, 290-297.

Grün, R. & Schwarcz, H.P. (1987). Some remarks on "ESR dating of bones". *Ancient TL*, 5(2), 1-9.

Grün, R., Schwarcz, H.P. & Zymela, S. (1987). Electron spin resonance dating of tooth enamel. *Canadian Journal of Earth Science* 224, 1022-1037.

Hennig, G.J., Geyh, M.A. & Grün, R. (1985). The first inter-laboratory ESR comparison project phase II: evaluation of equivalent doses (ED) in calcite. *Nuclear Tracks and Radiation Measurements* 10, 945-952.

Hütt, G., Molodkov, A., Kessel, H. & Raukas, A. (1985). Dating of the subfossil Holocene shells in Estonia. *Nuclear Tracks and Radiation Measurements* 10, 891-898.

Radtke, U., Mangini, A. & Grün, R. (1985). ESR dating of marine fossil shells. *Nuclear Tracks and Radiation Measurements* 10, 879-884.

Sales, K.D., Robins, G.V. & Oduwole, D. (1989). Electron Spin Resonance study of bones from the Palaeolithic Site at Zhoukoudian, China. *Advances in Chemistry Series* 220, (Archaeological Chemistry IV), 353-368.

Smith, B.W., Smart, P.L. & Symons, M.C.R. (1985). ESR signals in a variety of speleothem calcites and their suitability for dating. *Nuclear Tracks and Radiation Measurements* 10, 837-844.

Williams, R., Pollard, A.M. & Evans J.C. (1988). An Electron Spin Resonance study of mollusc shells from archaeological contexts. In: *Science and Archaeology*, Glasgow, 1987, ed. E.A. Slater & J.O. Tate, BAR 196, 635-645.

MAGNETIC SURVEY AND DATA PROCESSING APPLIED TO THE LOCATION OF ROMAN KILNS: METHODOLOGICAL EVALUATION AND ARCHAEOLOGICAL RESULTS

A. Casas[1], V. Pinto[1], J.M. Gurt[2], S. Riera[2] and L. Burés[2].

[1]Dpt. Geoquímica, Petrologia i Prospecció Geológica, Facultat de Geologia, Universitat de Barcelona, 08028 Barcelona, Spain.

[2]Dpt. Prehistòria, Història Antiga i Arqueologia, Facultat de Geografia i Història, Universitat de Barcelona, 08028 Barcelona, Spain.

Abstract

The results from a geophysical survey carried out to locate Roman kilns of 'Terra Sigillata Hispanica' at the Pla d'Abella, near the town of Naves (Lleida, Spain), are presented. The existence of these kilns was described in 1912, being the first ones of this kind found in the Iberian peninsula. Therefore, the relocation of these kilns has proved fruitless until now.

Magnetic prospection was used for the survey, since ceramic clays contain haematite particles, which when heated to a temperature above their Curie point and subsequently cooled in the presence of a magnetic field, acquire a stable thermoremanent magnetism.

The measurements were done with a proton magnetometer with a sensitivity of one nano Tesla (nT), and the survey followed a square pattern of one metre. Three significant anomalies were detected, and the residual anomalies, clearly dipolar due to the obliquity of the magnetization vector, displayed intensities of up to 200 nT. The application of the magnetic pole reduction to the experimental data, has enabled us to correct this effect and make the magnetic data relatively easy to interpret.

The results of the survey were confirmed by excavation, and three kilns have been found which have provided large quantities of archaeological data.

Introduction

Pla d'Abella is a small depression of about 2 km diameter in the Catalan Pre-Pyrenees at the meeting of the Cardener and Aiguadora rivers (Figure 1). Pla d'Abella is an important basin, from which many streams drain into the river.

Serra Vilaró started archaeological research on the area in 1912, and in 1925 excavated three Terra Sigillata kilns in the Pla d'Abella. This settlement, however, has never been found since this date.

The first excavations gave little data about the kilns location, structure, technical characteristics, historical context and chronological function.

The next series of exploration began in 1986 with a team of Roman archaeologists and geologists from the University of Barcelona. Their aims were as follows:

1. Archaeological prospection was undertaken of Pla d'Abella and the nearby hills in order to a) re-identify the kilns found in 1912, and b) find out about other archaeological sites in the area which could give information either about population of the region or historical information about the kilns.

2. A typological study was undertaken on the finds from the Serra Vilaró excavations.

3. An excavation of 'La Rectoria', a Roman settlement in the Naves (Solsones region) was undertaken. Cartography and

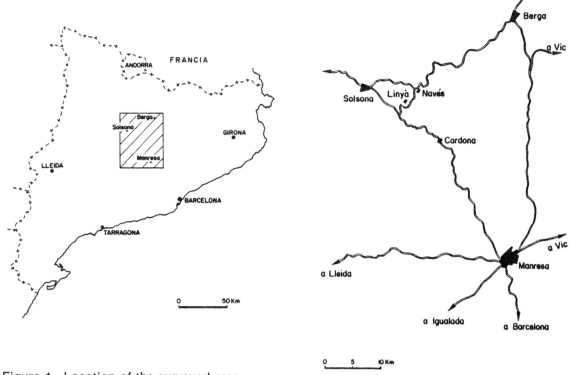

Figure 1. Location of the surveyed area.

Figure 2. Zones where field walking prospection provided important numbers of scattered pottery.

266

aerial photographs were examined, and field walking was conducted in the area surrounding Pla d'Abella.

From fieldwalking Iberic pottery was found in the nearby hills. In Pla d'Abella pottery was located in three different areas which corresponds to three farms in the area: Guingueta, Can Passavant and Pla d'Abella in the highest part of the depression (Figure 2). Iberic amphora and some Terra Sigillata fragments were found in Guingueta. In Can Passavant and Pla d'Abella only artefacts from the Roman period were found.

Magnetic survey

The need for an effective, fast and non-destructive method was essential for the detection of the ceramic kilns of Pla d'Abella, whose location was concealed as described above. Therefore a magnetic survey was carried out.

The reason for selecting magnetic prospection methods was also considered because of its simplicity of handling, easy and fast operation and the contrast between the susceptibilities of the refractory materials used in the kiln walls and the surrounding terrain. This susceptibility contrast is produced by the reorientation of the magnetic domains in the ferromagnetic minerals (generally haematites), which when heated and cooled down in the presence of a magnetic field align along that field. This is thermoremanent magnetism, which is very stable, and is therefore particularly useful for the prospection of refractory kilns and baked clays (Thellier & Thellier, 1951).

Once the geophysical method was decided upon, the choice of the most suitable grid size was based upon the size of the kilns already excavated by Serro Vilaró (1925). Thus a grid interval of 1 metre square was used throughout the survey area.

The magnetic measurements were carried out with a 1 nT sensitivity Geometrics G-816 proton magnetometer, placing the sensor over the terrain surface in order to enhance the detectability of the kilns.

The correction for the temporal variation of the earth's magnetic field was achieved both from successive measurements at the same place and from the data recorded at the nearest geomagnetic observatory. The results obtained by both methods allowed the diurnal correction, with a precision of 5 nT, to be made.

Magnetic interpretation

The magnetic survey detected significant anomalies on a piece of land situated eastwards of the Pla d'Abella field house. The distribution of the total magnetic anomalies, after correction for temporal variation of the geomagnetic field, and subtraction of the regional value, were estimated to be 44956 nT. Three intense anomalies and other low amplitude dispersed ones were evidenced (Figure 3).

Immediately after the location of the main anomaly the measurement network was concentrated to one point each 0.1 meter, in order to determine, with precision, the size of the kiln and to plan any future excavation. The residual anomalies found within this dense network depict a clear dipolar pattern, characteristic of magnetic anomalies, and are due to the inclination of the magnetic field (Figure 4).

In order to facilitate the interpretation of the geometric characteristics of the body responsible for this anomaly, a reduction to pole numerical transformation was applied to the data (Baranov, 1957). This is based on calculating the theoretical field obtained in the case when the inductor field would have an inclination of 90°, as in the geomagnetic poles. The method used to perform this transformation derives from the algorithm proposed by Battacharyya (1965), which expresses the intensity of the magnetic field reduced to pole as a double Fourier series on a regular grid.

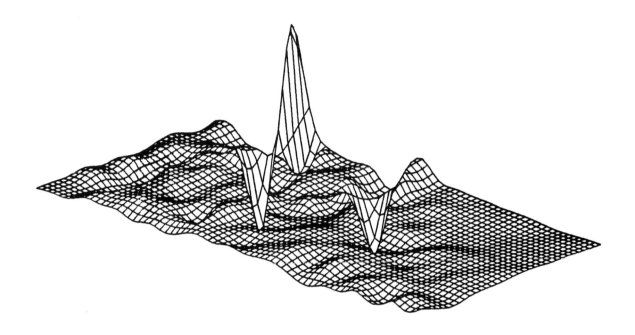

Figure 3. Perspective display of the main magnetic anomalies detected.

In this method the data are transformed first into the wave number domain by means of the Fast Fourier Transform (FFT). The Fourier coefficients are then multiplied by the wave number response of the appropriate digital filter. Finally the resulting Fourier coefficients are inversely transformed back into the space domain yielding the desired transformed data.

The Fourier integral transformation F(u,v) of a function f(u,v) is defined as

$$F(u,v) = \int_{-\infty}^{\infty} \int_{-\infty}^{\infty} f(u,v) \; e^{-2ni(ux+vy)} \; dx \; dy$$

Where u and v are the unit frequencies associated with the x (north) and y (east) coordinate axis, respectively. Since the total magnetic field f(x,y) is only known as discrete points (grid points) their transformation is obtained by using the discrete Fourier transformation:

268

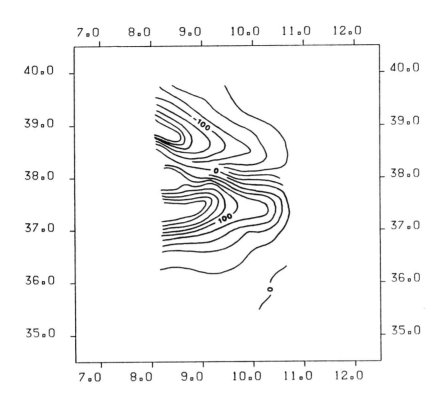

Figure 4. Detailed magnetic anomaly from a 0.1 m grid.

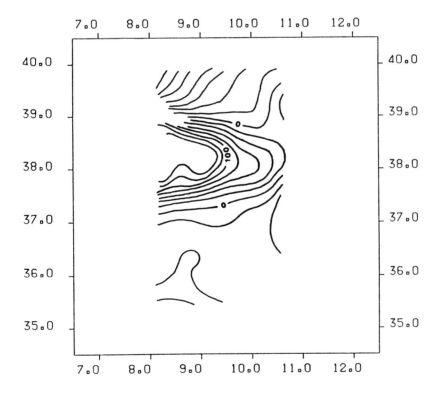

Figure 5. Detailed magnetic anomaly reduced to pole.

$$F(u,v) = \sum_{m=0}^{M} \sum_{n=1}^{N} f(u,v) \; e^{-2ni \left(\frac{Km}{M} + \frac{ln}{N} \right)}$$

where m and n are integers representing grid point locations in the space domain, k and l are integers representing grid point locations in the frequency domain, and M and N are the number of rows and columns, respectively.

Grid or two dimensional filtering may also be described by the convolution formula:

$$F(x,y) = h(x,y) * f(x,y)$$

which is the shorthand notation for the integral

$$F(x,y) = \int_{-\infty}^{\infty} \int_{-\infty}^{\infty} h(x,y) \; f(x-y,y-v) \; du \; dv$$

where

F(x,y) is the input data
f(x,y) is the filtered output
f(x,y) is the filtering function.

Inclination and declination parameters of the magnetic field have been taken such as those of the actual magnetic field in the studied area, ie 58 degrees for inclination and 3 degrees 40 minutes west for declination. The anomaly map obtained by this mathematical transformation shows an elongated monopolar anomaly with an intensity of up to 140 nT and steep gradients on both sides (Figure 5).

This anomaly was interpreted as being produced by one of the predicted kilns, therefore the excavation was started vertically under the anomaly, revealing a kiln at an approximate depth of 30 cm.

A detailed map of each fired clay refractory block from the kiln walls was recorded, using the same grid reference as that used in the magnetic survey. The comparison between the anomaly map reduced to pole, and the structure of the discovered kiln showed perfect correlation (Figure 6).

Furthermore, the magnetic susceptibility of thirteen samples analysed from the kiln walls gave a mean value of 5×10^{-4} cgs, and a range of between 0.3×10^{-4} and 12.5×10^{-4} cgs.

Archaeological results

As the efficiency of the geophysical survey was proved with a view to locating magnetic anomalies, the following objectives were achieved.

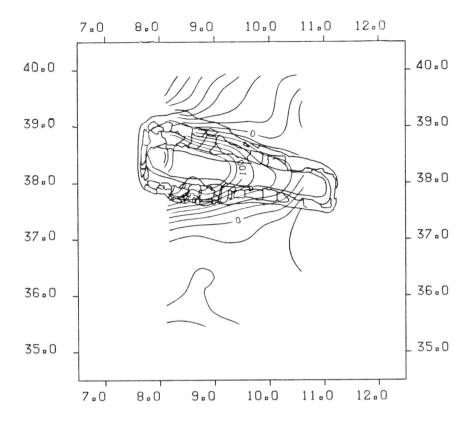

Figure 6. Comparison between the anomaly reduced to pole and the structure of the kiln.

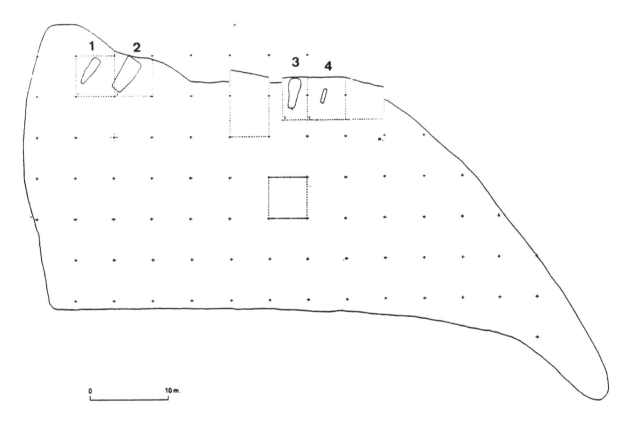

Figure 7. Location of the four excavated kilns.

271

1. Relocation of the kilns and their orientation in the field.

2. The production of detailed plans of the kiln structures to obtain information about their technology.

3. The discovery of future non-excavated kilns in order to provide further information about production characteristics, chronology and furnace structure and function.

4. Location of kiln slags with a view to identifying the stratigraphy which would allow the study of the evolution of the Abella workshop production.

Excavation followed the survey in 1987 and 1988. Four kilns were found, three of them with magnetic anomalies. None of these gave *in situ* finds because they had been previously excavated by Serra Vilaró. The fourth kiln produced no pottery because it was in a very bad condition.

From the excavations of the kilns the following conclusions were drawn:

1. The position of the kilns was corrected from those stated in the 1925 excavation, now being sited in a row near the edge of the hillside.

2. New data on the structure and technology were obtained.

3. Various building elements, grille pieces, clay cylinders relating to air conductance, were found, in addition to clay rings to hold the pottery in the kilns and some other, larger, unidentified rings.

4. From trenches throughout the site, more pottery has been recovered, although no kiln slag has been found.

5. The main pottery found from the kilns was Terra Sigillata Hispanica.

6. Recent studies on the kiln finds have been directed towards typological, chemical and physical, using X-ray fluorescence and X-ray diffraction techniques. It is hoped this characterisation will allow us to be able to identify the products of different pottery workshops, and look at diffusion, trade and organisation patterns. X-ray diffraction and Mossbauer spectroscopy may shed light on manufacturing processes.

7. The excavation in La Rectoria in Naves, has produced a new stratigraphy which indicates a date of the second to first half of the third century for the last period of kiln pottery production.

Conclusions

Magnetic prospection has enabled the relocation of three kilns of Roman pottery in Pla d'Abella. Magnetic reduction to pole has allowed identification of the structures responsible for magnetic anomalies. Detailed plans about the kiln structures has been determined and a hypothesis about their elements has been attempted.

References

Baranov, V. (1957). A new method for the interpretation of aeromagnetic maps: pseudo-gravimetric anomalies. *Geophysics* 22, 359-383.

Bhattacharyya, B.K. (1965). Two-dimensional harmonic analysis as a tool for magnetic interpretation. *Geophysics* 30, 829-857.

Serra Vilaró, J. (1925). Ceramica en Abella. Primer taller de 'Terra Sigillata' descubierto en España. *Memorias de la Junta Superior de Excavaciones y Antigüedades*, n. 73.

Thellier, E., y Thellier, O. (1951). Sur la direction du champ mangétique terrestre retrouvé sur les parois de fours des époques punique et romaine, à Carthage. *C.R.A.S.* 233, 1476-1478.

THE ARCHAEOLOGICAL PERCEPTION OF GEOPHYSICAL DATA

P.N. Cheetham[1], J.G.B. Haigh[2] and S.S. Ipson[3]

[1]Department of Archaeological Sciences, University of Bradford, Bradford, West Yorkshire, BD7 1DP.

[2]Department of Mathematics, University of Bradford, Bradford, West Yorkshire, BD7 1DP

[3]Department of Electrical Engineering, University of Bradford, Bradford, West Yorkshire, BD7 1DP.

Introduction

The application of geophysical survey to archaeological fieldwork involves the collection of data at a large number of stations distributed over a wide area of land. In order that the significance of so large a collection of results may be readily appreciated, it is necessary to use a form of graphical display which clearly illustrates the spatial distribution of the data. A variety of suitable methods of display have been discussed by Aspinall & Haigh (1988), who favoured those techniques where the physical readings are shown as a distribution of variations in intensity or colour over a plan of the area of the survey. The advantages of such techniques include the ease of location of particular features revealed by the survey, and the ability to relate the data to other known features of the site.

The recent rapid development of computer technology has greatly influenced the practical means by which displays can be produced. Only a few years ago the majority of workers had to rely on mainframe computers for their graphical results, which usually appeared in hardcopy form and took several days to produce. The advent of powerful microcomputers now allows workers to expect a high-quality screen display on the same day as results are collected, and one looks forward to the time when portable computers will allow the display to be formed simultaneously with the reading of data.

Clearly such developments can be beneficial only if the display gives the archaeologist a proper impression of his data. As well as information of archaeological interest, the data must contain signals of geological origin and from recent agricultural and industrial usage. There may also be noise arising from the survey instruments (although most modern equipment is sufficiently good to reduce true instrument noise to a negligible level), from interference by external electromagnetic devices, from occasional erroneous readings, and from minor variations in the soil surface. It is important that the archaeological features should be clearly perceived in contrast to other competing effects.

Haigh & Ipson (1989) have discussed the application of image processing techniques to visual displays of geophysical data. This paper is intended to specialise the discussion to one particular type of display, namely the video graphics array (VGA), which is a popular option in the range of PC-compatible computers. The significance of the VGA display is that it is sufficiently flexible to be used in connection with image processing, that it provides an attainable standard for a great number of computer users, and that it is likely to remain a standard for some time to come, albeit with steadily improving alternatives at its top end.

When geophysical data are displayed as a distribution of intensities on a high-quality screen, such as a VGA display, the results have a strong resemblance to a visual image, which provides a natural way of perceiving features revealed by the data. However, it is important to realise that the data set is not exactly equivalent to an image and, in particular, the method of measurement is entirely different from the mechanism of image formation. This paper will demonstrate clearly that some techniques of image processing are directly applicable to the enhancement of a display of geophysical data. It would, however, be wrong to assume that every imaging technique can be applied, and careful consideration of their relevance should be given before introducing more sophisticated methods.

The VGA display used with grey shades

In its standard form the VGA display offers an array of 640x480 pixels, each able to show any of sixteen shades, which may be selected from a palette of 262144 colours. All VGA cards offer at least one alternative resolution, as well as the resolutions associated with the older CGA and EGA cards. In this paper discussion will be mainly confined to the standard resolution, since this is available to the widest range of users; clearly there is some benefit to be gained by going to higher resolution, but it will result in increased times to create the display and may be useful to only a relatively small number of users.

If it is desired to display the data as a distribution of intensity, then it is natural to employ sixteen shades of grey, which may be selected from the 64 shades of grey available within the palette. Although the human eye is capable of distinguishing between many different shades of grey over an extended region (Gonzalez & Wintz, 1987), in fact it is quite difficult to choose sixteen which are clearly visible within a limited area and are perceived as uniformly graded. This is partly because of hardware problems with many video-monitors; the electron guns do not fire for low intensities and the response to increasing intensity is very non-linear. Hence the darkest shade of grey has to be set a fair way up the scale, and its neighbours have to set at wide intervals, gradually decreasing to the lighter end of the scale.

To create a display of regularly gridded data, the available area of the screen may be scaled to represent the overall area of the survey, or a smaller section within it. The screen is divided into squares, one corresponding to each of the geophysical readings of the survey, and each square is shaded with an intensity determined by the magnitude of its reading. Provided that the correspondence between the grey scale and the magnitude of the geophysical data is well chosen, the resulting display gives a clear representation of most of the discernible features. The division of the display into grey squares causes a hard straight edge between adjacent readings, which many observers find to be a distraction from their perception of delicate features within the display. Means of smoothing the variation between adjacent readings will be discussed in a later section.

The VGA scale with colour gradation

It is possible to represent the variation in the magnitude of the readings by colour as well as by grey-scale intensity. Very sharp changes of false colour are often introduced when processing visual images. These are intended to enhance local contrast, with the aim of emphasising particular details in the image. Such sharp contrasts are rarely appropriate for geophysical data; the aim here should be to provide a graduated enhancement of contrast over the whole range of intensities, so that two readings at opposite ends of the scale are more sharply contrasted than two readings of similar intensity.

One obvious way to introduce a graduated enhancement is to use a rainbow spectrum of colours ranging from violet-blue for low magnitudes, through greens and yellows for intermediate values, up to red for high magnitudes. The authors have attempted to combine this rainbow variation with intensity variation, so that the blue shades are made rather dark and the red shades rather light, verging on pink. The combined effect gives a very clear spread of contrast, and has proved to be very useful in highlighting certain aspects of the data.

A particular problem of geophysical survey is balancing readings from different parts of the survey area. On surveys conducted by the University of Bradford, the custom is to divide the overall site into square grids of 20x20 readings, and to survey one grid at a time. The need from time to time to relocate the fixed probes in resistance survey, or to recheck the zero of the instrument in magnetometer survey, often results in discrepancies between adjacent grids. Similar discrepancies are bound to arise, no matter what scheme of survey is employed. The rainbow scale has proved to be particularly useful in ascertaining the magnitude of the discrepancies, so that standard facilities within the program can be used to bring adjacent grids into acceptable balance.

Commercial packages often use a similar range of colours to give a spread of contrast, but cause the order of the colours to be reversed in the yellow and red region of the spectrum. The reason for this reversal is that, since the human eye does not observe very much contrast over the citrus hues (lime-green, yellow, orange), an artificial contrast is deliberately introduced into the middle.

Unfortunately, a direct transition from green to red can only be achieved by going through brown shades, which are far darker than any of the surrounding colours. Consequently a very dark band is introduced into the middle of the intensity scale, making a very sharp contrast with the surrounding regions. There are some occasions when this contrast will be of use in picking out particular features of the geophysical survey, but there are far more occasions when such an artificial contrast will distract from the perception of other features. Overall the alternative method of colour gradation has not been found to be helpful, and the authors have largely abandoned its use in favour of the rainbow scheme.

A third scheme of colour gradation was discussed by Aspinall & Haigh (1988), with a view to application in magnetometer survey. Here positive readings are shown as increasingly saturated shades of red, so that the highest readings are dark red and zero readings nearly white. Likewise negative readings are shown as increasingly saturated shades of blue. This scheme was intended to make a clear distinction between the positive and negative regions of the survey, since they are often associated with different types of feature - positive regions with ditches and

negative regions with baulks and other compacted features. It also emphasises the structure of large positive features, which often have a negative shadow on their northern edge.

An unexpected advantage of the blue-white-red scheme is that near-zero readings are emphasised as the white areas of the display, causing the surveyor to think carefully about the choice of the zero-point, which is often difficult to ascertain in the field. In many surveys it becomes apparent that there is an imbalance between positive and negative regions of the survey, and that the zero-point must be shifted as part of the operation of balancing the different grids. The scheme has also proved to be useful in optimising displays for resistance surveys, although here the balance is between regions of low and high resistance, rather than between negative and positive readings; this application will be discussed in a later section of the paper.

Another application of the blue-white-red scheme is in ensuring that the range of the displayed levels matches the range of readings in the original data. If the levels are adjusted so that the darkest shade of blue and the darkest shade of red are both discernible on the display, then it may be assumed that a good match has been achieved. In the terms of image processing, this operation is known as contrast stretching. Although many colour schemes can be helpful for contrast stretching, the blue-white-red scheme has proved to be particularly advantageous.

A great advantage of the VGA display is that different palettes may be set very quickly. Once the palettes have been selected to give the required range of colours, which may be a time-consuming operation, they may be incorporated into the program in such a manner that the user is able to flick between them almost instantaneously. Many experienced surveyors prefer the grey-scales as the best method of displaying information about a site overall, but it may still be advantageous to have access to the various coloured scales in order to fine-tune the display and to check on some detailed features.

Interpolation between readings

As mentioned earlier, there is a general consensus that the division of the display into grey squares, one for each reading, causes hard edges which may distract from the perception of delicate features. The distraction is probably exacerbated by the use of colour palettes to improve the contrast. It is therefore necessary to find some method of interpolation which gives a smooth gradation between neighbouring readings. Several interpolative formulae are commonly used in connection with regular square grids.

The simplest possible formula is a linear interpolation. However, this works best over a triangular cell, since its definition requires data values at three points, which may be taken as the vertices of the triangle. Thus each of the original square cells has to be divided into either two or four triangles. Experience shows that neither of these gives a particularly good solution, not least because all the contours are made up of linear segments and may have a markedly jagged appearance.

A better alternative is provided by bilinear interpolation, whose definition requires the data values at four neighbouring points, which may be taken to be the vertices of a square cell. The contours are smooth curves (rectangular hyperbolae) through each grid cell, but may not connect smoothly to adjoining curves at cell boundaries. One of the authors has considerable experience in the use of bilinear interpolation for contour plotting (Haigh & Kelly, 1987), and has demonstrated that it is a successful and useful technique; the irregularities at cell boundaries tend not to be very conspicuous.

A popular technique in image processing is bicubic interpolation, whose definition requires the data values at sixteen neighbouring points, which may be taken to be the vertices of the grid cell under consideration, together with the remaining vertices of the eight surrounding cells. The contours are smooth curves through each grid cell and join smoothly to the curves in adjacent grid cells. The technique is somewhat cumbersome to apply when plotting curvilinear contours, since it requires a much lengthier calculation than bilinear interpolation. It can, however, be applied conveniently at the intervening pixels between readings on a square grid. The resulting grey-scale display has an acceptably smooth appearance, having lost the hard edges between squares.

When a bicubic interpolation is displayed with a rainbow scheme of sixteen colours, the improved contrast shows up distinctly the boundary between adjacent intensities. In image processing such visible steps in intensity might be considered undesirable, and are referred to as 'false contouring'. For the display of geophysical data, however, such boundaries are entirely acceptable and represent exactly the same contours as were produced by the earlier methods of presentation. Since bicubic interpolation has a higher degree of continuity at cell boundaries, the new contours are more smoothly rounded than those produced by bilinear interpolation. The intermediate method, biquadratic interpolation, is rarely used for this purpose, since it would have to be defined over a region asymmetric with respect to a given cell.

Obtaining a hardcopy of the display

Once an acceptable display has been obtained as a range of 16 shades on screen, one needs some means of making a hardcopy of it, both as a record and for publication purposes. It is possible to obtain a good hardcopy by translating the display into a pattern of fine dots which can be output to a graphics laser printer.

Each pixel on the screen is interpreted as a group of dots in a 4x4 array on the hardcopy. Level 0, the darkest shade on a monochrome screen, has all 16 dots set, so as to give the appearance of almost solid black on the paper; level 15, the lightest shade, has no dots set and remains white on the paper. This technique of interpreting intensity in terms of a dot pattern is known as 'dithering'; here a systematic dither is in use, since a given intensity is always represented by the same pattern of dots. In fact 17 shades are available on the printer (from 0 to 16 dots), whereas only 16 are available on screen, but one of the printer levels is omitted to increase the contrast in a difficult part of the scale.

A laser printer with the standard resolution of 300 dots per inch allows approximately 560 pixels to be represented across the width of the paper. This is effectively the whole width of the screen (640 pixels) when allowance is made for a narrow column in which information can be displayed to guide the program user. Provided that the printer has a large enough memory, there is no problem in representing the 480 pixel depth of the VGA screen. The same technique may be used to transfer the display to modern high-resolution ink-jet printers.

It is unfortunate that the hardcopy from many laser printers shows a pattern of vertical and horizontal striations superimposed on the desired reproduction of the display. This problem appears to arise from the fact that laser printers do not give an entirely uniform distribution of dots over the whole surface of the paper. The non-uniformity is too small to be noticed in printed text, but is sufficient to be apparent against delicately contrasted shades of grey, particularly in the middle density region. In the authors' experience, the striations are yet more apparent on output from ink-jet printers.

The impact of the striations might be reduced by using random dither patterns for the dot distributions. On the other hand, if the dot pattern were sufficiently random to get rid of the striations, then it would probably remove significant detail from the image of the data. The hardcopy from the laser printer is superior to the popular dot-density presentation not only because it offers higher resolution, but also because its systematic structure allows detail to be preserved.

All the figures for this paper, with the exception of Figure 3, have been prepared by these techniques. Figure 1 shows the results of a resistance survey over an area with considerable variation of background values. Contrast stretching and bicubic interpolation have been applied and several contour curves are distinctly visible, although striations cause interference in some parts of the diagram.

Sampling frequency and sampling pattern

When laying out a standard 20x20 metre grid, there is a tendency to assume that the readings must be taken at one-metre intervals. Experience has shown that this is often a satisfactory sampling frequency, giving a reasonable compromise between speed of survey and resolution of detail within the results. When the data are examined by sophisticated methods, such as those discussed in this paper, it becomes apparent that the situation is not always as satisfactory as it might appear.

There are cases where a collection of apparently randomly varying readings begins to assemble itself into a systematic pattern as the imaging techniques are applied. The initial failure to observe the pattern is often caused by the interaction between the underlying data and the sampling pattern. Such an effect is an example of a phenomenon known in signal processing as 'aliasing'. Bicubic interpolation has some anti-aliasing properties and, in some circumstances, is capable of restoring the appearance of the underlying data.

The only certain way to avoid aliased artefacts is to increase the sampling frequency, which should be at least twice the highest frequency component present in the original data. In practice, it is difficult to predict what that highest frequency is likely to be in archaeological geophysics. Instead, the surveyor needs to receive some warning that the data contains information at higher frequencies than are reliably observable with the present sampling structure. On receipt of such a warning, a decision can be made as to whether it is worth increasing the sampling frequency. Likewise, if the data contains only low-frequency information, a decision could be made to decrease the sampling frequency, and hence to allow the available survey effort to be used to best effect.

It has been contended that aliasing can be avoided by the use of a randomised sampling pattern. It is difficult to discuss this point definitively, since the standard sampling theorems do not apply to random data and consequently the topic is not introduced into the majority of standard works on image processing. It may be that aliased effects are converted into a different and perhaps less obtrusive form. It is certainly true that the efficiency of information gathering will be reduced, since randomisation will cause certain areas to be oversampled and other areas to be undersampled. The experiments of Fletcher & Spicer (1988) do not indicate any overall improvement from random sampling.

A more important concern is that any systematic sampling pattern is likely to be subject to small positional errors. The effect of such errors is not normally taken into account, but is likely to be quite small provided that any positional error is substantially smaller than the sampling interval. Consequently surveyors who wish to improve the quality of their data by increasing their sampling intensity must also take steps to improve their positional accuracy. Hence there are many practical problems to be solved if the quality of field data is to be improved.

Non-linear scales

The majority of users find bicubic interpolation to be a natural method to display data, and tend to use it automatically in preference to the grey squares. There are, however, some problems with it, which can be troublesome if the user does not take care. These usually occur in regions of very rapid intensity change, as in the case of odd points with very high or very low readings sitting against a fairly uniform background. When bicubic interpolation is applied these odd points may appear to be surrounded by

haloes of contrasting values; this is best seen by using the blue-white-red colour scheme, when high readings will be seen as a red dot surrounded by a blue halo, and low readings as a blue point surrounded by a red halo. The halo is caused by the interpolant trying to cope with too rapid a change and overshooting as a result. It has no true relationship to the original data, and is known as an 'artefact' of the numerical method. Similar artefacts are liable to occur with almost any numerical method which is powerful enough to cope with a wide range of situations.

Several such haloes are clearly visible in the upper left-hand section of Figure 1. In most cases the odd readings which cause these problems are probably wildly incorrect, and may be regarded as random noise spikes. They may appear in a variety of ways: because resistance probes have not been inserted properly into the ground; by some local object, such as a stone or piece of metal, interfering with the reading; through some failure within the recording equipment. Such random noise can be smoothed away by the application of a gentle spike-removing routine to the data, but care must be taken to ensure that no new artefacts are introduced. In the case of Figure 1, the surveyor suspected that some of the anomalous readings are correct, and wished to retain them in the data. The spike-removing routine cannot then be applied, and other action must be taken to remove the halo artefacts.

In many surveys, both resistance and magnetometer, the number of low readings greatly exceeds the number of high readings. Consequently, when the data are shown with the blue-white-red intensity scale, the area shown blue is much greater than the area shown red, indicating that the range of display colours is not used to best effect. This situation can be remedied by using a non-linear scale of displayed intensities, with much smaller intervals at the low end of the range than at the high end of the range. Scollar et al. (1986) have suggested that the arc-tangent transformation is an appropriate means to introduce such a non-linear scale.

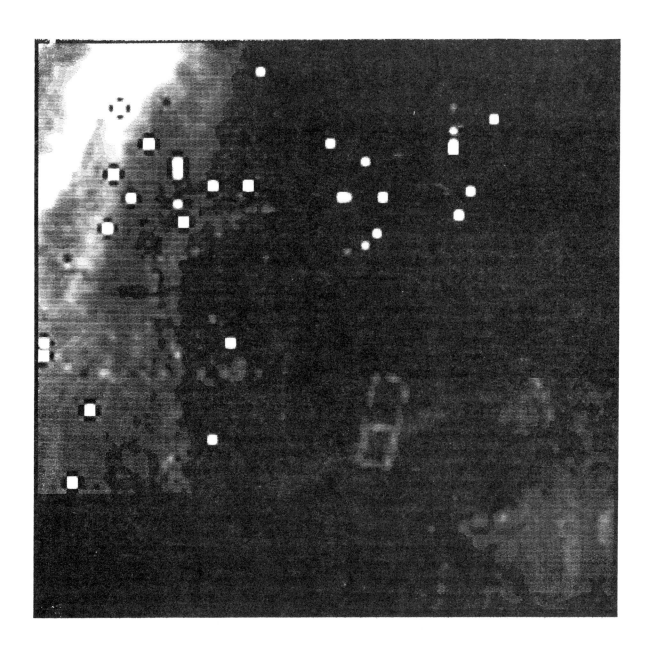

Figure 1. Results of a resistance survey at All Hallows Hill, Tickhill, South Yorkshire. The data are displayed on a linear grey-scale with bicubic interpolation. The overall extent of the survey is approximately 100 m by 100 m.

The arc-tangent transformation has a number of advantages over the more familiar logarithmic transformation. It is capable of handling zero values, which do not have to be shifted artificially. It can be applied to both positive and negative data in a symmetrical manner, as is essential for magnetometer surveys. The interval between levels for low values can be reduced by a ratio which is set as a user-definable parameter. These advantages have persuaded the authors to introduce the arc-tangent transformation into their program, although they do not apply it in quite the manner specified by Scollar et al.

It should be noted that application of the transformation does not alter the underlying data, but simply the manner in which they are displayed. Hence it can be used without any significant risk of introducing artefacts, except that the user should be aware of the possibility of emphasising certain types of feature at the expense of others. Bicubic interpolation is applied after the arc-tangent transformation, so that the variation in high values is apparently reduced because of the large interval between successive levels; as the reduction ratio for low values is increased, the halo round high values can be seen to weaken and eventually to disappear.

The more important effect of introducing the arc-tangent transformation can also be seen to best advantage on the blue-white-red display. As the strength of the non-linearity is gradually increased, the blue areas of the screen tend to decrease in size and the red areas to increase. If the non-linearity can be adjusted to make the blue and red areas more or less equal in size, then it can be said that good use is being made of all available screen colours, and that the data is displayed to best effect. This operation has a strong resemblance to a technique of image processing known as histogram equalisation, although the latter implies an attempt to achieve a balance between individual levels, not merely between the bottom and top halves of the scale.

This simple approach to histogram equalisation, combined with bicubic interpolation and with contrast stretching, has proved remarkably effective in bringing out the details of a large number of surveys. In many cases it has revealed patterns and structures which were not at all visible in more primitive forms of presentation. The results of its application to the data of Figure 1 are shown in Figure 2, where the following points should be noted: the random noise spikes are relatively inconspicuous, and their surrounding haloes have disappeared; there is a much more acceptable range of contrast over the whole area of survey; and features are clearly visible in both the high-resistance and low-resistance regions of the survey. It is interesting to note that there are two isolated low-readings in the high-resistance section of the survey; the introduction of a strongly non-linear scale has caused them to be surrounded by haloes in Figure 2.

Of course, there are plenty of cases where the prescription of the last paragraph cannot be expected to work too well. There are a few examples of resistance survey where large readings are more numerous than small values, so that the red part of the display dominates the blue part. Under these circumstances it is not appropriate to apply an arc-tangent transformation, but rather a tangent transformation, which will expand the intervals between lower lying levels and compress those between higher levels. As yet such a tangent transformation has not been introduced into the authors' program, but it is under active consideration.

In more complicated examples there are a large number of very low readings and a large number of very high readings, but relatively few readings with intermediate values. Such cases would require a tangent transformation with the transformed origin shifted into the intermediate values, corresponding more nearly with the approach of Scollar et al. (1986), rather than with author's approach. There is a strong case for taking another careful look at the application of tangent and arc-tangent transformations. Alternatively it may be worthwhile to seek a completely different approach to the problem.

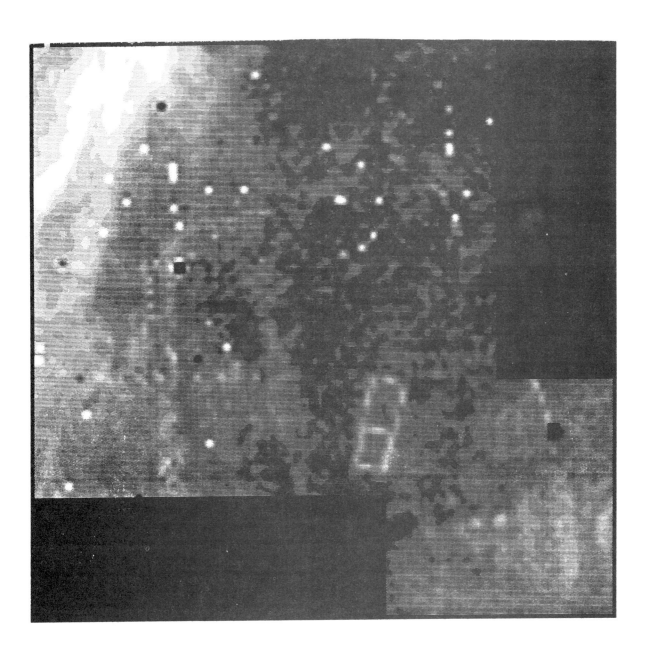

Figure 2. The data of Figure 1, displayed with a strongly non-linear grey-scale (arc-tangent transformation), applied before bicubic interpolation.

Figure 3. Dot density plan of a resistance survey around the guest house at Kirkstall Abbey; reproduced from Chamberlain & Haynes (1983).

A case study; the guest house at Kirkstall Abbey

Over the past decade extensive resistance surveys have been carried out in the grounds of Kirkstall Abbey, Leeds, and one of the authors (PNC) has been closely associated with the work. Although many of the data must remain confidential in advance of publication, the section of the work in the vicinity of the guest house has already been published (Chamberlain & Haynes, 1983); the dot-density plan from the original publication is reproduced with permission in Figure 3. The machine-readable versions of the original data seem to have been lost, but fortunately someone recorded the individual readings on a large sheet of sectional paper. It has been possible to transcribe this record into computer files, each containing the standard Bradford grid of 20x20 readings. The authors' program was then used to reassemble the data from the files, ready for display.

Figure 4 is a laser print showing the original data after contrast stretching and bicubic interpolation, but with a linear grey-scale, so that the data are represented with their natural intensities. Striations have caused some degradation of contrast in the final

Figure 4. The data of Figure 3, displayed with a linear grey-scale and bicubic interpolation.

283

print, and rather more detail was visible in the screen display. If Figure 4 is compared with Figure 3, it will be seen that much more detail is now visible, in spite of the degradation; it should be noted that the order of the grey-scales is reversed between the two figures, so that high-resistance regions are shown dark in Figure 3, but light in Figure 4.

Some of the features indicated by Chamberlain & Haynes, but not clearly apparent in Figure 3, are now distinctly visible; these include the modern trenches (indicated as broken lines numbered 7, 8, 18 and 20 in Figure 3), the Victorian bandstand (numbered 3), and Victorian paths (numbered 4, 5, 6 and 19). The mediaeval features are not as clearly visible as the later features in Figure 4; something shows of the suspected great barn towards the left-hand side of the diagram, of the kitchen to the lower left of the guest house, and of gardens and a range of buildings to the right.

Following the standard prescription, an attempt was then made to improve the display using the arc-tangent transformation to achieve a more balanced histogram. Some improvement was achieved in the locality of the suspected great barn, but the appearance of the buildings at the right-hand side degenerated slightly. The reason for this unevenness in the results is clearly apparent; the detailed evidence for the mediaeval structures is set against a strong variation in the background readings, probably due to geological features. When structures are set against a low-level background, the arc-tangent transformation will increase the contrast, and hence emphasise them; when they are set against a high-level background, it will decrease the contrast, and hence make them less obvious.

The circumstances where one wishes to enhance detail in both the high-level regions and in the low-level region are fairly unusual, but they are likely to occur sufficiently often to make it worthwhile seeking an alternative solution. A possible application of image processing is discussed in the next section.

The use of high-pass filters

In the present problem, it has been shown that the perception of the mediaeval structures is effectively a matter of discerning fine detail against a more slowly varying background. In terms of image processing, this might be described as emphasising the high-frequency information at the expense of the low-frequency information. Such an emphasis can be achieved by means of Fourier filtering techniques.

A fast Fourier transform is applied to the data set to produce the equivalent distribution in frequency space. To achieve such a transformation the data set must be augmented by null-values to make up a complete rectangular image. The low-frequency components of the distribution are eliminated through the application of a high-pass Butterworth filter, designed to give a very smooth cut-off at the low-frequency limit and hence to avoid resonant artefacts in the final image. An inverse Fourier transform is then applied to the filtered distribution to produce the filtered image, which is shown in Figure 5.

All the details that were observed in Figure 4 are more sharply visible in Figure 5, but set against a more uniform background. All the modern and Victorian features are defined with increased clarity, and the mediaeval features in particular can now be discerned with greater certainty. A major reason for the improved clarity is that the absence of strong variation in the background allows a greater degree of contrast stretching to be employed. A side-effect of emphasising small details is that larger uniform regions may be reduced to the average level. There is evidence that some garden features may have suffered from this effect in the present study. On the other hand, it is usually the smaller-scale features which are of archaeological significance.

One artefact of the filtering technique is that sharp edges may be replaced by double borders, with light and dark regions on opposite sides of the boundary. This problem is very apparent round the edges of the survey area, but does not appear to have had

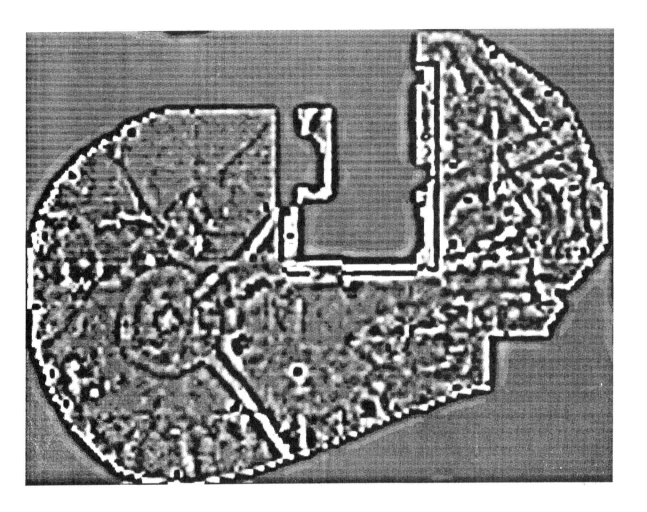

Figure 5. The data of Figure 3, following Fourier transformation and the application of a high-pass Butterworth filter; displayed with a linear grey-scale and bicubic interpolation.

much effect in the interior of the survey. Nevertheless it is important to watch out for such artefacts of the method. Overall the filtering technique has proved to be appropriate, and remarkably successful, for this particular site.

Conclusion

A grey-level display on a VGA monitor has been shown to be an extremely good means of examining the results of geophysical survey. The ability to obtain hard copies quickly and conveniently through a laser printer greatly enhances the utility of the procedure. Appropriate application of simple image-processing techniques - bicubic interpolation, contrast stretching, and histogram equalisation - can be made to improve the presentation of the results. Although the use of monochrome images is usually satisfactory for the presentation of the results, the discreet use of colour may aid the process of selecting the optimum presentation; the blue-white-red scale has proved to be particularly helpful in this respect.

There are cases where the standard prescription fails, and under those circumstances it is worth looking at more sophisticated techniques from image processing. The authors have presented an example where high-pass Fourier filtering has been entirely successful, and they have found this technique to be applicable on many other occasions. Unfortunately, because of its requirements for the storage of large arrays, this technique is not readily transferrable to PC-compatible computers running under MS-DOS, and hence must be regarded as rather more specialised. The next generation of microcomputers should have more flexible operating systems, and the filtering technique might then be made available to the general user.

Yet more sophisticated techniques of image processing exist, and particular emphasis has been placed on non-linear methods of enhancement, of which maximum entropy is the best known (Frieden, 1979). Although such methods have been devised to avoid visible artefacts of the types discussed in this paper, they are based on properties of visual images which may not be appropriate to all forms of geophysical data. Hence their range of applicability is uncertain at the present time.

Enhanced versions of the VGA display are currently being marketed, allowing 256 different colours to be displayed with a resolution of 640x480 pixels. Unfortunately only 64 shades of grey are available with the VGA palette, preventing the full flexibility of Super VGA being used for grey-scale images. To increase the number of grey levels from sixteen to 64 would give a noticeably smoother image, usually without visible contours, but should not be crucial in the analysis of the majority of data. It would, however, be possible to produce non-linear scales by selecting from the wider range of shades available. Consequently non-linear scales could be displayed almost instantaneously, rather than having to wait for an extensive calculation, as at present.

Another relevant development is in the increasing portability of powerful computers. Within the next few years, most of the techniques discussed in this paper will become available to the surveyors while they are still in the field, so that data may be inspected as they are collected. It will be interesting to see how the availability of sophisticated techniques of display will affect the quality of the raw data.

Acknowledgements

The authors are grateful to Mr A. Aspinall for making available the data which form the basis for Figures 1 and 2, and to Mr J.A. Pocock who released the original artwork for Figure 3.

References

Aspinall, A. & Haigh, J.G.B. (1988). A review of techniques for the graphical display of geophysical data. In: *Computer and Quantitative Methods in Archaeology, 1988*, International Series 446, ed. S.P.Q. Rahtz, pp. 295-307. British Archaeological Reports, Oxford.

Chamberlain, M.P. & Haynes, M. (1983). A resistivity survey at Kirkstall Abbey, Leeds, West Yorkshire. In: *Geophysical Surveys*

1982, Occasional Paper 3, ed. J.A. Pocock, pp 117-39. School of Archaeological Sciences, University of Bradford.

Fletcher, M. & Spicer, R. (1988). Clonehenge: an experiment with gridded and non-gridded survey data. In: *Computer and Quantitative Methods in Archaeology, 1988*, International Series 446, ed. S.P.Q. Rahtz, pp. 295-307. British Archaeological Reports, Oxford.

Frieden, B.R. (1979). Image enhancement and restoration. In: *Picture Processing and Digital Filtering*, 2nd edn., ed. T.S. Huang, pp 177-248. Springer-Verlag, Berlin.

Gonzalez, R.C. & Wintz, P. (1987). *Digital Image Processing*, 2nd edn. Adison-Wesley, Reading, Massaschusetts.

Haigh, J.G.B. & Ipson, S.S. (1989). Image processing in archaeological remote sensing. In: *Computer Applications in Archaeology, 1989*. British Archaeological Reports, International Series 548, Ed. S.P.Q. Rahtz & J. Richards, pp.99-109. BAR, Oxford.

Haigh, J.G.B. & Kelly, M.A. (1987). Contouring techniques for archaeological distributions. *Journal of Archaeological Science* 14, 231-41.

Scollar, I., Weidner, B. & Segeth, K. (1986). Display of archaeological magnetic data. *Geophysics* 51, 623-633.

AN EXPERIMENTAL STUDY OF GEOPHYSICAL METHODS

Yasushi Nishimura

Center for Archaeological Operations, Nara National Cultural Properties Research Institute, 2-9-1 Nijo-cho, Nara-shi 630, Japan.

In 1988, a hillside kiln site 13 x 18 m was to be excavated by the Aizu Wakamatsu City Board of Education in Fukushima Prefecture, Japan, as part of an attempt to determine the extent and nature of the ancient kilnworks. This site was known to be a kiln site by the scattering of wasters and kiln body fragments found on the surface. Professor Tatsumasa Douke of the Tokyo Institute of Technology viewed this excavation as an opportune time to test several surface prospecting methods for comparative use under his project, 'The Application and Evaluation of Physical Prospecting Methods in Archaeology', funded by the Ministry of Education. Application was made by Professor Douke to Aizu Wakamatsu to carry out comparative surveys by members of his project.

Prior to the excavation, the site area was surveyed with four different varieties of prospecting equipment: magnetometer, resistivity, electro-magnetic and ground radar. Two magnetometers were employed for comparison with each other: the FM18 fluxgate gradient magnetometer produced by Geoscan Research in England, and the G-856 proton total intensity magnetometer made by Geometrics in the United States. The equipment utilised for the resistivity survey was the RM-4 resistance meter also made by Geoscan Research. The electro-magnetic survey was conducted with the EM38 made by Geonics in Canada; and the Japanese KDS-3AM produced by the Koden Company was employed for the ground radar survey.

Subsequent to the surface surveys, the area was excavated to reveal an 8th century tunnel kiln for firing Sue stoneware in the left centre of the survey area running perpendicularly upslope. One portion of the kiln roof remained, but its absence in the upper half revealed the stepped floor of the firing chamber. The kiln feature measured 4.5 m long and 1 m wide.

Magnetometer Survey

The total intensity survey with the G-856 utilised a measuring unit of 0.1 gamma and a grid unit of one metre; it has a measuring range of 120 gamma. The gradient magnetometer survey with the FM18 was conducted using a measuring unit of 1 nT and a grid unit of one metre; the range of the FM18 is only 60 nT (Fig. 1).

Both magnetometers produced two sets of high and low anomalies in the approximate centre and the lower right-hand corner of the target area. On the basis of these survey data alone, we hypothesized the existence of two kilns in the survey area. But as we later learned, this evidence was misleading.

In comparing the results of the two kinds of magnetometers, minor differences were noted in the way anomalies were formed and how they related to the archaeological feature itself. However, the source of the differences are not yet known and cannot yet be evaluated for their significance. The clarification of these differences needs further research.

Resistivity survey

For the resistivity survey using the RM-4 resistance meter, we employed a frame on which five probes were positioned 30 cm apart. By switching individual probes on and off, we can adjust the number of probes from two to five, and the probe spacing from 30 to 120 cm. For this survey, we use two probes with the interval set at 90 cm.

MH33 G856

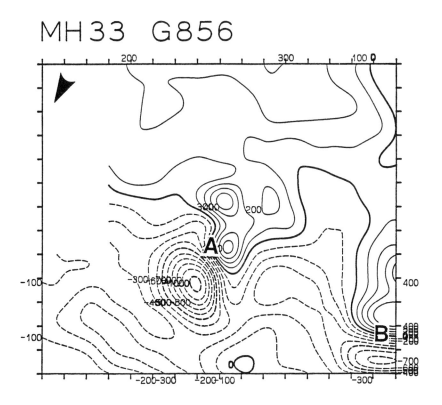

Figure 1a. MH33 G856 (Proton) Total Intensity

MH33 FM18

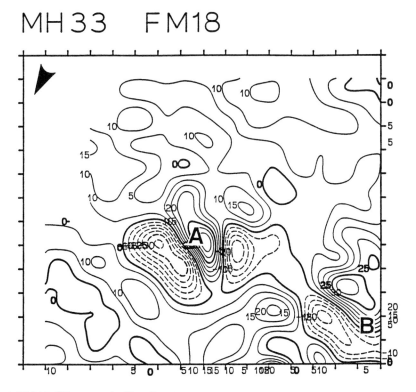

Figure 1b. MH33 FM18 (Fluxgate) Gradient

The resulting patterning of resistivity levels over the site was very different from the patterning of results in the magnetometer survey. The site was divided into left and right halves by a diagonal juncture between low resistivity values in the lower left and high resistivity values in the upper right. Thus what looked like two similar structures in centre and lower right in the magnetometer survey were shown to correspond to two very different structures in the resistivity survey. From experience, these two areas were thought to be an area of weathered bedrock sand (high resistivity) such as would be found in sterile subsoil located at shallow depths, and an area of moist soil (low resistivity) such as would be found in thick topsoil (Fig. 2).

Ground radar

In the KDS-3AM made by the Koden company, the transmitting and receiving antennae are contained in one unit, and the frequency is 167 mHZ. This is the lowest frequency unit available among the products of three companies in Japan; consequently, it can penetrate very deeply.

The antennae are extremely heavy due to their multi-layered shields for preventing leakage of the signal. On a hill slope where wheels could slip or drag because of the weight, there was the danger of not obtaining a correct reading of the signal since the distances were measured by wheel turns. This is one disadvantage of using this machine on sloped ground.

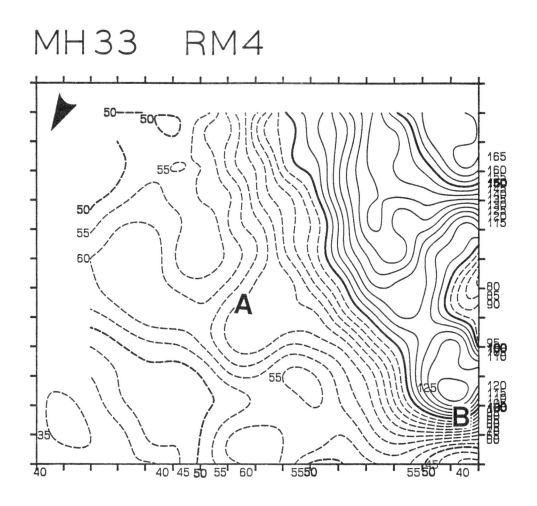

Figure 2. MH33 RM4 (Probe Spacing 90 cm).

We believe that effective depth of the signal at this site was 1.5 m. The plotted results of the radar survey show a stratigraphic section with signal disturbances between 0.5 and 1.10 m depth in the left centre area and parallel strata rising to the right-hand edge. Between these two patterns occurred a vertical area of broken signal which we associate from experience with soils of heavy water content. Three stratigraphic units could be perceived: the undisturbed subsoil, an old disturbed surface, and the topsoil (Fig 3). Within the disturbed area occurred some arched signal traces that were interpreted as belonging to the kiln structure. This identification, however, was confirmed from the subsequent excavation of the site, but it would have been very difficult from the radar data alone to perceive the presence of a kiln without extra knowledge of the site. The signal from the kiln was lower than expected, and it is now understood that a kiln is a difficult target to detect in a radar survey.

The rising parallel layers in the right-hand portion of the target area correspond to the area of high resistivity in the resistivity survey and to one of the anomalies in the magnetometer survey.

Electro-magnetic Survey

The electro-magnetic method has been applied unsuccessfully in many case studies, but it was decided to try it again since the objective (the kiln) was shallowly placed.

In this survey, the EM38 was placed directly on the ground and measurements were made in both vertical and horizontal modes. The distance between coils on the instrument was one metre, the coils being positioned on both ends of the bar. The unit of measurement was mho = 1/ohm, with the vertical mode penetrating to 1.5 m depth and the horizontal mode penetrating to half that depth in measuring the degree of conductivity.

The results of the vertical mode showed that conductivity values were high in the downward slope portion of the survey area and low in the upslope portion of the survey area. The survey area thus seemed to be cut horizontally into areas of high and low resistance rather than diagonally as known from the resistivity survey. Although this division was slightly different, the upper right and lower left portions of the target area produced comparable results in both the resistivity and electro-magnetic surveys.

However, the distribution of relative conductivity at shallow depths, measured by the EM38 in its horizontal position, was completely different from that of the vertical mode. These results cut the target area vertically into left and right halves with high conductivity on the right and low conductivity on the left. Not only did these results differ from those described for the vertical mode, they are completely contradictory to the resistivity results. It is not known why these results are so different from the rest (Fig. 4).

Comparisons

In comparing the results of the above surveys with each other and with subsequently excavated data, several points are clear. Excavation proved that the bedrock did rise up to the surface in the right-hand portion of the survey area. The resistivity meter did distinguish between the soil in the kiln area and the bedrock but did not successfully identify the kiln structure itself. The excavation data also made it clear that the magnetometer was producing the same signal for the kiln and the rising bedrock. Thus, although the magnetometers pinpointed the location of the kiln structure very accurately, they could not distinguish it from the location of subsurface bedrock since both were represented by similar dipole anomalies. The shallow bedrock was very clearly recovered in the ground radar survey, and the distribution of water-laden soils in the kiln area was illustrated. The presence of these soils as shown in the ground radar survey then allow us to interpret the reasons for the low resistivity results in the left-hand portion of the target area as measured by the resistivity meter.

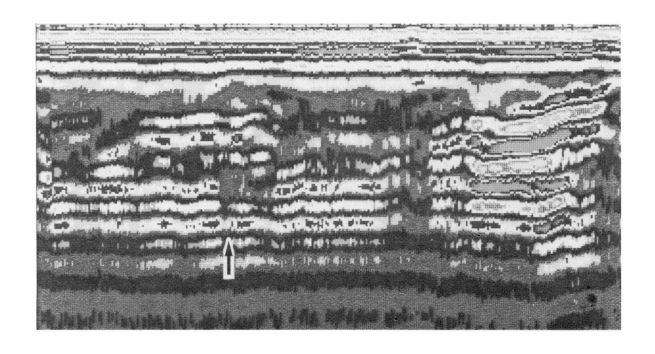

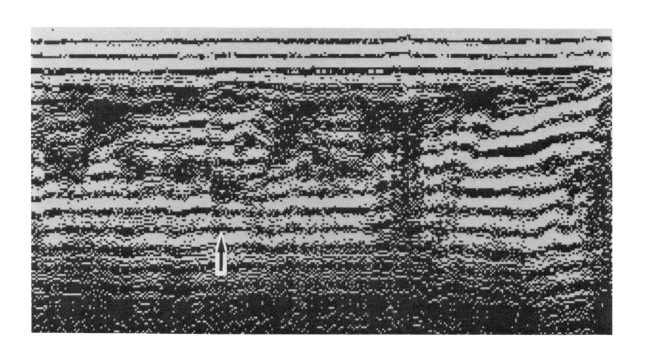

Figure 3. Ground radar results.

MH33 EM38-V

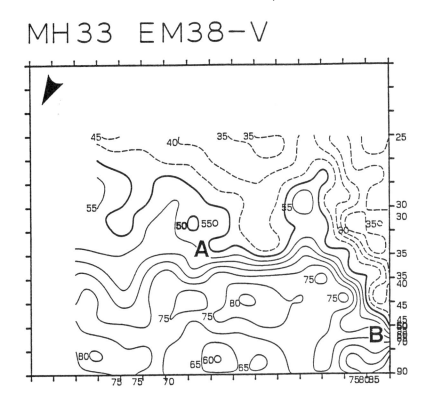

Figure 4a. EM38 Vertical Mode

MH33 EM38-H

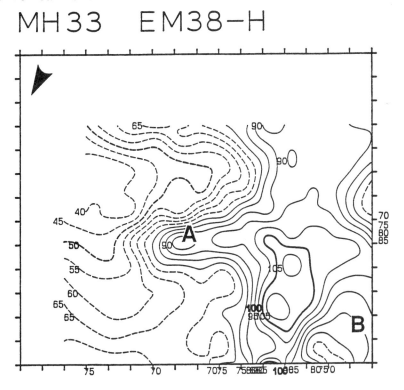

Figure 4b. Horizontal Mode

From these results, it should be emphasised that the use of multiple survey methods with different capacities and different interpretational frameworks is very advantageous in investigating archaeological sites from the surface. The fact that no single method gave unequivocal results is important to recognise. Except for the electro-magnetic survey, each method had its individual contribution to make to the overall interpretation and thus could not have been dispensed with. On the other hand, even with multiple forms of data, the nature of this site could not have been fully known without the information from the surface artefact collection and excavation. We are not yet at a point where surface geophysical methods can inform us completely on the subsurface structure. More correlated data are necessary to continue to build up the interpretational framework for geophysical survey in archaeology.

Some specific points are as follows:

1.   There are differences in the kind of results obtainable from the total intensity and gradient measures in magnetometry survey. In particular, the plotted results of the former usually occur in graded bands of values whereas the results of the latter occur in scattered spots of differing values. It is surmised that the graded values of the total intensity measures possibly reflect changes in the subsurface stratigraphy more than the value distributions of the fluxgate results.

2.   Although resistivity survey did not seem to be efficient in detecting this buried kiln, given drier conditions the method might be useful in detecting contrasts between the feature and the surrounding soil. We regret using only the twin-probe array in this investigation and believe it would have been better to use the Wenner array under these wet conditions.

3.   As we supposed, the electro-magnetic method was not as useful in obtaining information on the stratigraphy as the resistivity method. Also, since this methods cannot be used in urban areas because of noise interference, its sphere of usefulness is extremely limited. As elsewhere, we use this equipment as a metal detector in many cases.

4.   The ability of the radar survey to show the actual kiln feature was worse than we expected because of water retention in the soil. It is recommended that a resistivity survey be conducted prior to the radar survey in order to determine the water content distribution in soil.

IMAGING ARCHAEOLOGY BY ELECTRICAL RESISTIVITY TOMOGRAPHY: A PRELIMINARY STUDY

Mark Noel[1] & Roger Walker[2]

[1]Department of Archaeology, University of Durham, 46 Saddler Street, Durham DH1 3NU.

[2]Geoscan Research, Heather Brae, Chrisharben Park, Clayton, Bradford BD14 6AE.

Summary

This paper presents the latest results from a series of numerical and field experiments aimed at developing a 3-D prospection system based on a multielectrode resistivity array. Image sections through archaeological structures are now constructed tomographically using a traversing array. Results from numerical simulations are encouraging and two resistivity sections have been obtained through buried features at Fountains Abbey. The success of the field and theoretical studies have stimulated the construction of a new resistance meter which will automate the collection of resistivity tomography data.

Introduction

Geophysical surveys now play an important role in archaeological resource assessment and excavation planning. The usual aim is to produce an 'area' map' showing lateral variations in some physical property from which buried archaeological features can be located and possibly identified. The first geophysical surveys in archaeology involved measurements of ground magnetism, (ie. surface magnetic fields) and electrical conductivity and these remain the principal prospecting methods in use today (Aitken, 1974).

Modern instruments, incorporating recent advances in electronics and data processing, now enable rapid area surveys of geomagnetic field gradient or ground resistivity (Aspinall & Walker, 1975; Papamarinopoulos *et al.*, 1986). Nevertheless, there remain serious constraints in an archaeological interpretation of the resulting data. The major problem centres on the lack of depth information which becomes particularly important when surveying multiperiod sites with complex stratification. In such cases, the geophysical 'map' represents the influence of buried structures at various depths on the instrument to an extent which depends on the magnetic sensor height or resistivity electrode spacing. In the majority of surveys a compromise must therefore be established between spatial resolution and 'depth of investigation'. Consequently, if burial depth varies, it will be difficult to optimise the geophysical appearance of all archaeological structures within the survey area.

In geological resistivity prospecting, the method of 'electric drilling' provides approximate stratigraphic information and a partial solution to this problem (Telford *et al.*, 1976). By expanding a 4-electrode array in stages, current penetration is increased and apparent resistivity values become biased towards deeper strata. If this procedure is repeated while displacing the array, the result is a 'pseudo earth section' revealing both lateral and vertical changes in resistivity (Griffiths & King, 1981; 101). Prototype instruments which automate this process have been described but these have yet to be evaluated for archaeological prospecting (Griffiths & Turnbull, 1985; Van Overmeeren & Ritsema, 1988).

This paper outlines a new approach to resistivity surveying in which area maps and vertical sections can be simultaneously reconstructed from potential data gathered on a multielectrode array. Resistivity sections beneath the array are synthesised tomographically using an algorithm adapted from a medical scanner system. The general

ELECTRODE NUMBER

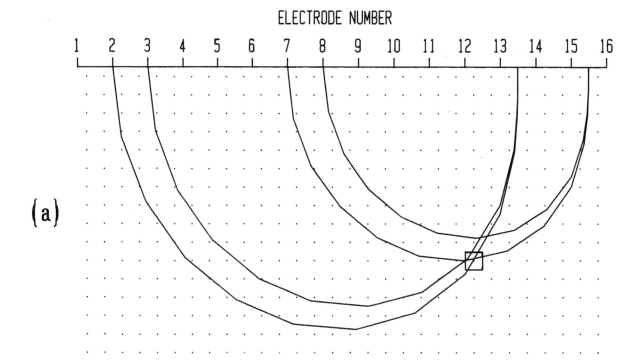

(a)

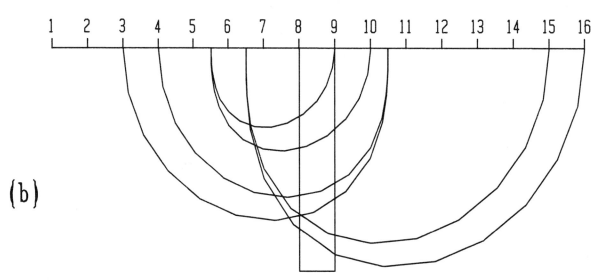

(b)

Figure 1. (a) The method of back-projection used to generate resistivity sections. Two stages in the image reconstruction are shown for a single pixel (small square) beneath an array of 16 electrodes. The potential measured between electrodes 2 and 3 when currents are passed between 13 and 14 is compared to the potential expected for a homogeneous region. The pixel density is adjusted according to this ratio and the procedure repeated as a sum for measurements: 7, 8, 15 and 16 and all other projections which intersect the pixel. (b) Method for producing a resistivity section beneath a traversing electrode array. Here back-projection is used to reconstruct resistivity values only in the column beneath the mid-point of the array where current isotropy is a maximum. As the array moves forward, a complete section is formed from adjacent columns and artefacts in the image are thereby reduced.

feasibility of this method has been examined in numerical simulation experiments and field trials which have led to the design of a new meter for multielectrode surveying.

Electrical Resistivity Tomography

The aim of this research is to develop a new geophysical system capable of producing both vertical archaeological sections and area maps from surface potential measurements. Of relevance here is the recent development of the medical imaging technique of 'Applied Potential Tomography' or APT (Brown, 1986). This method computes anatomical resistivity cross-sections by backprojecting measurements of skin potential on an array of 16 electrodes (Barber & Seager, 1987). Preliminary laboratory studies by Powell, Barber and Freeston (1987) suggested that the medical APT technique could be adapted for use with a linear array and this configuration was further explored numerically and in field trials at Verulamium by Noel (in press, (a)). The method for reconstructing a resistivity section from multielectrode data is explained with reference to Fig. 1a. The primary objective is to obtain all possible surface measurements of apparent resistivity in order to achieve an optimum description of the underlying structure. Thus, with N electrodes, there exist N(N-3)/2 possible independent measurements, assuming that the current and potential probe pairs are always discrete. These are referred to here as the 'Data Set'.

The potentials V recorded at each point in the Data Set are used to modify the resistivities of pixels (cells) lying between equipotentials extending from the measurement electrodes (Fig. 1a). If V_u is the corresponding potential which would be recorded for a region of uniform resistivity, ρ_u, them the simplest modification to an individual pixel is to change the resistivity by the ratio $V_m{:}V_u$ to derive a new value ρ_n, where

$$\frac{\rho_n}{\rho_u} = \frac{V_m}{V_u}$$

(Powell *et al.*, 1987). Projecting the entire Data Set through all pixels in the subsurface yields the following first approximation to the resistivity section:

$$\rho_{x,y} = \frac{\rho_u}{\Sigma\,\omega} \cdot \sum_{\rho=1}^{N(N-3)/2} \left(\frac{V_m}{V_u}\right)_\rho \cdot \omega\,(x,y)$$

where $\omega(x,y)_\rho$ is a weighting term (Noel, in press (b)). The accuracy of the reconstruction could be further improved by iterative backprojection (Griffiths & Ahmed, 1987) or by using this initial resistivity section as an input to finite element or other inversion models (Yorkey & Webster, 1987).

The electrical resistivity tomography (ERT) Data Set is larger than that generated by classical pseudosection profiling; eg. 104 measurements for a 16 electrode array as opposed to only 35 expanding Wenner data. The result, therefore, is an improved electrical description of the subsurface and the basis for a more constrained inversion. Moreover, since the ERT Data Set is complete, it contains all the information required to synthesise a conventional area survey corresponding, for example, to the twin-electrode array. Finally, an ERT survey stimulates a more isotropic distribution of current vectors at each point in the subsurface than is the case for the simple expanding Wenner scheme thus reducing the sensitivity of the survey to boundaries with specific orientations. These and other advantages of the ERT method have been discussed by Noel (in press (b)).

Resistivity Tomography surveying has been evaluated by numerical modelling experiments and field trials at Verulamium (Noel, in press (a)). As a result, a simple extension to the basic method has been proposed aimed at improving the accuracy of the resistivity

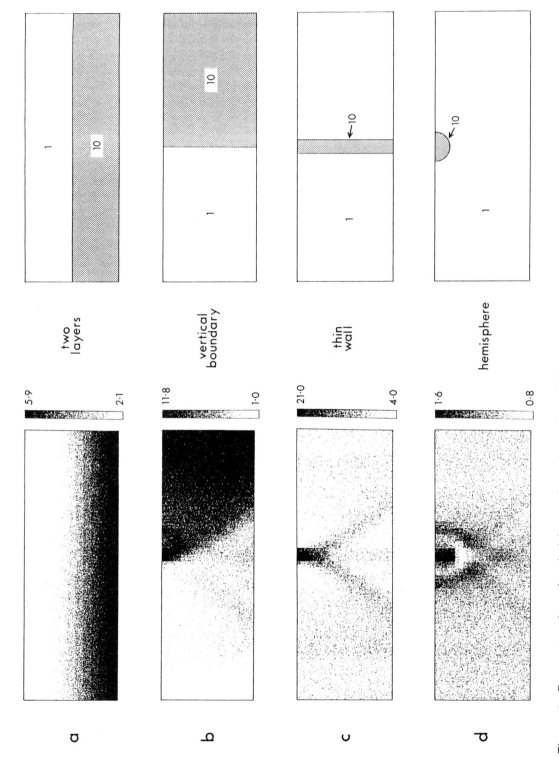

Figure 2. Reconstruction of resistivity sections through model structures based on the traversing back-projection method of Fig. 1b. (a) Two layers of contrasting resistivity. (b) A vertical boundary between two regions of contrasting resistivity. (c) A thin vertical wall of higher resistivity. (d) A small, more highly resistive hemisphere lying in the surface. Horizontal and vertical scales are identical.

298

inversion (Fig. 1b). In the revised scheme, backprojection is used to compute resistivities only in adjacent central columns beneath the traversing array thus maximising current isotropy in the image and significantly reducing abberations. Forward movement of the array can be simply achieved by leap-frogging the rear electrode. The traversing ERT survey method has again been simulated numerically for a number of simple two-dimensional structures and the results are shown in Fig. 2.

Field Trials

Preliminary trials of the basic ERT method at Verulamium succeeded in producing approximate sections through unexcavated Roman walls. Encouraged by the results of the numerical simulations (Fig. 2), a second series of experiments were designed to evaluate the traversing ERT method over known archaeological features with simple geometry. A suitable test site was chosen at Fountains Abbey (Fig. 3). Two 16.5 m long ERT traverses were made along parallel lines extending west from the Cloisters and crossing an unexcavated wall that had been located on the basis of crop marks which appear in dry weather. The survey employed an Abem Terrameter SAS300 signal-averaging meter with a sensitivity of 0.0005 ohms together with an array of 16 steel electrodes spaced at 75 cm intervals.

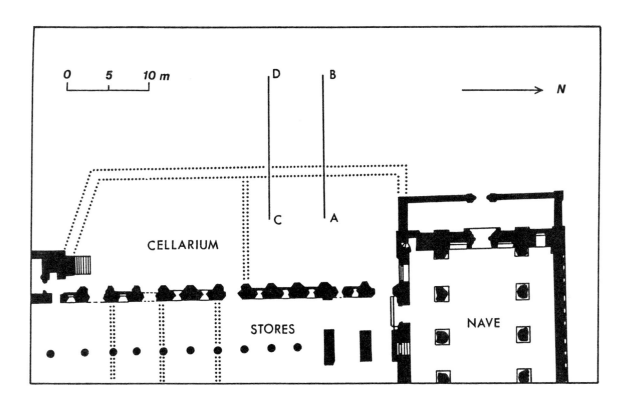

Figure 3.   The Cloisters and Cellarium area at Fountains Abbey. The array of 16 electrodes was traversed along two parallel lines, A-B and C-D, crossing an unexcavated wall (dotted). From the resistivity measurements, two 16.5 m long tomographic sections could then be computed.

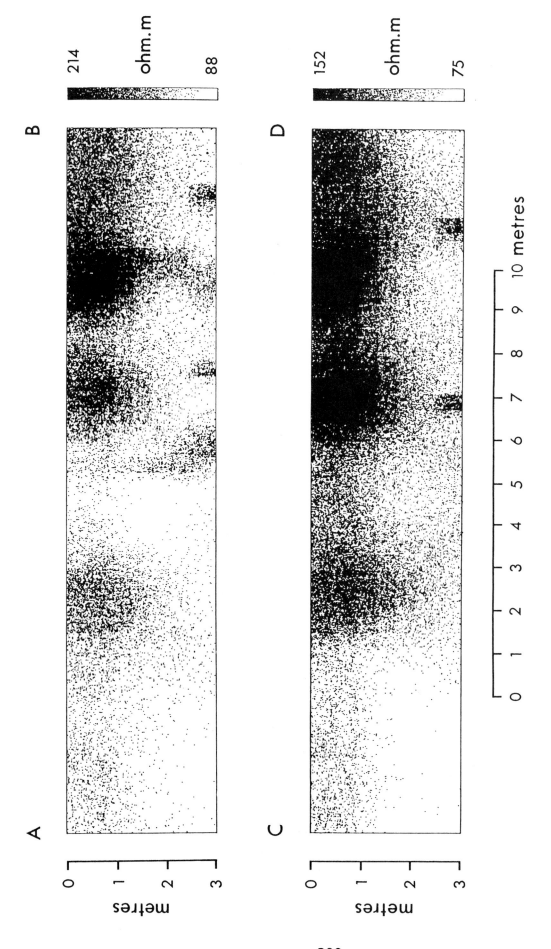

Figure 4. The two parallel ERT sections across the courtyard wall at Fountains Abbey. Letters correspond to the end-points shown in Fig. 3. Three shallow features of high resistance have been located, one of which matches the estimated position of the known wall: the others may be rubble-filled ditches or additional walls. The high resistance features at the base of each section and the semi-circular arcs are artefacts.

At the start of each traverse, the meter was used to gather a full Data Set of 104 readings from which the first column in the resistivity section could be computed (Fig. 1b). The array was then moved forward in 75 cm steps by leapfrogging the last electrode and only the N(N-3)/8 new readings that were required to generate the next full dataset were collected. In this way, each traverse line required a total of 650 resistivity measurements. Electrode connections were changed by hand.

The two ERT sections are very similar and they provide evidence for three, rather than one, high resistivity features which appear to be inserted into a horizontal, more resistive layer about 1 m thick (Fig. 4). The most western structure in both sections has dimensions of ~1 x 1 m and is in a position

consistent with the conjectured wall. It is possible that the remaining two features record unknown rubble-filled ditches or walls cut into the upper layer.

The New Resistivity Meter

Experience with using conventional equipment in the preliminary field trials indicated that a practical ERT system for archaeology will require a new resistivity meter, multiplexed to the electrode array and capable of gathering and storing an accurate Data Set in a very short time. A new instrument, the Geoscan RM15, has been designed to meet these criteria and will form the basis of an ERT system for imaging archaeological structures (Fig. 5).

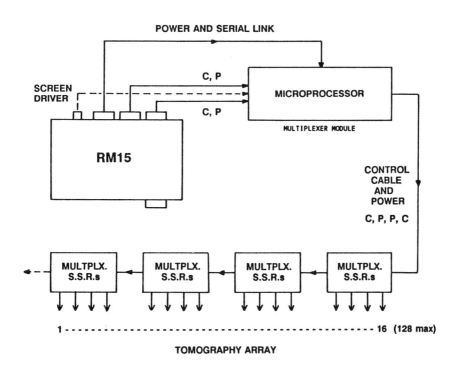

TOMOGRAPHY SYSTEM 2
(ALSO GENERAL PURPOSE PSEUDO-SECTION SYSTEM)

Figure 5. Design for the second tomography system based on the new RM15 resistivity meter. Up to 128 electrodes can be connected, in groups of four, each controlled by a multiplexer with solid state relays. The entire array is then linked to the RM15 via a processor module and multicore cable.

For a linear array of N equally spaced electrodes inserted in homogenous ground, potentials in a Data Set with inter-electrode spacing = 1 will have a dynamic range of

$$\frac{0.5-1/(N-2)+1/(N-3)}{1/(N-1)-2/(N-2)+1/(N-3)} : 1$$

This equates to 690:1 for a 16 electrode array and this wide dynamic variation of potentials is illustrated by the results of field trials with the Abem Terrameter (Fig. 6). If we aim to measure the smallest reading to 1% then this represents an instrument dynamic range exceeding 100 000:1. Clearly, very high accuracy is required. To achieve this, three constant current ranges will be provided (0.1 mA, 1 mA and 10 mA) along with a precision A/D converter and in addition, it will be possible to stack the lowest value readings from 4 to 512 times. A constant current source with better than 0.02% accuracy, regardless of probe contact resistance, has already been developed.

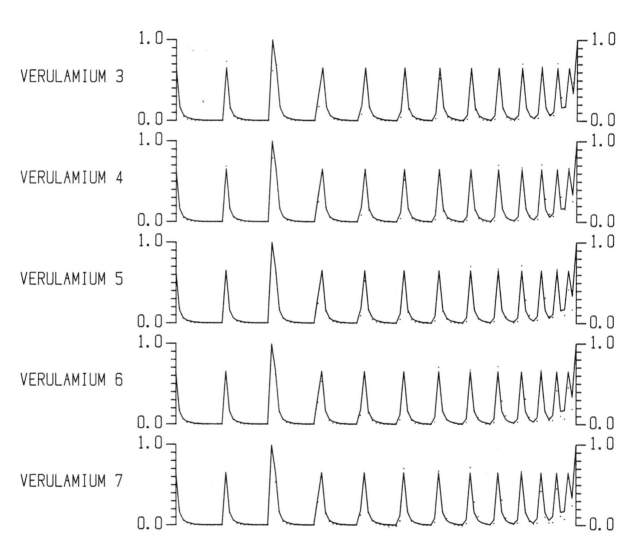

Figure 6. Normalised potentials for the Data Set gathered with a 16 electrode array and an Abem Terrameter at Verulamium (104 points). Data Sets 3-7 correspond to successive displacements of the array by one electrode spacing of 75 cm and the solid line represents the best-fit potentials for a homogeneous region. These results illustrate the wide dynamic range of the measured potentials and the small perturbations corresponding to archaeological anomalies.

The choice of operating frequency will be a compromise between speed of use and keeping capacitative and inductive errors low and will probably be in the range 20-100 Hz. If the A/D converter samples at 10 readings per second, and no stacking is required, then each Data Set will take about 11 seconds to capture. A 20 m profile, using a probe spacing of 1 m, will require 170 readings which, if stacking is used, will probably require about 3.7 minutes to complete.

In addition to connections for the current and potential probes, the new meter will incorporate a logger and a serial port for transmitting data to an external processor. This port will also enable the RM15 to receive control commands from a portable microcomputer. An LCD display will show the current survey position and instrument status and a menu system will permit configuration of the meter from a keyboard.

An expansion port will be used to control a variety of multiprobe systems via external multiplexer modules (Fig. 5). These modules will derive their power and control signals from the expansion port. Conventional relays are not suitable for this application because of their high power consumption and slow speed. Instead, the multiplexers will incorporate high performance solid state relays with exceptionally good isolation and low leakage currents.

The tomography system is being developed in two stages. The first simple system will use a multiplexer module that can control 16 probes, each of which will be connected directly to the module. It will contain its own microprocessor that is in turn controlled by the resistance meter. Different multiplexing patterns can then be programmed into the RM15 and down-loaded into the multiplexer module. This system will be versatile enough to be used as a conventional pseudosection profiler based, for example, on the Wenner array.

The second tomography system will use a multiplexer module that drives a single multicore cable, thereby connecting the array more conveniently and allowing for up to 128 electrodes. The multiplexer will now control several sub-modules via the cable, each containing a multiplexer and solid state relays which select four probes. Thus a large array can be constructed simply by linking the required number of modules. Furthermore, as one sub-module become redundant, because its part in the permutation of probes has been completed, it can be leap-frogged ahead ready to be included in the next Data Set. This is thought to be a more practical and economic arrangement than installing multiplexer and relay circuits in the handle of each probe, although the architecture of the system leaves this possibility open should it later prove desirable.

We anticipate that a major problem with such a multicore system will be capacitative coupling effects, especially between the current and potential cables. Following the example of medical tomography systems (Brown & Seager, 1987) we propose to overcome this problem by employing cables with a driven screen. This will allow relatively high frequencies to be used, in the range 20-100 Hz.

Discussion

Preliminary experiments at Verulamium provided the first indication that Electrical Resistivity Tomography might be a useful prospecting tool for archaeology. Numerical simulation studies now show that the accuracy of the resistivity inversion can be significantly improved by modifying the Data Set to include only back projections beneath the midpoint of the array. Although the results of field trials have been encouraging, there remains considerable scope for improving the quality of the resistivity sections and thus their archaeological value. Removal of image artefacts could probably be achieved by employing more rigorous inversion schemes such as iterative back-projection or finite element methods (eg. Dey & Morrison, 1979; Murai & Kagawa, 1985). However, a simple back-projection may be retained as a fast, one-step inversion for use in field-portable microcomputers. Undoubtedly, the prime factor of archaeological interest will be the attainable image resolution; this parameter is primarily governed by the size of the Data Set which in turn is related to the number of array electrodes and the data capture time. The new RM15 meter will permit much larger Data Sets to be gathered in a realistic time and the interaction of these parameters can then be explored in detail.

Acknowledgements

This research is supported by a awards from the Science and Engineering Research Council and The Royal Society in collaboration with Geoscan Research. We are grateful to J. Roebuck of English Heritage for giving permission to survey at Fountains Abbey and for commenting on the geophysical interpretation. D McCann and G. Ripin of the British Geological Survey loaned the Abem Terrameter to which C. Batt kindly acted as multiplexer. We also appreciate the loan of a PC5541 graphics computer from the Sharp Corporation.

References

Aitken, M.J. (1974). *Physics and Archaeology*, Oxford University Press.

Aspinall, A. & Walker, A.R. (1975). The earth resistivity instrument and its application to shallow earth surveys. *Underground Services*, 3, 12-15.

Barber, D.C. & Seager, A.D. (1987). Fast reconstruction of resistance images, *Clinical Phys. Physiol. Meas.*, 8, 47-54.

Brown, B.H. (1986). Applied potential tomography, *Phys. Bull*, 37.

Brown, B.H. & Seager, A.D. 91987). The Sheffield data collection system, *Clinical Phys. Physiol. Meas.*, 8, 91-98.

Dey, A. & Morrison, H.F. (1979). Resistivity modelling for arbitrarily shaped two dimensional structures, *Geophys. Prosp.*, 27, 106-136.

Griffiths, D.H. & King, R.F. (1981). *Applied Geophysics for Geologists and Engineers*, Pergamon, Oxford.

Griffiths, D.H. & Turnbull, J. (1985). A multi-electrode array for resistivity surveying, *First Break*, 3, 16-20.

Murai, T. & Kagawa, Y. (1985). Electrical impedance computer tomography based on a finite-element model, *IEEE Trans. Biomed. Eng.* BME-32, 177-184.

Noel, M. (in press (a)). Multi electrode tomography for imaging archaeology, *British Archaeological Reports*.

Noel, M. (in press (b)). Archaeological prospecting in 3-D by electrical resistivity tomography, *Geophys. J. R. Astr. Soc.*

Powell, H.M., Barber, D.C. & Freeston, I.L. (1987). Impedance imaging using linear electrode arrays, *Clinical Phys. Physiol. Meas.*, 8, 109-118.

Papamarinopoulos, S.P., Tsokas, G.N. & Williams, H. (1986). Electric resistance and resistivity measurements of the archaeological relics on the castle of Mytilene, *Boll. Geog. Teor. Appl.*, 77, 23-34.

Telford, W.M., Geldart, L.P., Sheriff, R.E. & Keys, D.A. (1976). *Applied Geophysics*, Cambridge University Press.

Van Overmeeren, R.A. & Ritsema, I.L. (1988). Continuous vertical electric sounding, *First Break*, 6, 313-324.

Yorkey, T.J. & Webster, J.G. (1987). A comparison of impedance tomographic reconstruction algorithms, *Clinical Phys. Physiol. Meas.*, 8, 55-62.

FREQUENCY MODE INDUCED POLARIZATION STUDIES FOR GEOPHYSICAL EXPLORATION IN ARCHAEOLOGY

S.M. Ovenden & A. Aspinall

Archaeological Sciences, University of Bradford, Bradford, West Yorkshire, BD7 1DP.

Introduction

Geophysics is now widely used and accepted in archaeology. Magnetometer and resistivity surveys are the most widely used and are continually being updated by improving speed, accuracy, and the data processing and presentation.

This project is concerned with a relatively new method - Induced Polarization (I.P.). I.P. is widely used in geological applications, but its use in archaeology has been limited. In the late 1960's the use of time domain I.P. on archaeological sites was tested (Lynam, 1970) with encouraging results, but the method was found to be time consuming and cumbersome. Most of the problems encountered can, however, be overcome by studying I.P. in the frequency domain; the present research project is concerned with a study of this technique.

Brief Theory

As this is a relatively unknown method the basic principles and theory will be outlined. By studying the potential difference associated with a current applied to the ground, it is clear that the voltage does not build up or decay instantaneously when the current is switched on or off. This results from storage of charge as the ground effectively acts as a capacitor, and becomes polarized. Because of the heterogeneous nature of the subsurface there will be varying localised concentrations of this stored charge. These variations may occur because of changes in chemical composition, as a result of physical controls, or as a combination of both electrical and physical phenomena.

There are two distinct forms of induced polarization described as electrode and membrane. Electrode polarization, also termed metallic or electronic polarization, is caused by variations in the ionic and electronic conduction through a material. A potential difference exists across the interface between a metal electrode and an electrolyte, even when no current flows across that interface. When a current is passed through the ground there is a loss of equilibrium and a build-up of ions at the interface results because of changes from ionic to electronic conduction and back to ionic conduction. The interface becomes polarized, as represented in Figure 1. Charge is transferred leading to a secondary voltage which opposes the primary current flow, and retards the instantaneous primary voltage build up.

Membrane polarization is superficially similar to electrode polarization but it occurs where no metallic or electronic conductors are present, and no chemical reaction takes place. It results from variations in the mobility of fluids through the material. Unsatisfied charges in the material attract a diffuse cloud of positive ions. In the presence of an electric field the mobile positive ions are attracted to the negative grain surfaces and a double layer forms. If the pores are of the same order of magnitude as the double layer, or if the positive charge zone extends into the pores, blockages occur causing the area to become polarized, as shown in Figure 2. A secondary voltage develops which opposes the primary field and the primary voltage build up is retarded.

For both electrode and membrane polarization the build up of charge dissipates and the polarizations equilibrate when the current is terminated. The two types of polarization can not be distinguished in the field and they often occur together. However, membrane polarization is usually much smaller in

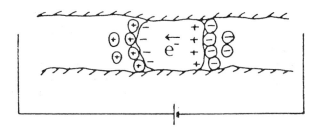

Figure 1. Creation of electrode polarization.

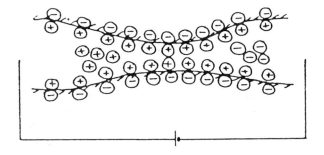

Figure 2. Creation of membrane polarization.

magnitude and is almost always present to some degree, and as a result, is often regarded as noise.

Application of I.P. to Archaeology

Like the resistivity method, the method should be able to detect massive bodies that alter the bulk resistivity of the subsurface, but it has one main advantage. Because of the nature of I.P. its magnitude is controlled by, among other factors, the surface area of the material creating the phenomena. It should, therefore, detect disseminated material relatively easily.

It is hoped that I.P. surveys can be used in archaeology to find buried structures, infilled ditches, disseminated features such as slag heaps or infilled mineshafts/tunnels and detect subtle disturbances in the ground, including changes in soil properties indicative of past human habitation.

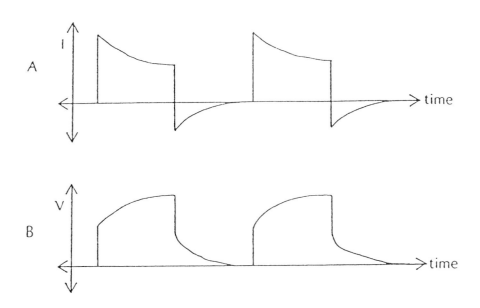

Figure 3. Comparison of current and voltage forms: a) Current through sample, b) Voltage across sample.

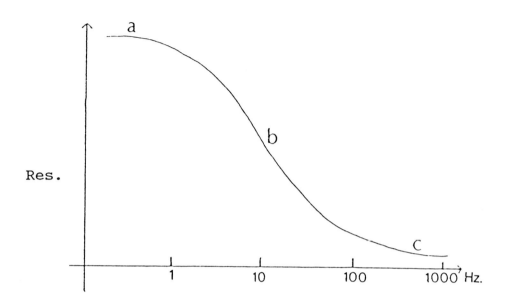

Figure 4. Variation of resistivity with frequency: a) Resistive region, b) Warburg region, c) Region of electromagnetic induction.

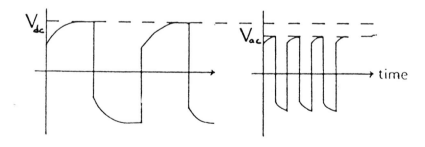

Figure 5. Variation of voltage with frequency.

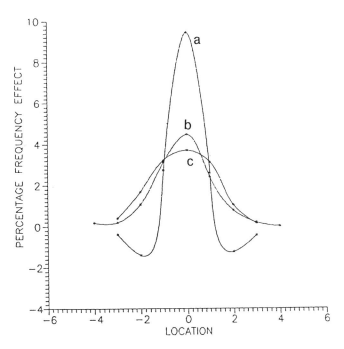

Figure 6. Responses for different arrays. a) Dipole-dipole, b) Twin probe, c) Wenner.

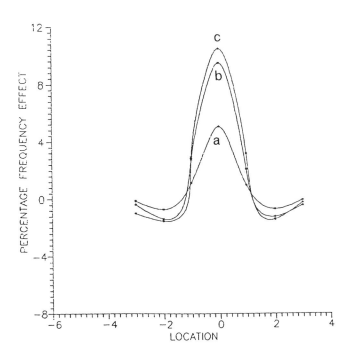

Figure 7. Responses for different sample size.

309

Measurement of Induced Polarization

There are many different methods of measuring I.P. effects but only that relevant to this research is discussed here. The presence of I.P. is indicated by a retardation in the build up and decay of voltage when an alternating current is passed through a medium, as shown in Figure 3. The initial build up is instantaneous followed by a slow non-linear increase of voltage until saturation is reached. This takes an infinite time to occur. Therefore the voltage attained depends, together with other factors, on the length of time the current flows. The longer the current flows the higher the voltage reached. So resistivity is a function of the frequency of the alternating current. This relationship is only true between 0.1 and 10 Hz. At very high and very low frequencies resistivity is independent of the current frequency, as can be seen in Figure 4.

Given this dependence on frequency consider the following. If a material is not very polarizable, saturation is reached almost instantaneously because the polarization effects are negligible. As a result, a longer period of current flow will not affect the voltage attained. If however the material is polarizable, the effect of frequency variations will be significant, as indicated in Figure 5. At a low frequency the instaneous build up is followed by the non-linear increase in voltage. At a higher frequency the non-linear time dependent build up of the voltage is curtailed by the short period of the current flow. As a result there will be a marked difference in the two voltages. The greater the I.P. the greater the difference in the voltages. The standard frequency domain measurement is the Percentage Frequency Effect:

$$P.F.E. = \frac{(Vdc - Vac)}{Vac} \times 100$$

The survey method is basically a resistivity survey at two frequencies and thus the two field survey techniques are almost identical except that now two reading are taken at each station. Also the AC data can be used to give a resistance plot, so one is in fact getting two sets of data from one survey.

Instrumentation

Prototype laboratory and field systems have been designed and built. Both are essentially the same although the laboratory equipment is more versatile to enable detailed theoretical studies to be made.

The transmitted current signal is a constant amplitude bipolar square wave. Laboratory studies have indicated that a current of 1 mA and frequencies of 1 and 10 Hz give optimum results. The resulting voltage signal is measured using a synchronous detector. Any of the accepted resistivity electrode arrays can be used although the twin probe array and the dipole-dipole array give the best results.

Examples of Laboratory Studies

In the laboratory a KCl solution tank has been used. Initial studies were undertaken to test the response profiles. As one would expect, the profile responses for simple models are the same as one would predict for a resistivity profile over a conducting body.

Figure 6 shows responses for different arrays over a wooden cylinder coated with highly conductive paint. The profiles indicate that the dipole-dipole array gives the best object definition followed by the twin probe array, and finally the Wenner array. However the twin probe array has a better depth response and is far easier to manage in the field, and has therefore been used in the field surveys. Figure 7 illustrates the effect of target size on the anomaly shape using a dipole-dipole array. The targets used were wooden cylinders coated with conductive paint, with diameters of a) a/2, b) a, and c) 2a where a is the electrode separation (15mm).

310

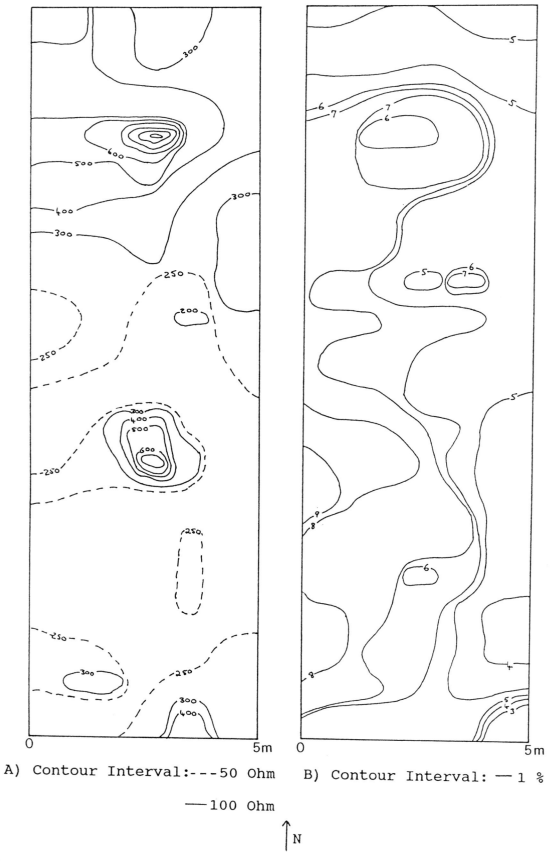

A) Contour Interval:---50 Ohm   B) Contour Interval: — 1 %

— 100 Ohm

↑N

Figure 8. a) Resistance, Manor Vale, b) I.P., Manor Vale.

Contour Interval: ——50 Ohm

N

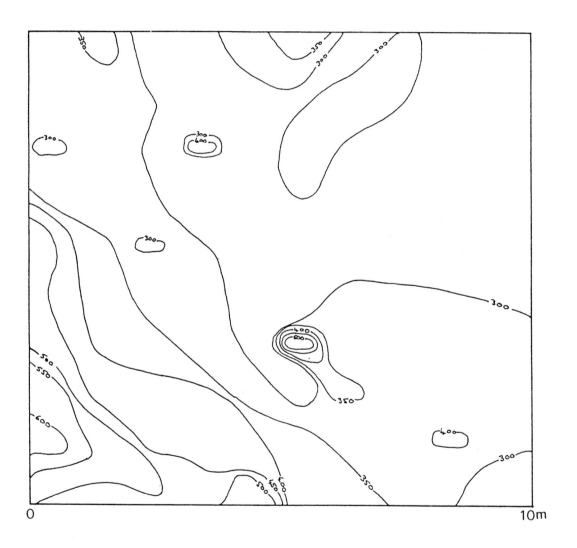

Figure 9. Resistance, Fountains Abbey.

Contour Interval: --- 0.5 %

——— 1 %

N

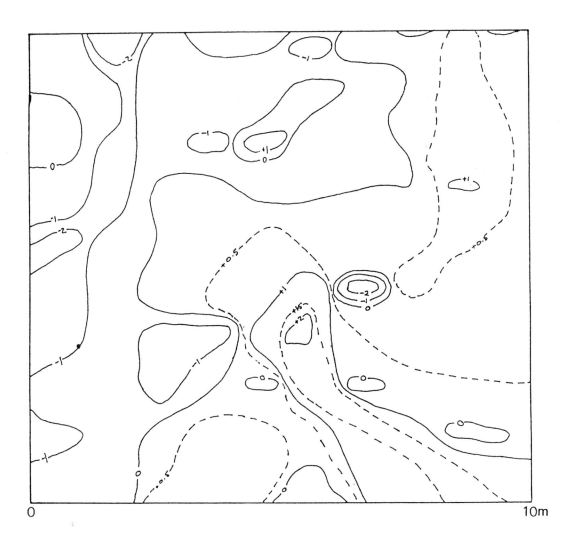

0   10m

Figure 10. Induced polarization, Fountains Abbey.

Field Testing of the Method

When undertaking an I.P. survey one is also carrying out two resistivity surveys, one at 1 Hz and one at 10 Hz, thus providing two different sets of data. As there is very little difference in the 1 and 10 Hz resistance plots, only the 10 Hz plots have been reproduced. To achieve a good signal the voltage readings have an arbitrary gain of x 5.275 relative to true resistance. The values shown on the resistance contour maps have not been adjusted, therefore to compare the values with other resistance data, the data must be divided by 5.275.

The first area to be surveyed (Manor Vale Lawn) was chosen because it has been surveyed frequently and therefore provides a good control. Figure 8a is a contour map of the 10 Hz resistance data. In the north an area of clinker infill shows clearly as a high resistance feature. In the centre region there is a lower more confined high resistance feature representing a known bricked feature. From the south, leading to this feature, there is a relatively high linear resistance representing a drain leading to the bricked feature. A similar line runs from this southern anomaly to the west, indicating another drain. There is also another possible pipe or drain running east - west in the area between the clinker infill and the bricked feature.

The I.P. data is shown in Figure 8b. The background of approximately 5% indicates a clay cover. There is some indication of the clinker infill the north, but it appears to be detecting the clay/clinker transition rather than the clinker itself. The east - west orientated pipe seen in the resistance plot is again visible. The most noticeable feature of this survey is the marked increase in percentage frequency effect to the west from approximately 4% to 9%. The order of magnitude and the distribution of the data suggests an increase in the clay cover to the west, which is in keeping with consolidation of the area against sloping ground to the west.

The second field test was carried out in the grounds of Fountains Abbey, North Yorkshire, in the vicinity of a known drainage pipe.

Figure 9 shows the 10 Hz resistance contour map. The high resistance region in the south west is almost certainly related to changes in soil cover. There is a relatively high resistance linear feature running from the south east towards the north west. This coincides with the expected location of the pipe. While the data is not distinct enough to be readily interpreted as a pipe, it suggests that the pipe is either earthenware or coated metal.

However the P.F.E contour map, Figure 10, shows a relatively high feature in the south east corner. Again the magnitude indicates a membrane (clay) effect. Given that the ground slopes down towards the south east this area of I.P. is probably caused by clay being washed down and building up against the pipe. One noticeable difference between this data and the Manor Vale data set are the background readings. Here the background ranges from 0% to -2%. The lower magnitude is caused by the sandy soil as apposed to the clay background of Manor Vale. The negative readings are caused by the lack of ground cover on this site. In such cases if the probes are in direct contact with, even slightly, polarizable material negative readings will result.

These negative readings are informative rather than a hindrance as they can provide information about the topsoil and the location and orientation of shallow targets.

Conclusions

The work is still in its early stages but shows definite promise. In all surveys carried out, the I.P. data has detected variations not represented in resistivity and/or magnetometer plots. While many variations are subtle and environmental rather than archaeological, they can be significant and aid interpretation. Obviously a wide variety of sites will have to be surveyed before a meaningful assessment of the method can be made.

Acknowledgements

We wish to thank the following collaborators:

Peter Dale, University of Bradford, for his assistance in the building of the laboratory and field equipment. Roger Walker, Geoscan Research, for assistance on the design of the equipment. One of us (S.M. Ovenden) acknowledges the assistance of a S.E.R.C Studentship.

References

Bertin, J. & Loeb, J. (1976). *Experimental and Theoretical Aspects of Induced Polarization*, Volume 1 & 2, Gebruder Borntraeger, Berlin.

Kearey, P. & Brooks, M. (1984). *An Introduction to Geophysical Exploration*, Blackwell Scientific Publications.

Lynam, J. (1970). *Techniques of Geophysical Prospection as Applied to Near Surface Structure Determination*, unpublished Ph.D. Thesis, Bradford University.

Sumner, J.S. (1976). *Principles of Induced Polarization for Exploration Geophysics*, Elsevier, Amsterdam.

INORGANIC ANALYSIS OF ORGANIC RESIDUES AT SUTTON HOO

P.H. Bethell

Department of Biochemistry, University of Liverpool, PO Box 147, Liverpool, L69 3BX.

This paper is essentially a short resumé of the results produced by a series of experiments described in detail elsewhere (Bethell, 1989; Bethell & Smith, 1989). It is somewhat in the nature of an interim description of those results, in advance of fuller publication, and will serve as a summary for readers not familiar with this work.

The experiments in question centre around the analysis of the degraded remains of human burials from the Anglo-Saxon cemetery at Sutton Hoo in Suffolk, England. The peculiar nature of these burials is well described elsewhere (eg. Bethell & Carver, 1987), and has parallels from other sites such as Snape, Suffolk; Mucking, Essex; and Burrow Hill, Suffolk. The degradation is a function of the soil type, namely acid sands and gravels. Under the prevailing acidic, free-draining conditions, bodies are reduced to a 'stain' of varying shape and intensity, consisting of residual products of the breakdown of the organic parts of the body. In most cases the mineral portion of the skeletons has partially or wholly dissolved. In the case of Sutton Hoo, the recently excavated interments were recovered as 3-dimensional 'sandmen', reflecting the original contours of the buried cadaver.

Much controversy surrounded the original ship-burial in 1939, because no trace of a body could be found. Whether it had totally decayed, or not been present was a matter for some argument. It was hoped to find a method of chemical analysis to resolve that question should it arise again.

To this end, the known burials at Sutton Hoo were sampled, and analysed for their elemental composition. The resulting chemical 'signature', if found where little or no visible trace of a body was present, should hopefully indicate the presence of body decay products, and hence of a burial.

The analytical technique used was Inductively-Coupled Plasma Emission Spectrometry (ICP). The technical details of this are well-documented elsewhere, but it was selected as being particularly appropriate for an elemental and trace-elemental 'scan' of a soil-sample's composition. ICP can measure the concentrations of 25-30 major and trace elements in a machine time of c. 1 min per sample. This makes it relatively quick and cheap compared to other analytical methods.

The full experimental details are described elsewhere (Bethell, 1989; Bethell & Smith, 1989). In summary a clear pattern of enhancement and diminution in the levels of various elements in the identifiable body residue was detectable, relative to the values of those elements found in the background soil. This produced the chemical 'signature' for a decayed body under those particular soil conditions.

The next phase of the experiment involved the sampling of the burial-chamber floor in Mound 2 at Sutton Hoo. This large mound had been very heavily robbed, but showed evidence of a ship or boat being buried there, in the form of iron ship-rivets. Other finds included fragments of glass, gold, silver and bronze, all indicating that this had been a very richly furnished burial before the various robbing episodes, and thus comparable to that in Mound 1. The actual burial chamber was empty on excavation, and the only visible traces of occupation were a number of 'object stances' probably related to grave-goods since disappeared or removed (Carver et al., 1989, Fig.4). The sampling was undertaken from the clean soil directly below the floor of the burial chamber, perhaps 2-3 cm beneath the original positions of any grave goods or bodies.

In this phase of the procedure, 490 samples on a 10 cm grid were taken, dried, and

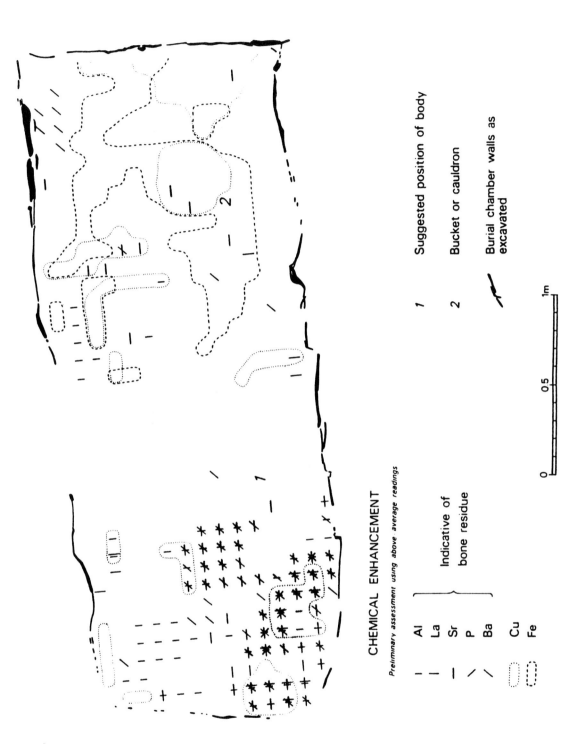

CHEMICAL ENHANCEMENT

Preliminary assessment using above average readings

Al
La   } Indicative of
Sr   } bone residue
P
Ba

Cu
Fe

1   Suggested position of body

2   Bucket or cauldron

Burial chamber walls as excavated

0   0.5   1m

Chemical microsurvey in Mound 2 burial chamber (Royle/Carver/Bethell).

Figure 1: Simple plot of positive anomalies of certain elements in Mound 2 burial chamber, Sutton Hoo. (Royle/Carver/Bethell).

analysed using ICP. In the analysis of the known body, The elements showing the largest positive anomaly over the background were Aluminium (Al), Barium (Ba), Calcium (Ca), Cerium (Ce), Chromium (Cr), Cobalt (Co), Lanthanum (La), Molybdenum (Mo), Phosphorus (P), Scandium (Sc), Strontium (Sr), and Yttrium (Y). Copper (Cu) showed a strong negative anomaly, and Iron (Fe) showed no appreciable variation. The diagram (Figure 1) shows the areas of positive anomaly of five of the 'body signature' elements (Al, Ba, La, P and Sr) and Cu and Fe.

There was a definite concentration of higher-than-background levels for the 'body' elements in the SW corner of the burial chamber; the obvious inference being that this was the site of deposition of an actual body which had subsequently decayed in situ. It could be argued that the residual anomalies could be due to an animal or possibly organic refuse left during one of the robbing episodes, but this seems unlikely. The rate of initial decay in the Sutton Hoo soil is extremely fast, so the 'fallout' from a buried corpse would probably be fixed in the adjacent soil fairly early on in the burial episode, and thus remain even if the 'sandman' itself was dug away. A further pointer towards the interpretation of these chemical traces as residues of a body lies in the distribution of the positive copper and iron anomalies. These were spread over the E end of the chamber, indicating that the larger grave goods were piled away from the body. This pattern can be seen in Mound 1 (Bruce-Mitford, 1975).

The conclusion of this experiment is that even very faint traces of buried organic material can be retained in the soil and detected. It is unlikely that such traces would travel far from the original source, so it can be inferred that the finding of these residual anomalies would indicate not just the presence of the original material, but its original location as well. In the case of Sutton Hoo, this is quite an important question, and could certainly be applied to other grave vs. cenotaph controversies.

The usefulness of this type of chemical tool in archaeology is, hopefully, demonstrated by the results of this experiment, and is something to be explored more fully, in other soil types and searching for the residues of other materials.

Bibliography

Bethell, P.H. (1989). Chemical analysis of shadow burials. In: *Burial Archaeology: Current Research, Methods and Developments*. Eds. C.A. Roberts, F. Lee & J. Bintliff. BAR British Series 211, 205-214.

Bethell, P.H. & Carver, M.O.H. (1987). Detection and enhancement of decayed inhumations at Sutton Hoo. In: *Death, Decay and Reconstruction*. Eds. A. Boddington, N. Garland & R. Janaway, pp 10-21. Manchester University Press, Manchester.

Bethell, P.H. & Smith, J.U. (1989). Trace element analysis of an inhumation from Sutton Hoo, using inductively-coupled plasma emission spectrometry (ICP): an evaluation of the technique applied to analysis of organic residues. *Journal of Archaeological Science* 16, 47-55.

Bruce-Mitford, R.L.S. (1975). *The Sutton Hoo Ship Burial*, Volume I. British Museum, London.

Carver, M.O.H., Glazebrook, J. & Royle, C.L. (eds). *Bulletin of the Sutton Hoo Research Committee* No. 6. Sutton Hoo Research Trust, London.

SOME QUANTITATIVE POLLEN EXTRACTION TESTS LEADING TO A MODIFIED TECHNIQUE FOR CAVE SEDIMENTS

D.N. Hale & M.J. Noel

Department of Archaeology, University of Durham, 46 Saddler Street, Durham, DH1 3NU, U.K.

Summary

Clastic cave sediments and speleothems have the potential for providing palaeoenvironmental data for periods considerably longer than the 15 ka available for most open-air sites. Work in the caves at Creswell Crags shows that sedimentation has occurred since the Ipswichian, but also that the pollen concentration of the sediments is low. There is a need to establish a more efficient pollen extraction procedure before reconstructing vegetation histories and discussing their anthropogenic influences. This article quantitatively describes some extraction techniques and puts forward a non-destructive, modified heavy liquid separation procedure.

Introduction

Pollen analysis was first recognised as a valuable palaeoenvironmental tool at the beginning of this century and has since been used throughout the world. In Britain, analyses of many lacustrine and peat deposits have enabled the construction of detailed late- and post-glacial climatic and vegetational histories. Clastic cave sediments and speleothems, however, have the potential for providing such data for periods considerably longer than the 15 ka available at most open-air sites. Work in the present study area, the Creswell Crags caves, East Midlands, has shown that sedimentation has occurred since the Ipswichian interglacial (125-75 ka) and produced evidence for intermittent occupation during and since the Mousterian period (Jenkinson, 1984). The sediments of the Creswell caves are generally fine sands and silts and a pilot study has shown that their pollen concentration is typically between 150-200 grains per 20 ml (Coles, 1987), using the swirling and sieving separation process of Hunt (1985). The low pollen concentration emphasises the need to establish a more efficient extraction procedure before reconstructing vegetational histories and discussing their anthropogenic influences. The resulting sequences are being complemented with palaeomagnetic and uranium series dating (Rowe et al., 1989) and by studies of faunal remains (Jenkinson, 1984). Twenty tests are described below which start to quantitatively investigate pollen separation processes.

Laboratory methods

To evaluate extraction techniques two control suspensions were established. By dissolving two spore tablets in 2 ml 0.1 N HCl + 23 ml filtered water the first suspension had a calculated concentration of 966 spores ml^{-1}, because each tablet is supplied containing an average of 12 077 spores (Batch no. 414831, Quaternary Biology Laboratory, Tornavagen). Tests 1-10, shown in Table 1, involved mounting and counting 1 ml aliquots of the standard suspension. The mean of the counted spore totals was 798 spores ml^{-1}. Although the number of spores in each tablet is known to vary (Coles, 1987), the average difference found between the expected and counted totals was a loss of 17.4%. This loss exceeds the manufacturer's variance in spore tablet totals ($\pm 3.1\%$) and seems likely to represent loss during transfer of suspension from sample tube to slides. Spores may be adhering to the sides of the transfer pipette or sample tube by electrostatic forces and so the use of a de-ionising agent will be evaluated in later trials.

The second control suspension contained 2700 Lycopodium spores ml^{-1}, since by using more tablets the effect of the different spore contents of the tablets would be lessened.

For this project we require a technique which effectively separates the sediment minerogenic and organic components without damaging any organic microfossils of environmental interest (including fungi and algae). Methods for the extraction of sediment pollen and spores are generally designed as concentration processes which aim to disintegrate and remove the minerogenic fraction and concentrate the pollen. The complexity of each method depends on the original proportion of the microfossils and the nature of the matrix. If the concentration of mineral material is high, as in most cave sediments, then it becomes necessary to remove it.

Established techniques in pollen separation include:

1) Hydrochloric acid treatment to dissolve free calcium carbonate,

2) Potassium hydroxide digestion to diaggregate the components of the sediment,

3) Acetolysis, often used with organic sediments since it effectively removes cellulose. Problems with the method involve selective destruction of certain pollen taxa, dissolution of many algal and fungal groups and the pseudo-fossilisation of modern contaminant grains, all of which might confuse later interpretation (Erdtman, 1952; Faegri & Iversen, 1974; Girard & Renault-Miskovsky, 1969; Leroi-Gourhan & Leroi-Gourhan, 1964; Wenner, 1942).

4) HF (hydrofluoric acid) maceration to dissolve silica, though the method requires special, costly equipment and there are inherent dangers in handling the liquid.

5) Heavy liquid separation. This is discussed below.

Due to the problems of acetolysis and HF, safer and less destructive tests have been carried out, as described here and outlined in Table 1.

The following ten tests (except 11 and 17) involved adding 1 ml aliquots of the suspension to 40 ml sediment samples and 10% potassium hydroxide (KOH). After boiling for ten minutes, the swirling and sieving method of Hunt (1985) was employed. The principle behind the swirling, carried out like panning, is that the organic-walled microfossils will remain in suspension while the denser, inorganic particles collect at the bottom of the swirling dish. The sieves used in this study were of 100 μm and 10 μm apertures, material in the pollen-sized fraction being rinsed from the larger to the smaller sieve where it is retained. Although more pollen will be extracted from a larger sample it is also likely that more pollen will become mechanically trapped within the sediment load, emphasising the need for thorough rinsing of the sediment.

Tests 12, 14, 15, 18 and 19 were processed as above, with 40 ml of sediment. They produced a mean Lycopodium recovery rate of 9.8%. Tests 11 and 17 were similar but without sediment in order to provide an estimate of the percentage loss of spores due to trapping within the sediment. The spore recovery of test 11, 10.4%, was low. Test 17, boiled in a microwave, gave a 41.4% recovery, giving a mean value of only 25.9%. These figures are much lower than expected and point to problems in the technique; when a how are spores being lost? The figures suggest that entrainment may be responsible for about 16% of the pollen loss during processing. UV fluorescence will later be used to see if spores are being caught in the sieve mesh.

Test 16 evaluated a simple technique whereby after initial boiling in KOH the beaker was magnetically stirred and the pollen suspension pipetted off the sediment. The sediment was repeatedly agitated to free trapped spores while more filtered water was being added. The pipetted liquid was sieved and the residue stained. The 1.9% recovery shows this technique to be totally unsuitable for pollen and spore extraction.

Table 1. Results of spore separation methods.

Sample No.	Method	Grains recovered	% Recovery
1	(aliquot)	838/966	86.75
2	"	1018/ "	105.38
3	"	856/ "	88.61
4	"	840/ "	86.96
5	"	695/ "	71.95
6	"	907/ "	93.89
7	"	668/ "	69.15
8	"	706/ "	73.08
9	"	750/ "	77.64
10	"	706/ "	73.08
11	KOH+sv/sw	100/ "	10.35
12	KOH+sv/sw+40ml	74/ "	7.66
13	KOH+sv/sw+40ml+ZnCl$_2$	145/ "	15.01
14	KOH+sv/sw+40ml	107/ "	11.08
15	KOH+sv/sw+40ml	150/ "	15.53
16	KOH+sv/sw+40ml+mag.st.	18/ "	1.86
17	KOH(m'wave)+sv/sw	400/ "	41.41
18	KOH(m'wave)+sv/sw+40ml	92/ "	9.52
19	KOH+sv/sw+40ml RHX	48/ "	4.97
20	KOH+sv/sw+40ml RHX+ZnCl$_2$	197/ "	20.37

Heavy liquid separation

To try to improve on the low recovery rate using these methods a heavy liquid process was employed after HCl, KOH dissolution, swirling and sieving. Heavy liquid separation was developed primarily to separate minerals on the basis of their relative densities (Pryor, 1965). The method involves raising the density of a liquid to a level between the relative densities of the two materials to be separated. The 'feed' is then added to the liquid. Matter with a relative density less than that of the heavy liquid rises to the top, while matter with a relative density greater than that of the liquid sinks to the bottom of the bath. The efficiency of the method will be a function of the liquid viscosity and particle diameter (ie. mobility) and the distribution of densities in the organic and inorganic functions.

Heavy liquid separation has been used in palynology both as an alternative and a supplement to HF maceration, acetolysis and ultrasonic vibration (Brown, 1960). Knox

(1942) and Frey (1951, 1955) report success using a bromoform-acetone mixture of density 2.3 while Funkhouser & Evitt (1959) describe using zinc chloride (ZnCl$_2$) solution of density 1.96. The same technique is described by Doher (1980) using zinc bromide (ZnBr$_2$). These zinc halides to not oxidise organic matter, are stable and easy to obtain and prepare. Doroganievka et al. (1952) have investigated the use of zinc halides and also mixtures of potassium iodide (KI) and cadmium iodide (CdI$_2$). The CdI$_2$+KI liquid is often referred to as Thoulet's solution.

Leroi-Gourhan (1967) and Girard & Renault-Miskovsky (1969) report analyses of cave and rock shelter sediments using standard (HF, HCl) procedures and later, managed to increase pollen recoveries by following standard procedures with CdI$_2$+KI separation. Van Zinderen Bakker (1982) describes successful pollen extraction of the Wonderwerk Cave deposits, South Africa, using HCl, followed by HF, KOH and ZnCl$_2$

1. HCl TREATMENT

40ml sediment
+HCl
+1ml Lycopodium suspension

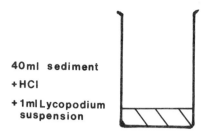

2. KOH TREATMENT

+100 ml KOH

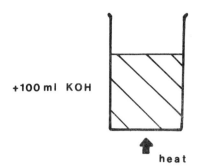

heat

3. SWIRLING AND SIEVING

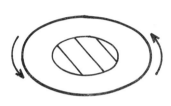

100 μ
10 μ
sieves

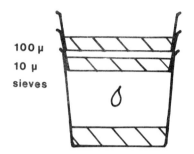

4. H₂O CENTRIFUGE

residue
+HCl
+H₂O

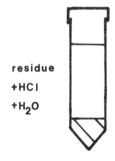

5. ZnCl₂ CENTRIFUGE

pollen etc
rd > 1.9

ZnCl₂ +HCl

residue
rd < 1.9

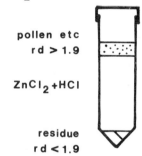

6. RINSE AND STAIN

10 μ

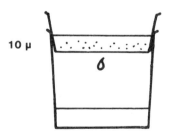

7. MOUNT

GG 89
240

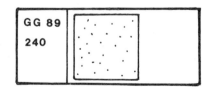

Figure 1: Modified zinc chloride separation process.

(r.d. = 2), sometimes further followed by acetolysis.

The modified process described below has the advantage of: (1) being less time consuming; (2) involving less stages of potential pollen loss; (3) being intrinsically safer; (4) causing no damage or pseudo-fossilisation of pollen grains and (5) being non destructive of other types of organic-walled microfossils. Briefly the process involves treatments with HCl, KOH and $ZnCl_2$ (Figure 1).

In tests 13 and 20, 40 ml of sediment were treated with HCl, KOH and initial swirling and sieving until about 10 ml of sediment remained. This residue was acidified with a little dilute HCl (to avoid $Zn(OH)_2$ precipitation), centrifuged in filtered water for two minutes at 2000 rpm, left 12 hours and centrifuged again. The supernatant was discarded and about 25 ml of the heavy liquid were added to the acidified residue. In these tests the liquid was a solution of $ZnCl_2$, with a density of 1.9, since this falls between the relative densities of organic particles (1.3- 1.7) and the lowest mineral grain density (>2.2). After centrifuging for ten minutes at 2000 rpm the heavy liquid and floats were rinsed on the 10 μm sieve, the floats then being stained and stored. The tests showed recoveries of 15.0% and 20.4% respectively, giving a mean of 17.7%. The results of the above tests are being replicated by further controlled experiments.

Conclusions

The quantitative tests described above demonstrate the need for more efficient pollen concentration methods to be developed. In situations where pollen concentrations in sediments are expected to be low, such as in cave sediments, the validity of interpretations will be supported by the extraction of greater numbers of grains. By using the modified zinc chloride technique the spore recovery rate is almost doubled. It is also worth noting that of the two tests that involve unstratified sediment actually from Robin Hood's Cave (19 and 20) the method without ZnCl2 recovered 5% of the added Lycopodium spores, while the test using ZnCl2 recovered 20%. Samples of similar sediment from nearby Steetley Cave have been processed with the above $ZnCl_2$ process yielding recoveries of 30-35%.

Other tests are being carried out to experiment with solutions of greater and lower densities. For example, repeated centrifugations with $ZnCl_2$ or using a higher relative density (for slower settling of sediment) may be expected to improve the spore recovery rate. The authors are also developing a froth flotation technique for pollen and spore extraction.

Acknowledgements

DNH thanks Sheffield's Town Trustees for funding his research and both authors would like to thank the Universities of Sheffield and Durham for their support.

References

Brown, C.A. (1960). *Palynological Techniques*. Baton Rouge, Louisiana, privately published.

Coles, G.M. (1987). Aspects of the application of palynology to cave deposits in the Magnesian limestone region of North Nottinghamshire. Unpublished PhD thesis, University of Sheffield.

Doher, L.I. (1980). Palynomorph preparation procedures currently used in the paleontology and stratigraphy laboratories, United States Geological Survey. *Geological Survey Circular* 830.

Doroganievka, E.A., Chenfinke, I.E. & Gritchouk, V.P. (1952). Nouveau liquide dense pour l'analyse pollinique. *Iza. Akad. Nauk. SSSR. Sed. Geog.* 4, 73-4. Translation no. 762, Bureau CEDP, Paris.

Erdtman, G. (1952). Pollen morphology and plant taxonomy. *Angiosperms: an introduction to palynology*. Almquist & Wiksell, Stockholm.

Faegri, K. & Iversen, J. (1974). *Textbook of pollen analysis*. 3rd edn. Blackwells, Oxford.

Frey, G.D. (1951). Pollen succession in the sediments of Singletary Lake, North Carolina. *Ecology*, 31(3), 518-33.

Frey, G.D. (1955). A differential flotation technique for recovering microfossils from inorganic sediments. *New Phytologist* 54(2), 257-8.

Funkhouser, J.W. & Evitt, W.R. (1959). Preparation techniques for acid insoluble microfossils. *Micropaleontology* 5(2), 369-75.

Girard, M. & Renault-Miskovsky, J. (1969). Nouvelles techniques de préparation en palynologie appliquées à trois sédiments du Quaternaire final de l'qabri Cornille (Istres-Bouches-du-Rhône). *Bulletin de l'Association Francaise pour l'étude du Quaternaire*, 4, 275-84.

Hunt, C.O. (1985). Recent advances in pollen extraction techniques: a brief review. In: *Palaeobiological Investigations, Research Design, Methods and Data Analysis*. British Archaeological Reports, International Series, 266, 181-7.

Jenkinson, R.D.S. (1984). *Creswell Crags: Late Pleistocene Sites in the English East Midlands*. British Archaeological Reports, British Series, 122.

Knox, A.S. (1942). The use of bromoform in the separation of non- calcareous sediments. *Science* 95 (2464), 307-8.

Leroi-Gourhan, A. (1967). Analyse pollinique des niveaux paléolithiques de l'abri Fritsch. *Review of Paleobotany and Palynology* 4, 81-6.

Leroi-Gourhan, A. & Leroi-Gourhan, A. (1964). Chronologie des grottes d'Arcy-sur-Cure (Yonne). *Gallia Préhistoire* 7, 1-33.

Pryor, E.J. (1965). *Mineral processing*, 3rd edn. Elsevier, London.

Rowe, P.J., Atkinson, T.C. & Jenkinson, R.D.S. (1989). Uranium-series dating of cave deposits at Creswell Crags Gorge, England. *Cave Science* 16(1), 3-17.

Van Zinderen-Bakker Sr, E.M. (1982). Pollen analytical studies of the Wonderwerk Cave, South Africa. *Pollen et Sores*, XXIV, 235-50.

Wenner, C-G. (1947). Pollen diagrams from Labrador. Thesis, reprinted from *Geografisker Annaler* 1-241.

GLUE, DISINFECTANT AND 'CHEWING GUM' IN PREHISTORY

C. Heron[1a], R.P. Evershed[1], B. Chapman[2] & A.M. Pollard[3.]

[1]Department of Biochemistry, University of Liverpool, P.O. Box 147, Liverpool L69 3BX.

[2]Department of Archaeological Sciences, University of Bradford, Bradford, West Yorkshire BD7 1DP.

[3]Department of Chemistry, University of Wales College of Cardiff, Cardiff CF1 3TB.

Introduction

Finds of tar or resin are quite common on well-preserved or waterlogged prehistoric sites, particularly peat bogs. In Northern and Central Europe there is plentiful but fragmentary evidence that tar distilled from the bark of the silver birch tree (*Betula pendula*) was commonly and extensively employed for a wide range of activities. Pliny (*Naturalis Historia*, 16.75) referred to the people of Gaul who extracted 'bitumen' from the bark of the 'white' birch tree by 'boiling' (Loeb Classical Library translation). Vogt (1949) surveyed the use of the birch tree as a source of raw material in prehistory and highlighted two important properties. Firstly, the bark could be detached from the tree in thin layers and easily worked and sewn like fine leather for use in roofing and for making bark containers and other utensils. The bark is also resistant to fungal attack (O'Connell *et al.*, 1988). According to Dimbleby (1978), maple syrup could be preserved for a whole year in vessels made from birch bark. The second property of the birch tree is that tar obtained from the bark could have been used for many activities, notably as an adhesive. Evidence for pottery decoration using cut-out patterns of birch-bark stuck by means of bark tar to the outer surfaces of the darker coloured fabric was uncovered at the Neolithic waterlogged sites of Lake Neuchatel and Egolzwil, Switzerland. Birch-bark tar was thought to have still been in use in the La Tène period.

Evidence for the use of birch-bark tar is known from a number of prehistoric sites, particularly the Swiss and South German lake dwellings (Schlichtherle & Wahlster, 1986). Excavation of the Mesolithic seasonal settlement at Star Carr, Yorkshire, U.K. led to the recovery of numerous tightly wound rolls of birch bark, argued by Clark (1954) to be used primarily as a source of tar. A microlithic arrowhead was found at the site embedded in resin or tar, as well as two barbed points with tarry traces on their tangs. A number of thin, flat cakes of an unidentified resinous substance were also found.

Previous studies have proposed the identification of birch-bark tar from a wide range of prehistoric artefacts. The principal techniques used include infra-red spectroscopy (Sandermann, 1965; Sauter, 1967; Schoknecht & Schwarze, 1967; Hadzi & Orel, 1978; Rottländer, 1983) and thin-layer chromatography (Funke, 1969; Rottländer, 1981; Rottländer, 1986). More recently, results using ^{1}H and ^{13}C-nuclear magnetic resonance spectroscopy have been briefly reported (Sauter *et al.*, 1987).

A regular feature of the reported identifications of birch-bark tar is that little information regarding the chemistry of the tar is given. The largest extractive component of fresh birch bark is the triterpene betulin (lup-20(29)-ene-3ß,28-diol, **1**;Ukkonen & Erä, 1979; Ekman, 1983; O'Connell *et al*, 1988). Triterpene compounds are found in the resins from many genera of broad-leaved trees (Mills & White, 1987), although betulin is very restricted in its distribution in plants.

Betulin is a white crystalline compound and it is the presence of this compound in such large quantities that actually gives silver birch bark its characteristic white coloration. In addition to betulin, other constituents including lupeol(3ß-hydroxy-20(29)-lupene, **2**) and betulinic acid(3ß-hydroxy-20(29)-lupaene-

[a]Current address: Dept. of Archaeological Sciences, University of Bradford, Bradford BD7 1DP.

28-oic acid, **3**) have been identified in the bark of the tree (Ukkonen & Erä, 1979).

The production of birch-bark tar in prehistory represents one of the earliest chemical-technological processes. Rajewski

1 Betulin   R = CH2OH

2 Lupeol   R = CH3

3 Betulinic acid   R = COOH

(1970) has reviewed the methods of production in the early Medieval and later periods (9th century onwards) in Poland. The tar was extracted by dry-distillation in *pechgruber*, literally 'pitch-graves'. According to Rajewski, many production areas go unnoticed due to the lack of attention or misinterpretation. At the site of Biskupin, a number of *pechgruber* of different shapes and dimensions have been found. Experimental investigations demonstrated how the tar was obtained. The hollow was separated into an upper and lower portion by means of a sieve. The top portion, with an approximate capacity of 5 litres, was filled with fresh birch-bark and sealed with loam. By strongly heating the bark without direct interference from the fire or air, the tar was obtained in the lower portion of the 'grave'. Later designs were larger and allowed the collection of tar in many containers simultaneously. An alternative method involved the heating of fresh birch-bark in a ceramic vessel. The tar was collected from the walls of a second pot which had been upturned on the heated one.

This paper is concerned with the analysis of birch-bark tar residues on pot sherds from an assemblage of Neolithic Altheim pottery found

at the site of Ergolding Fischergasse, Germany (Ottaway & Hill, 1985). The sherds were excavated from a partially waterlogged site. (The water table was lowered just prior to and during the excavation.) Resinous deposits from the site have been found mainly in association with pot sherds although deposits have also been found observed on bone and lithic material. A more comprehensive study of all artefact residues from the site will be published elsewhere (Heron *et al.*, in preparation). The deposits have a distinctive appearance as shiny black tar/pitch traces found either on the inner wall or outer wall of the vessels. Those on the inner wall are often uniform layers whilst those on the outer wall consist of a vertical flow line or smear of tar running down the vessel. The majority of sherds with residue were large bodied thick walled pots, either in the shape of storage vessels or bowls. The sherds, mostly medium or coarse ware, were well fired, probably to between 700°C and 900°C. A full report on the analysis of the ceramics will be published elsewhere (Chapman, in preparation).

Materials and Methods

Small samples (100-200 mgs) of each deposit were removed from the sherd surface with fresh scalpel blades. Each sample was ground to a fine powder using an agate pestle and mortar and transferred to a 50 ml round-bottomed flask to which 30 mls dichloromethane and methanol (2:1 v/v) were added. The solutions were refluxed for 20 minutes, cooled and gravity filtered into fresh flasks in order to separate the solvent-soluble residue from carbonised material and pot grains. The solvent was evacuated under water- pump vacuum at 50°C and the residue dried in a vacuum dessicator.

Portions of each residue were examined by infra-red (IR) spectroscopy using a Perkin Elmer 783 Spectrometer. The residues were deposited in solvent onto sodium chloride discs and run as melt films. Prior to gas chromatography (GC) and combined GC/mass spectrometry (GC/MS), the residues were treated with ethereal diazomethane to covert free carboxyl groups into methyl ester derivatives (Schlenk & Gellerman, 1960). GC analyses were performed on a Hewlett Packard 5890A microprocessor-controlled gas chromatograph equipped with flame ionisation detection. An on-column injector system was employed and the GC oven was temperature programmed from 50-330°C at 10°C per minute. An SGE 12 m x 0.22 mm i.d. aluminium clad BP-1 coated capillary column (OV-1 equivalent; 0.1 μm film thickness) was used for all the analyses with helium as the carrier gas.

Combined GC/MS analyses were carried out using a Pye Unicam 204 linked to a VG 7070H double-focusing magnetic sector mass spectrometer. Data acquisition and processing were controlled by a Finnigan INCOS 2300 data system. Scanning was from m/z 40 to 700 in a total cycle time of 3 s at an accelerating voltage of 4 kV. A direct mode of connecting a capillary column into the MS ion source was employed and detection was by electron ionisation (EI, 70 eV). The peaks were identified by computerised library searching and by interpretations based on the known fragmentations of organic compounds.

Results

The residues from the tarry deposits produced largely similar infra-red spectra. Spectra of possible aged birch-bark tar have been recorded by Sandermann (1965), Sauter (1967), Hadzi & Orel (1978) and Rottländer (1983). The spectra of two examples from the present investigation shown in Figure 1 closely match previously published examples. Also shown in Figure 1 is the spectrum of fresh laboratory birch-bark tar distilled at 300-350°C. Each sample exhibits prominent C-H stretching (2840-2950 cm^{-1}) and a broad carbonyl (C=O) absorption from 1700-1750 cm^{-1}. In the fingerprint region of the spectrum the absorptions generated by the fresh tar at 1250, 1025 and 750 cm^{-1} are more clearly defined than in the Neolithic samples. The absorption at 1025 cm^{-1} arises from carbon-oxygen (C-O) bonding.

GC and GC/MS analyses of the Neolithic residues revealed a complex composition. The total ion current (TIC) chromatogram obtained by GC/MS analysis of the methyl ester derivatives of sample EF 219939 is shown in Figure 2. The major component in the trace was identified as betulin (1; Figure 2, peak 4). Lupeol (2; Figure 2, peak 1) and an isomer of betulin, possibly allobetulin (Ukkonen & Erä, 1979; Figure 2, peak 3) were clearly evident. Betulinic acid (3; Figure 2, peak 2) was also present as the methyl ester derivative. Minor components eluting before the triterpenes were identified as long-chain alkanes, not unexpected in the extracts of higher plant tissues. The EI mass spectrum of betulin (1) is shown in Figure 3. The molecular ion (M$^+$)is clearly visible at m/z 442. The fragment ion at m/z 411 results from the loss of the alcohol group (-CH$_2$OH) from the intact molecular species. Betulin was identified as the major component in all the extracts including the sample of freshly prepared tar, although slight differences existed in the relative proportions of the constituents.

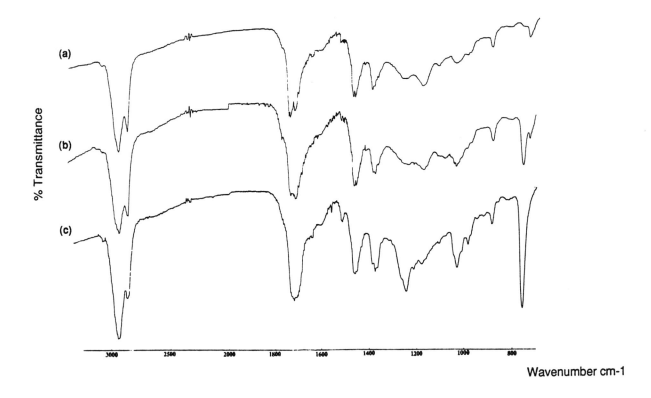

Figure 1. Infra-red absorption spectra of solvent extracts from (a) EF 218839; (b) EF 327; (c) freshly prepared birch-bark tar.

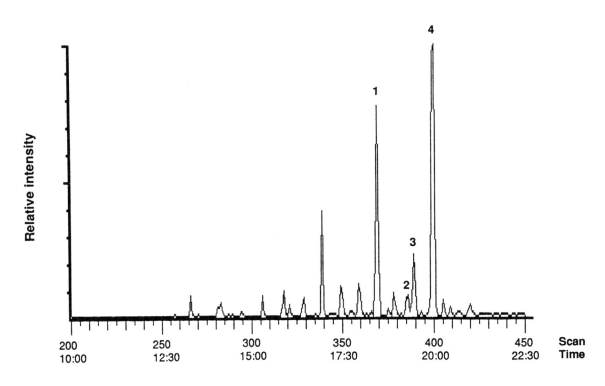

Figure 2. Total ion current (TIC) chromatogram given by sample EF 218839. Peak identities are: 1 = lupeol; 2 = betulinic acid; 3 = betulin isomer; 4 = betulin.

328

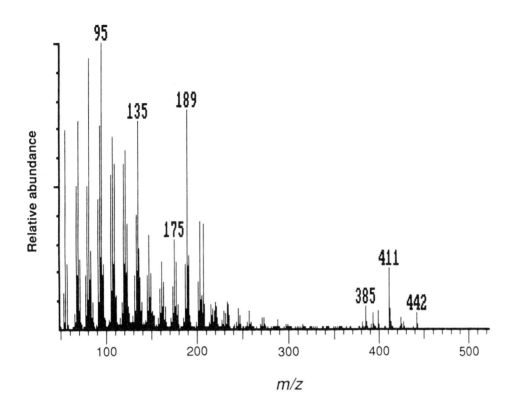

Figure 3. EI (70 ev) mass spectrum of betulin from a sample of Neolithic birch-bark tar.

The prominent carbonyl absorption in the IR spectra of the ancient and freshly prepared tars is unlikely to arise from the low abundance of betulinic acid present in the residues. Furthermore, the ratio of betulin to lupeol in the heated samples is considerably reduced from the quantitative analyses calculated from fresh bark extracts (Ekman, 1983; O'Connell et al., 1988). This may suggest some loss of betulin during distillation of the bark to produce the tar. The appearance of a strong carbonyl absorption suggests that betulin may be oxidised and polymerised during distillation.

Discussion

The identification of betulin and other triterpenes from the solvent extracts of the tarry deposits provides unambiguous confirmation of the use of birch-bark tar at Ergolding Fischergasse. Raw natural products such as birch-bark tar would have been of considerable importance to prehistoric communities. According to Hadzi & Orel (1978), finds of tar are documented as grave goods in Northern Yugoslavia. The principal use of tar would have undoubtedly been as a universal adhesive for fixing flint implements to hafts of wood or antler (Clark, 1954; Funke, 1969; Hadzi & Orel, 1978; Larsson, 1983), for mending broken pots and fixing birch bark decoration to the outside walls of vessels (Vogt, 1949) and as an internal sealant for porous pottery vessels (Hadzi & Orel, 1978; Rottländer, 1986). The uses of tar by prehistoric communities can be supplemented by an historical survey of birch-bark tar in Medieval Poland by Rajewski

(1970). These include a number of practical uses such as the waterproofing of hunting and fishing nets, the preserving of leather objects, the disinfection of fermented beverages and as an ingredient of medicinal preparations. Furthermore, thin layers of tar spread over the ground became useful traps for catching small birds.

Perhaps the most novel interpretation on the use of birch-bark tar in prehistory belongs to Rottländer's study of nine lumps of tar with human tooth impressions from the Neolithic lake-dwelling settlement at Hornstaad-Hornle I in South Germany (Rottländer, 1981; Schlichtherle & Wahlster, 1986). Quoting parallels referring to tobacco, fir-resin and betel chewing, it was considered that chewing birch-bark tar released small quantities of disinfectants during mastication that may have stimulated respiration. Finds of tar displaying obvious chewing marks are known from other sites (Larsson, 1983).

Birch-bark tar has yet to be unambiguously characterised from a prehistoric site in Britain. Evidence that the tar was in use during the Romano-British period has been confirmed by the finding of a residue enveloping the rim of a coarse-ware vessel (North Kent Shelly Ware) used for the transport of salt. It is likely that the tar was applied to the rim of the vessel to hold down a hide or animal skin and so protect the contents from spoilage (Heron, unpublished results).

Conclusion

The study represents the first GC/MS analysis of birch-bark tar residues from prehistoric artefacts. Whilst IR spectroscopy is useful in order to provide preliminary data on the nature and origin of aged resinous deposits, the use of GC and GC/MS to separate and characterise the individual molecular species in the samples is desirable. Chemical analysis of amorphous tarry deposits can potentially assign samples to genus or even species origin. The deposits themselves produce important information on technology, artefact use and other practical aspects of prehistoric societies.

Acknowledgements

The authors would like to thank Dr B.S. Ottaway and also Dr B. Engelhart and S. Aichison for their permission to examine the ceramic samples and Mr M.C. Prescott for running the GC/MS analysis.

References

Clark, J.G.D. (1954). *Excavations at Star Carr*, pp.166-7, Cambridge University Press, Cambridge.

Chapman, B.C.G. (in preparation). The Altheim pottery from Ergolding Fischergasser. In: B.S. Ottaway. *Eine Altheimer Feuchtbodensiedlung in Niederbayern.* Materialhefte zur Bayerichen Vorgeschichte Reihe A.

Dimbleby, G.W. (1978). *Plants and Archaeology*, p.36, 2nd Edition, John Baker, London.

Ekman, R. (1983). The suberin monomers and triterpenoids from the outer bark of Betula verrucosa Ehrh., *Holzforschung* 37, 205-11.

Funke, H. (1969). Chemische-analytische Untersuchungen verschiedener archäologische Funde, Dissertation, Hamburg.

Hadzi, D. & Orel, B. (1978). Spektrometricne raziskave izvora jantarja in smol iz Prazgodovinskih Najdisc na Slovenskem, *Vestn. Slov. Kem, Drus.* 25, 51-63.

Heron, C.P., Evershed, R.P., Chapman, B. & Pollard, A.M. (in preparation). The analysis of ceramic and lithic residues from Ergolding Fischergasse, Germany.

Larsson, L. (1983). Ageröd V: An Atlantic bog site in central Scania, pp.75-6, *Acta Archaeologica Lundensia 12*, Lund, Sweden.

Mills, J.S. & White, R. (1987). *The Organic Chemistry of Museum Objects*, pp. 92-5, Butterworths, London.

O'Connell, M.M., Bentley, M.D., Campbell, C.S. & Cole, B.J.W. (1988). Betulin and lupeol in bark from four white-barked birches, *Phytochemistry* 27(7), 2175-2176.

Ottaway, B.S. & Hill, P.H. (1985). Vorbericht der ausgrabung 1983 in Ergolding Fischergasse, *Archäologische Denkmalpflege in Niederbayern*, pp.38-45, Arbeit 26 Des Bayerischen Landesamtes für Denkmalpflege.

Rajewski, Z. (1970). Pech und teer bei den Slawen, *Z. für Archäologie* 4, 46-53.

Rottländer, R.C.A. (1981). A Neolithic chewing gum, unpublished poster presentation, *21st Archaeometry Symposium*, Brookhaven, U.S.A. •

Rottländer, R.C.A. (1983). *Naturwissenschaftliche Methoden in Der Archäologie*, p.214, Archaeologica Venatoria 6, Tubingen.

Rottländer, R.C.A. (1986). Chemical investigation of potsherds of the Heuneberg, Upper Danube. In: *Proceedings of the 24th International Archaeometry Symposium*, eds. J.S. Olin & M.J. Blackman, pp.403-6, Washington.

Sandermann, W. (1965). Untersuchung vorgeschichlicher 'Graßerharze' und Kitte, *Tech. Beitrage Z. Archaologie II*, 58-73.

Sauter, F. (1967). Chemische untersuchung von 'Härzuberzugen' auf halstaffzeitlicher Keramik, *Arch. Austriaca* 41, 25-36.

Sauter, F., Hayek, E.W.H., Moche, W. & Jordis, U. (1987). Betulin aus archäologischen Schwelteer, *Z. für Naturforsch* 42c (11-12), 1151-2.

Schlenk, H. & Gellerman, J.L. (1960). Esterification of fatty acids with diazomethane on a small scale, *Analytical Chem.* 32(11), 1412-4.

Schlichtherle, H. & Wahlster, B. (1986). *Archäologie in Seen und Mooren - Den Pfahlbauten auf der Spur*, Konrad Theiss Verlag, Stuttgart.

Schoknecht, U. & Schwarze, E. (1967). Hinwiese zur Pechbereitung in frühslawischer zeit, *Ausgrabung Funde* 12(4), 205-10.

Ukkonen, K. & Erä, V. (1979). Birch-bark extractives, *Kem, Kemi*, 6(5), 217-20.

Vogt, E. (1949). The birch as a source of raw material during the Stone Age, *Proc. Pre. Soc* 5, 50-1.

NEW APPROACHES TO THE ANALYSIS OF ORGANIC RESIDUES FROM ARCHAEOLOGICAL REMAINS

C. Heron[1a], R.P. Evershed[1], L.J. Goad[1] & V. Denham[2]

[1]Department of Biochemistry, University of Liverpool, P.O. Box 147, Liverpool, L69 3BX.

[2]Raunds Area Project Co-ordinator, Central Excavation Unit, Historic Buildings and Monuments Commission for England, Fort Cumberland, Fort Cumberland Road, Portsmouth, PO4 94D.

Introduction

There is a understandable bias in archaeological science towards the analysis of organic materials such as ceramics and metals, and the anatomical study of ecofacts in which morphology is well preserved as in the case of the more durable components of some flora and fauna (Carbone & Keel, 1985). This has arisen through a combination of the greater likelihood of survival or inorganic materials and the long established relationship between archaeology and microscopy (Biek, 1963). The investigation of amorphous organic residues has yet to be developed to a comparable level although there has been a marked expansion in recent years in the application of analytical organic chemistry to archaeology. The relatively few studies undertaken have revealed that organic residues can be preserved in a wide range of microenvironments.

The use of the term 'organic residue' is one that may apply to a wide range of diverse substances - for example, the remnant silhouettes of human bodies in some soils (Bethell & Smith, 1989), a caulking agent from boat timbers (Evershed et al., 1985) or a possible blood stain from the surface of a lithic artefact (Loy, 1983). It has been demonstrated that pottery vessels may contain recognisable organic components which have come into contact with the vessel during its time of use, retained within the porous microstructure of the fired clay wall of preserved in visible surface residues (Heron, 1989 and references therein). Chemical analysis of organic residues associated with ceramics can afford valuable information regarding vessel contents and usage. Fatty carboxylic acids are the most commonly studied compound class, with the compositional information usually being provided by gas chromatography (GC). Although fatty acid distributions may be affected by oxidative and microbial degradation (den Dooren de Jong, 1961; Morgan et al., 1984; Evershed, this volume), there is evidence to suggest that in ceramic matrices and charred residues these processes can be retarded (Condamin et al., 1976; Needham & Evans, 1987).

The sterol content of a lipid extract may be a useful indicator of the origin of the residue. Sterols are present in fats and oils as the major component of the so-called non-saponifiable fraction which usually makes up less than 1% of a fat or oil. Whereas animals produce an abundance of cholesterol (cholest-5-en-3ß-ol,1), higher plants produce sterols alkylated at the C-24 position such as sitosterol (sigmast-5-en-3ß-ol,2) which predominates in many species. Much of the work on sterols in archaeological contexts has so far concentrated on faecal material (eg. Knights et al., 1983) and preserved bog body tissues (Evershed, this volume).

Other isoprenoids also offer potential for classifying residues. Generic origin has been clearly defined in the case of pine tars, pitches and resins from a range of amphoras, owing to the presence of alkyl-substituted diterpenoids based on the abietane and pimarane skeletons (Robinson et al., 1987; Heron & Pollard, 1988; Beck et al., 1989). The positive detection of triterpenoid compounds from *Pistacia* sp. has also confirmed the identify of the resins contained in the amphoras recovered from the Late Bronze Age shipwreck at Ulu Burun in Southern Turkey (Mills & White, 1989).

[a]Current address: Dept. of Archaeological Sciences, University of Bradford, Bradford BD7 1DP.

1

Cholesterol

2

Sitosterol

Similarly, the characterisation of triterpenoids found in the bark of the silver birch tree (*B. pendula*) has confirmed the widespread use of birch-bark tar by prehistoric communities (Heron *et al.*, this volume). In these latter investigations, detailed compositional information was acquired through GC/mass spectrometry (GC/MS).

In other studies, cooking vessels and associated residues (chars) have been the focus of analysis. Charring of food to the inside wall of the vessel may entrap organic components within a carbonised matrix. Preservation is promoted since carbon is not digested by most microorganisms. The principal aim of these investigations has been to identify foods prepared in pottery containers and provide highly specific data on the diet of early man. Indications are that a wide range of compound classes are retained in the char (Needham & Evans, 1987). Charred deposits found on cooking vessels have also been investigated by X-ray fluorescence techniques (Bush & Zubrow, 1986) and stable isotope analysis (Hastdorf & DeNiro, 1985). In the latter study, the ratios of $\delta^{13}C$ and $\delta^{15}N$ were used to characterise plant remains and determine the use of leguminous plants in cooking.

Hitherto, most residue work has concentrated on single vessels of specific interest or groups of vessels of particular form or function (notably amphoras and cooking vessels). Since the former type of study on a single vessel cannot demonstrate whether the organic residue found in a particular vessel is 'typical or atypical' of its type, date or context, the information may be of limited archaeological value. As a result of sampling inconsistencies or biases, residue analysis has often been seen as a novel adjunct to more mainstream ceramic studies. It is therefore proposed that the significance of the use of one vessel or a series of vessels can only be judged against a background of knowledge of general use. To this end, we have embarked upon the systematic study of large numbers of vessels from well-defined ceramic assemblages.

Sampling of sherds

The source of ceramics is the Raunds Area Project (RAP) - the study of an area of 40 km^2 in the Nene Valley, Northamptonshire (Dix, 1986-7). Investigations are concentrating on two specific sites. The 1st century Romano-British villa settlement at Stanwick and the Saxon/Medieval hamlet at West Cotton where samples have so far been

333

collected. On-going excavation has allowed the implementation in the field of specific, project designed, sampling strategies including handling and storage. Care is taken not to remove possible surface residues and burial soil is collected from around the sherd to enable comparative analyses. All pottery submitted for residue analysis is thoroughly examined and given a full contextual description.

Chemical analysis of organic residues

Prior to extraction of any organic constituents from the samples, simply demonstrating the presence of significant levels of organic matter provides an important first step in classifying vessel use. This can be established by organic elemental analysis (C,H & N). Whilst the level of organic matter in the clay wall is too low to give reliable assessments, the method can be used to distinguish visible surface residues such as charred food remains, sooting and inorganic deposits or accretions that have built up on the vessel during burial.

Preliminary investigations have been largely concerned with extractable lipid residues (derived from fats, oils, waxes, resins, etc.), although the presence and diagnostic potential of other major compound classes such as carbohydrates and proteins is also being investigated. Lipids tend to survive archaeological conditions better than other classes of compound and are mostly amenable to the sensitive methods of analysis such as GC and GC/MS.

For statistical purposes the analysis of several hundred different vessels is envisaged. To achieve this end, consideration has been given to streamlining the extraction procedures for releasing organic residue from the inorganic ceramic matrix. Thus far, 2 gram samples of the potsherd have been found to be sufficient for yielding readily recognisable organic constituents. In the case of visible surface residues (usually weighing considerably less than 1 gram), about one half of the sample is removed for study. The remainder can be saved for alternative

methods of study (eg. pollen analysis or scanning electron microscopy). A small portion of the sherd is ground to a fine powder and the lipids extracted by solvent washing (2:1 v/v chloroform/methanol; 30 minutes, ultrasonication). The samples are then centrifuged and the supernatant decanted into a clean vessel and evaporated (50°C; *in vacuo*).

In order to achieve our numerical objectives of examining several hundred vessels, it is essential to employ a rapid and uncomplicated analytical protocol. Furthermore, the relatively small amounts of organic residue available (typically a few milligrams/sherd) render conventional approaches employing adsorption chromatographic fractionation wholly impractical. Fractionations are lengthy procedures and without the careful use of internal standards important quantitative data and 'fingerprinting' potential may be obscured or lost.

Preliminary screening of the crude extracts uses thin-layer chromatography (TLC). Visualisation with appropriate spray reagents allows clear assessment of the major compound classes present. It must be stressed that TLC is limited in its application to the analysis of trace organic residues from sherds. Degradation can produce complex patterns that may differ significantly from the pattern given by the original material. Interpretations of specific organic products from aged material must be treated with caution if not supplemented with more rigorous analysis using more selective techniques such as GC and GC/MS.

GC and GC/MS hold the key to residue classification through characterisation or individual molecular species and 'fingerprint' profiles. This approach permits the rapid analysis of relatively crude extracts from vessels. Identification of each component by GC relies on retention time comparisons with authentic compounds. Combined GC/MS gives molecular information on each component in the mixture. Identifications by GC/MS are achieved by computer library searching and by interpretations based on the known fragmentations of organic compounds. The GC/MS data is also stored on a dedicated microcomputer for further processing. Ultimately the compositional data derived from GC/MS analyses will be integrated with conventional ceramic and contextual data, as well as floral and faunal information, for the purpose of data-basing and statistical analyses.

An approach has been adopted which attempts to resolve and characterise all the major classes of lipid in a given residue in a single GC or GC/MS run. To do this we have taken advantage of the potential offered by modern high-temperature stable immobilised GC phases in order to simultaneously determine free fatty acids, sterols, wax esters, monoacylglycerols, diacylglycerols, triacylglycerols (triglycerides), etc. The ability to detect the full suite of lipids in a single GC run renders arduous fractionation unnecessary. Furthermore, the addition of an internal standard to each sample allows comparison of yields not only from sherd to sherd but also from samples removed from different parts of the same vessel (eg. rim and base) and from soil samples taken off the sherd surface. Chromatographic resolution is improved by blocking protic sites present in free fatty acids, sterols etc. by trimethylsilylation.

The lipid profiles can be interpreted by calculating the relative proportions of fatty acids in the residues, by determining the type of sterol present and by investigating the origin of other lipid components such as hydrocarbons, ketones, alcohols and wax esters. The presence of triacylglycerols allows an assessment of the preservation of the organic material.

By way of an example, Figure 1 shows the total ion current (TIC) chromatogram observed from the extract of a Saxon sherd from the Furnells Manor site, Raunds, and serves to demonstrate the advantage of employing high resolution capillary GC to resolve the total lipid extract in a single analytical run. The presence of high levels of oleic acid (Z-9-octadecenoic acid; peak 3) along with the detection of intact triacylglycerols

suggests excellent preservation of the residue in the vessel matrix. The major sterol component in the residue (peak 7) was identified by its mass spectrum as sitosterol, the most widespread sterol in the plant kingdom. The high proportion of oleic acid and the presence of sitosterol is strongly indicative of a vegetable origin such as seed oil.

Classification of organic residues is based on the known chemosystematics of biogenic matter while taking into account variability arising from the differing nature of residues and environmental factors. In circumstances where there is evidence for the contents or function of a particular vessel, it is important to test for an appropriate chemical signature. Figure 2 shows the GC/MS analysis of the lipid extract from a Saxon lamp found in the North Raunds area (unstratified find). Saturated fatty acids are the dominant constituents. However, the expanded portion highlights the steroidal components, unaltered cholesterol (peak 4) and two oxidation products, cholest-5-en-3ß,7ξ-diol (peak 5) and cholest-5-en-7-one-3ß-ol; (peak 6). The absence of intact acyl glycerols also suggests degradation, perhaps as a result of heating. These results strongly indicate that the fuel for the lamp was derived from an animal source (eg. tallow derived from an animal fat).

The chemical data is also being supplemented with reference data on the composition of foodstuffs, the evaluation of known degradative pathways and the undertaking of cooking simulations and burial experiments.

Areas of investigation

The long-term goal of this work is to test whether or not organic residue analysis can be used diagnostically to define a specific or general pattern of vessel use and to provide a means of classification that will augment conventional typological, petrological, manufacturing and contextual information and so assist archaeologists in constructing models of economic activity. The investigation must take into account a number

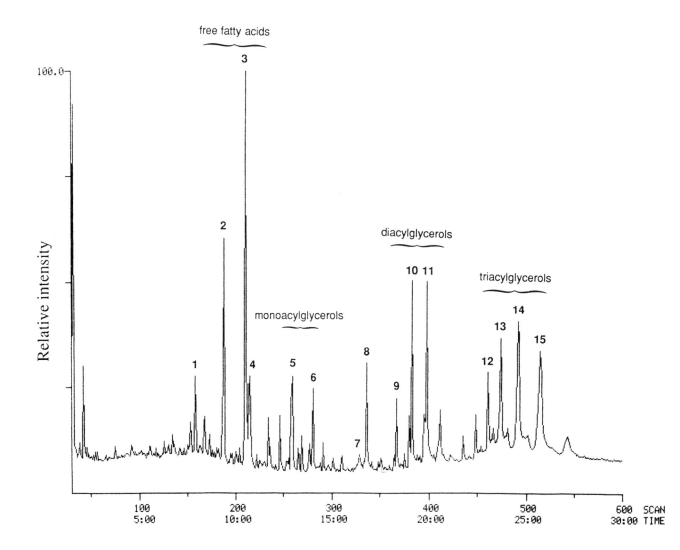

Figure 1. Capillary total ion current (TIC) chromatogram of lipid extract from a Saxon vessel, Furnells Manor Site, Raunds showing the major compound classes found in the residue. Peak identities are; 1 = tetradecanoic acid (C14:0); 2 = hexadecanoic acid (C16:1); 3 = octadecenoic acid (C18:1); 4 = octadecanoic acid (C18:0); 5 = monopalmitin; 6 = monostearin; 7 = sitosterol; 8 = unnassigned triterpene; 9 = palmitomyristin; 10 = dipalmitin; 11 = palmitostearin; 12 = dipalmitomyristin (PPM); 13 = tripalmitin (PPP); 14 = dipalmitostearin (PPS); 15 = distearopalmitin (SSP). Full experiemental details are presented elsewhere (Evershed *et al.*, in press).

336

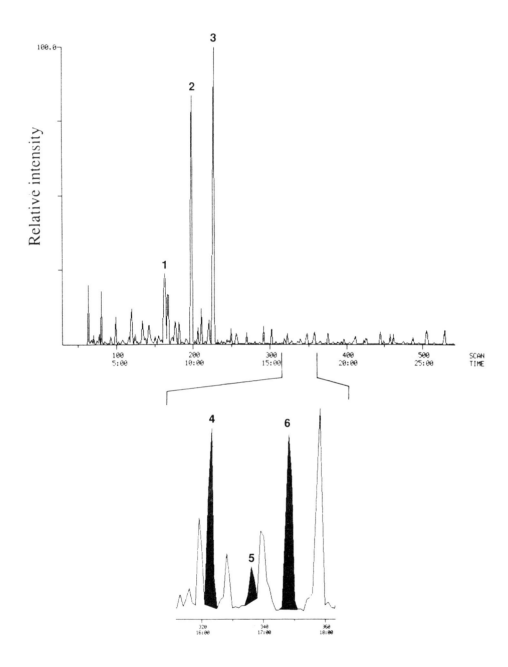

Figure 2. Capillary TIC chromatogram of lamp extract with expanded portion demonstrating the presence of animal sterols. Peak identities are: 1 = tetradecanoic acid; 2 = hexadecanoic acid; 3 = octadecanoic acid; 4 = cholesterol; 5 = cholest-5-en-3ß-7ξ-diol; 6 = cholest-5-en-7-one-3ß-ol.

of natural and cultural factors that may affect the recovery and interpretation of organic archaeological residue. These include:

- the composition of, and possible contamination from, soil organic matter (Heron, *et al.*, in preparation).

- microbial alteration or reworking (bacterial or fungal) of the original substances).

- possible effects arising from naturally occurring organic matter residual in the clay after firing (Johnson *et al.*, 1988).

- multiple use of single vessels.

- potential biases in data recovery (eg. wet as opposed to dry storage).

Conclusions

The aim of this paper is to propose a specific methodology governing the sampling, extraction and analysis of organic residues found in pottery vessels. The results presented here serve to illustrate cogent points arising from the discussion. Successful discrimination and identification of food and other organic substances found in pottery vessels would enable significant advances to be made in the study of food preparation and consumption as well as general pottery usage. However, there is a need for more rigour in the analysis of trace organic residues. The simple equation of organic constituents from samples with highly specific foodstuffs appears to be made consistently and uncritically. Furthermore, the interpretations are frequently derived from a rather fragile foundation using methods such as TLC and IR spectroscopy; techniques that usually provide broad classifications and whose interpretive potential is limited when analysing degraded substances. Only passing attention in the literature has been given to control or comparative sampling, for example, from soil adhering to sherds or handles not expected to have come into contact with the contents of a particular vessel. At present the confusion arising from such apparent inconsistencies will limit wider and more effective application of organic chemical and biochemical studies in archaeology.

Acknowledgements

The authors would like to thank the SERC Science-Based Archaeology Committee and English Heritage (HBMC-England) for funding.

References

Beck, C.W., Smart, C.J. & Ossenkop, D.J. (1989). Residues and linings in ancient Mediterranean transport amphoras. In: *Archaeological Chemistry IV*, Advances in Chemistry Series 220, Ed. R.O. Allen, pp. 369-80. American Chemical Society, Washington DC.

Bethell, P.H. & Smith, J.U. (1989). Trace element analysis of an inhumation from Sutton Hoo using inductively coupled plasma emission spectrometry: An Evaluation of the technique applied to the analysis of organic residues, *J. Archaeol. Sci.* 16(1), 47-56.

Biek. L. (1963). *Archaeology and the Microscope*. Praeger Press, London.

Bush, P. & Zubrow, E.B.W. (1986). The art and science of eating, *Science and Archaeology* 28, 38-43.

Carbone, V.A. & Keel, B.C. (1985). Preservation of plant and animal remains. In: *The Analysis of Prehistoric Diets*, eds. R.I. Gilbert, Jr. & J.H. Mielke, pp. 1-19, Academic Press, London.

Condamin, J., Formenti, F., Metais, M.O., Michel, M. & Blond, P. (1976). The application of gas chromatography to the tracing of oil in ancient amphorae, *Archaeometry* 18(2), 195-201.

den Dooren De Jong, L.E. (1961). On the formation of adipocere from fats, *Antonie von Leuwenhoek J. Microbiol. & Serol.* 27, 337-61.

338

Dix, B. (1986-7). The Raunds Area Project: Second interim report, *Northamptonshire Archaeology* 21, 3-29.

Evershed, R.P., Jerman, K. & Eglinton, G. (1985). Pine wood origin for pitch from the Mary Rose, *Nature* 314, 528-30.

Evershed, R.P., Heron, C.P. & Goad, L.J. (in press). Analysis of organic residues of archaeological origin by high-temperature gas chromatography and gas chromatography/mass spectrometry. *Analyst* in press.

Hastdorf, C.A. & DeNiro, M.J. (1985). Reconstruction of prehistoric plant production and cooking practices by a new isotopic method, *Nature* 315, 489-91.

Heron, C.P. & Pollard, A.M. (1988). The analysis of natural resinous material from Roman amphoras. In: *Science and Archaeology - Glasgow 1987*. Proceedings of a conference on the application of scientific techniques to archaeology, eds. E.A. Slater & J.O. Tate, pp.429-46, BAR British Series 196, Oxford.

Heron, C.P. (1989). The Analysis of Organic Residues from Archaeological Ceramics. Unpublished PhD Thesis, University of Wales College of Cardiff.

Heron, C., Evershed, R.P. & Goad, L.J. (in press). Effects of migration on organic residues associated with buried potsherds. *J. Archaeol. Sci.* in press.

Johnson, J.S., Clark, J., Miller-Antonio, S., Robins, D., Schiffer, M.B. & Skibo, J.M. (1988). Effects of temperature on the fate of naturally occurring organic matter in clays, *J. Archaeol. Sci.* 15, 403-14.

Knights, B.A., Dickson, C.A., Dickson, J.H. & Breeze, D.J. (1983). Evidence concerning the Roman military diet at Bearsden, Scotland in the 2nd Century AD., *J. Archaeol. Sci.* 10, 139-52.

Loy, T.H. (1983). Prehistoric blood residues: Detection on tool surfaces and identification of species of origin, *Science* 220, 1269-71.

Mills, J.S. & White, R. (1989). The identity of the resins from the Late Bronze Age shipwreck at Ulu Burun (Kas), *Archaeometry* 31(1), 37-44.

Morgan, E.D., Titus, L., Small, R.J. & Edwards, C. (1984). Gas chromatographic analysis of fatty material from a Thule midden, *Archaeometry* 26(1), 43-8.

Needham, S.P. & Evans, J. (1987). Honey and dripping: Neolithic food residues from Runnymede Bridge, *Oxford J. Arch.* 6(1), 21-8.

Robinson, N., Evershed, R.P., Higgs, W.J., Jerman, K. & Eglinton, G. (1987). Proof of a pine wood origin for pitch from Tudor (Mary Rose) and Etrurian shipwrecks: Application of analytical organic chemistry in archaeology, *Analyst* 112, 637-44.

THE SURVIVAL OF WAXES IN COPROLITES: THE ARCHAEOLOGICAL POTENTIAL

S. Wales[1], J. Evans[2] and A.R. Leeds[3]

[1]Department of Human Environment, Institute of Archaeology, University College London, London, WC1 OPY.

[2]Environmental Sciences Division, Department of Biosciences, Polytechnic of East London, Romford Road, London, E15 4LZ.

[3]Department of Food and Nutritional Sciences, Kings' College London, Campden Hill Road, London, W8 7AH.

This study originated with fragment of material recovered from the hearth areas of the Epipalaeolithic levels of Abu Hureyra in Syria. They are thought to be fragments of infant faeces preserved by charring of their outer layers. Similar fragments have also been recovered from the palaeolithic site of Wadi Kubbanyia in Egypt (Hillman, 1989). Tell Abu Hureyra stood at the edge of the Euphrates valley in northern Syria and was occupied during the Epipalaeolithic and Neolithic periods, and so represents a period when human diet underwent a drastic change as the hunter/gatherer economy was superseded by that of an early agricultural community. A wide variety of plant remains have been recovered as charred remains from the epipalaeolithic levels, suggesting a diverse plant diet prior to the beginnings of agriculture (Hillman et al., 1989). Fragments of possible coprolites from this period are both rare and potentially very rewarding as they can give direct evidence of diet. Infant coprolites are especially interesting as we have no knowledge of the type of food fed to infants at this period.

The usual procedure for studying dietary information from coprolites involves their rehydration in a suitable chemical, usually trisodium phosphate. Once the coprolite has disaggregated it is washed through sieves of various sizes and the identifiable macrofossils are recovered and identified (Callen & Cameron, 1960; Callen, 1969; Bryant, 1975). These methods, however, are not applicable to these fragments which appear to consist of finely ground material containing no fragments of plant material large enough to identify on morphological grounds (Hillman, 1989). It was, therefore, necessary to look for chemical evidence of the plant food consumed (Wales & Evans, 1988). In considering what substance, or substances, are likely to provide this evidence there are three criteria that must be fulfilled. Firstly, it must survive digestion in a chemically recognisable form. Secondly, it must be capable of surviving in an archaeological context in a chemically recognisable form. Thirdly, it must be able to indicate the nature of the original food.

Protein was an obvious consideration. Although a likely candidate the survival of protein after burial for a long period is rather uncertain. Any protein that does survive is likely to be contaminated from other protein sources, such as bacteria and fungi, and the separation of protein from different sources is quite complex.

Lipids were our next consideration. This was especially hopeful as the survival of lipids is likely to be good in arid areas where the Abu Hureyra fragments, and many other coprolites, are preserved. Triglycerides often survive in archaeological contexts and a number have been recovered from organic residues in pottery (Rottlander & Hartke, 1983; Evans, forthcoming). Unfortunately the processing of tryglycerides by the digestive system means that the triglycerides present in faeces do not have any direct correlation to those taken in as food. Waxes, however, appear to be likely to fulfil our three criteria.

Waxes often survive well in archaeological situations. Beeswax, for example, has been found in a number of contexts, including in pottery from the British Neolithic (Needham & Evans, 1987). Wax is present as the outer protective covering of most plant parts (Kolattukudy, 1975 & 1980). The amount of wax can be up to 15% of the dry weight of plant tissue (Eglinton & Hamilton, 1967), and the percentage of wax is likely to be high in plants growing in arid areas (Smith, 1978). Studies in chemical taxonomy suggest that wax is a useful taxonomic indicator.

The question whether wax survives human digestion in a chemically recognisable form does not appear to have been studied. As it was necessary to answer this question a dietary study was devised and carried out at the Department of Food and Nutritional Sciences, Kings' College London.

Dietary Study

Beeswax was chosen for this study as it was easy to recognise and it is familiar to the authors. It is also safe for human consumption. The beeswax used in this study was provided by Mr C. de Bruyn of Writtle Agricultural College. It came from bees whose source of nectar was free of the use of chemicals. This removed danger of contamination with pesticides or fungicides that might complicate analysis, or indeed be harmful to the volunteers.

The study involved five volunteers over a twenty day period. During the period of the study they maintained a low-wax background diet. This involved avoiding eating the skins of meat, seafood, vegetables and fruit (including fruit skins in jam preserves), and honey of any sort. The study was divided into four periods of five days each. At the beginning of each period each volunteer took a small number of small radio-opaque plastic markers. At the beginning of periods 1 and 4 the markers were square, at the beginning of period 2 they consisted of small rings and at the beginning of period 3 they were large rings.

In the first period no wax was added to their diet. During the second period they consumed 2.5 g beeswax per day and during the third period they consumed 5 g beeswax per day. The wax was ground in a Moulinex food processor and then incorporated into prepared snacks of Jordans' special recipe muesli, Waitrose carrot and nut salad or Waitrose brown rice salad. During the fourth period the volunteers consumed no additional wax and resumed their normal diet.

All stools were collected and labelled with time and date, then X-rayed to locate the markers. This enabled them to be divided into groups to correspond to the four periods. Each group was pooled, homogenised and a sample freeze-dried. Volunteers also provided duplicate food samples from each period. These were also pooled, homogenised and a sample freeze-dried. The study was approved by the ethical committee of Kings' College London.

Analysis

Analysis was carried out at the Polytechnic of East London. A sub-sample of each stool and food sample, weighing between 0.6 g and 1.2 g, was extracted with hexane in a soxhlet apparatus. Each extraction continued for at least 6 hours and was carried out in a fume cupboard. At this stage wax was clearly visible in some of the extracts. The concentrated extracts were examined by Infrared spectroscopy using a Perkin Elmer 781 Infrared spectrometer with data station.

All the food samples showed a very similar common lipid spectra, as expected. The extracts from the stool samples from the second and third periods, however, showed a very clear wax spectra. The wax in the faecal extract was so strong that it masked other faecal components. The spectra was compared with known wax spectra and it correlated well with the spectra of the beeswax used in this study (Figure 1), indicating the survival of the beeswax through the digestive tract. A few samples were examined by Gas Liquid Chromatography and

341

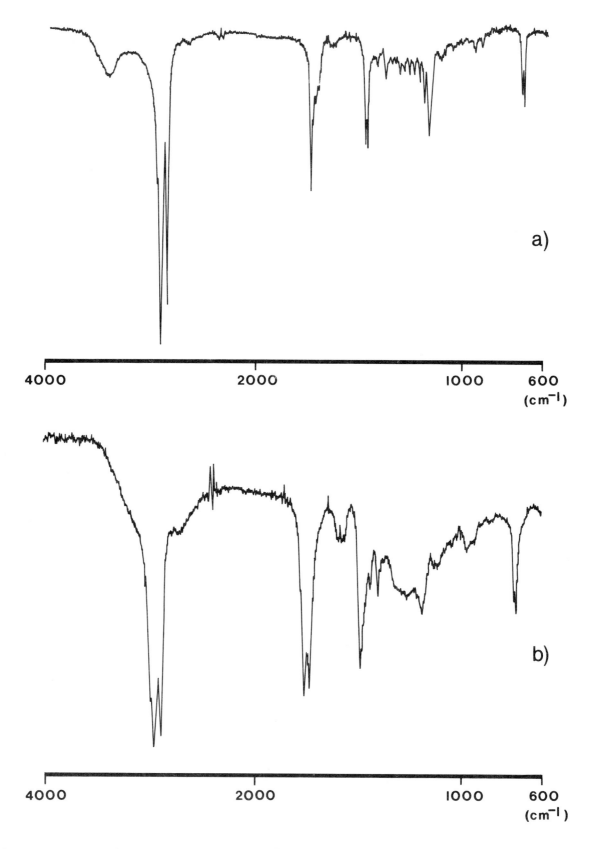

Figure 1. Infrared Spectra a) beeswax, b) stool B.015

342

this confirmed the presence of beeswax in the stools.

Conclusion

This work has shown that waxes survive human digestion in a chemically recognisable form. There is good reason to believe, therefore, that waxes may also be present in coprolites. This study is being continued with the investigation of some North American coprolites from an area where it is believed cactus pads were one of the staple foods. If this proves successful it will open up a number of possibilities. Most macrofossils recovered from coprolites consist of those items that are resistant to digestion, such as seeds and bone. Although these provide valuable information on diet they are unlikely to provide the complete picture. Items of food that do not survive in a physically recognisable form, however, may survive in a chemically recognisable form. This work will give us the potential to recover evidence of the consumption of a variety of leafy foods, immature pods of legumes and stems. These are foods which may have made up to a third of the diet of ancient peoples (Stahl, 1984), and for which comparatively little evidence is available at the present time.

Acknowledgements

The authors would like to thank all those who gave them help and encouragement, especially Gordon Hillman and Don Brothwell of the Department of Human Environment, Institute of Archaeology, University College London; and also Cheng Ong at Kings' College London for her work on the dietary study.

References

Bryant, V.M. Jnr. and Williams-Dean, G. (1975). The coprolites of man. *Scientific American* 232(1), 100-109.

Callen, E.O. (1969). Diet as revealed in coprolites. In: *Science and Archaeology*, ed. D.R. Brothwell & E.S. Higgs. Thames and Hudson, London.

Callen, E.O. & Cameron, T.W.M. (1960). A prehistoric diet revealed in coprolites. *New Scientist*, 7 July.

Eglinton, G. & Hamilton, R.J. (1967). Leaf epicuticular waxes. *Science* 156, 1322-1335.

Evans, J. (forthcoming) Analysis of organic compounds from archaeological sources. In: *Archaeological Chemistry* 6, ed. M. Hughes. Wiley, New York.

Hillman, G.C. (1989). Wild plant food economy and seasonality at Wadi Kubbanyia: riverine subsistence hyper-arid North Africa in the Late Palaeolithic. In: *Foraging and Farming*, eds. D.R. Harris & G.C. Hillman. Unwin Hyman, London.

Hillman, G.C., College, S.M. & Harris, D.R. (1986). Plant-food economy during the Epi-palaeolithic period at Tell Abu Hureyra, Syria: dietary diversity, seasonality and modes of exploitation. In: *Foraging and Farming*, eds. D.R. Harris & G.C. Hillman. Unwin Hyman, London.

Kolattukudy, P.E. (1975). Biochemistry of cutin, suberin and waxes - the lipid barriers on plants. In: *Recent Advances in the Chemistry and Biochemistry of Plant Lipids*, eds. T. Galliard & E.I. Mercer. Academic Press, London and New York.

Kolattukudy, P.E. (1980). Cutin, suberin and waxes. In: *The Biochemistry of Plants*, Vol.4, Lipids: Structure and Function, ed. P.K. Stumpf. Academic Press, London and New York.

Needham, S. & Evans, J. (1987). Honey and dripping: Neolithic food residues from Runnymede Bridge. *Oxford Journal of Archaeology* 6(1), 21-8.

Rottlander, R.G. & Hartke, I. (1983). New results of food identification by fat analysis. In: *Proceedings of 22nd Symposium on Archaeometry*, eds. A. Aspinall & S.E. Warren, pp. 218-23.

Smith, P.M. (1978). Chemical Evidence in plant taxonomy. In: *Essays in Plant taxonomy*, ed. H.E. Street. Academic Press, London and New York.

Stahl, A.B. (1984). Hominid dietary selection before fire. *Current Anthropology* 25, 151-168.

Wales, S. & Evans, J. (1988). New possibilities of obtaining archaeological information from coprolites and similar materials. In: *Science and Archaeology*, Glasgow 1987, eds. E.A. Slater & J.O. Tate. BAR British Series 196, Oxford.

BACKSCATTERED ELECTRON IMAGING OF DIAGENETIC CHANGES TO TEETH AND RELATED ALVEOLAR BONE

Lynne S. Bell, Sheila J. Jones & Alan Boyde

The Hard Tissue Research Unit, Department of Anatomy and Developmental Biology, University College London, Gower Street, London WC1E 6BT, UK.

Introduction

Diagenetic changes to bone and teeth have largely been investigated using routine light microscopic methods (Garland, 1989; Hackett, 1981; Poole and Tratman, 1978; Falin, 1961; Sognnaes, 1955; Wedl, 1864). Such methods have allowed a good understanding of the general distribution of diagenetic change and enabled its consequent morphologies within bone and teeth to be assessed. Backscattered imaging extends this body of work to demonstrate the relative density of compositional changes within hard tissue microstructure caused by the effects of diagenesis.

Backscattered electron (BSE) imaging using a scanning electron microscope (SEM) is a technique already utilised in the study of bones and teeth in basic medical and dental research. Work by Boyde et al. (1986) and Boyde and Jones (1983a) showed to good effect the density variations produced by initial mineralization and later maturation during growth and normal bone-turnover, and also of impaired mineralization during pathological conditions such as Paget's disease, fluorosis, osteomalacia and osteogenesis imperfecta. Dental and related tissues (and encrustations) ie. enamel, dentine, cementum, Sharpey fibre alveolar bone and calculus have also been studied to advantage (Boyde and Jones, 1983b); they exhibit incremental and remodelling mineral phase changes in common with the atomic number (Z) compositional changes present in bone - with the mean densities of the dental tissues generally higher than those of bone. Similarly, dental tissue can be affected by pathological interference during development and after teeth have erupted into the oral cavity (Jones, 1987; Jones and Boyde, 1987).

SEM/BSE imaging has been applied little to bones and teeth in the study of archaeological materials. To date most work in the archaeological field has been concerned with the Z composition of metals, ceramics and glassware (Meeks, 1988), even though scanning electron microscopy has been used extensively to study surfaces on many different artefacts (see Olsen, 1988). Recently, we have used BSE imaging to investigate diagenetic effects in adult human femora and tibiae with particular reference to archaeological normal and pathological bone (Bell, 1990); whilst Dobney and Brothwell (1986) applied the same technique to the morphology of calculus in archaeological samples. Other workers have employed the technique in wider studies, and in particular, to tooth enamel structure during primate evolution (Martin et al., 1988).

The study presented here examines the use of SEM/BSE imaging mode in the SEM to assess the effects and distribution of diagenesis in teeth retained in situ of their supporting bone.

Methodology

Soil-buried, archaeological, human mandibular and maxillary specimens with teeth still in situ, with good microscopic preservation, were selected for this study. A single tooth and accompanying socket was isolated from the mandible or maxilla by thick sectioning the entire tooth and socket free, using a wet, diamond-edged circular saw. Each section was then rinsed in tepid tap water and allowed to air dry. These sections were put in polymethylmethacrylate (PMMA), placed in an oven set at 37°C and removed after the PMMA had solidified. The methacrylate had been prepared by the flash distil method described by Boyde et al. (1986). The embedded specimens were then cut longitudinally buccolingually using an Isomet-11-1180 circular saw, polished using graded abrasives and finished with a water-dispersed 1 μm diamond abrasive on a rotary lap. Each block face received a coating of carbon and was then mounted on an aluminium stub.

The specimens were examined using a Cambridge Stereoscan S4-10 SEM, operated in BSE mode working at 20 kV beam voltage. The BSE flat section images were created using a solid state BSE detector in a four-segmented ring configuration and the topographical images were obtained by subtracting the east and south quadrants simultaneously. The topographical images were necessary for the identification of bubbles, holes, edges and scratches which may have contributed to the overall signal artifactually; whilst the flat section images gave qualitative relative assessments of density changes within the specimen alongside any morphological changes.

Results and Discussion

All the hard tissues observed in this study, excepting enamel, had undergone quite radical diagenetic change.

Bone

The characteristic morphology of adult bone includes circumferential lamellae, secondary osteons and areas of interstitial lamellae. A radiating system of osteocyte lacunae and interconnecting canaliculae is present throughout the tissue (Figs. 1 and 2). Within this general framework the process of remodelling results in areas of bone which appear relatively denser (lighter in BSE images) and areas which appear relatively less dense (darker in BSE images), with holes appearing as completely black (Fig. 1). This difference in density is a direct indication of the relative ages of packets of bone, the progress of mineralization being extended by a relatively long maturation phase. Thus, due to the incremental structure of bone, where time and density are interconnected during

growth and the normal bone turnover that occurs throughout life, BSE imaging should be an ideal method to observe such changes, even in archaeological material.

To interpret prepared samples it was necessary to check topographic detail assiduously (as previously mentioned). BSE imaging detects high energy electrons released from the specimens surface, to an approximate depth of 1 μm (Boyde and Jones, 1983a), with back-scattering increasing steeply with increasing Z values. Hence, any local variations in Z composition gave variations in intensity and so image, due to modulations in the back- scattering electron signal (Boyde and Jones, 1983a&b; Boyde et al., 1986; Watt, 1985). However, topography can alter this Z related signal artifactually, causing an erroneous count with the scan. Such artefacts are usually products of embedding and polishing, and can include bubbles, holes, edges or scratches. With biological specimens, where density can vary, it is difficult to avoid creating such topography, since less dense areas will be abraded much faster than relatively more dense areas. Even with very fine abrasives one must accept the existence of such artefacts, keep them to a minimum, and record topography for each BSE image taken. Interpretation can then be cross-checked from flat section image to topographic image, thus avoiding false conclusions.

Topographic detail due to polishing was observed in the normal non-archaeological specimen (Fig. 2). This was due to three main factors. Firstly, bone wears less than PMMA; secondly, collagen orientated parallel to the surface wears more than that vertical to the surface (Reid, 1986); and thirdly, more highly mineralized bone wears less than less well mineralized bone (Boyde and Jones, 1987). Thus the sites of osteonal canals and osteocyte lacunae were marked by slight depressions; younger less well mineralized osteons, or last formed regions of osteons, were less proud than older, more mineralized zones; and interstitial lamellae were generally proud of the mean surface level. Lamellar topography depended on collagen orientation and was enhanced in less well mineralized

346

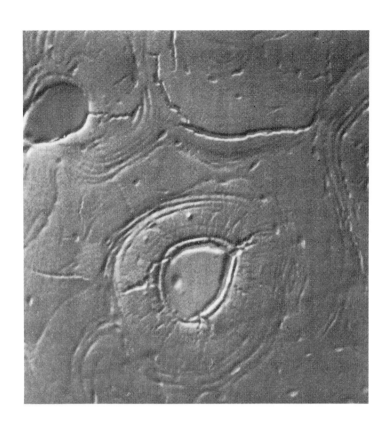

Figure 2. SEM/BSE topographical image of Figure 1. The haversian canals and osteocyte lacunae are here seen as depressions within the specimen. The more highly mineralized areas appear smoother; whilst areas which are relatively less well mineralized and lamellae with collagen orientated parallel to the block face have abraded faster, producing small depressions in the flat surface section. FW 410 μm.

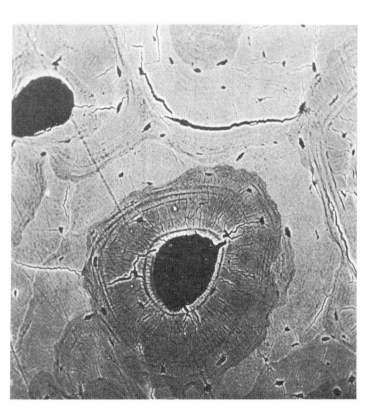

Figure 1. SEM/BSE image. A PMMA-embedded transverse section, lateral aspect, of a laboratory reference adult tibia. The field shows the characteristic morphology of cortical bone remodelling, with younger secondary osteons appearing as slightly less dense (darker) than relatively older ones. The areas of interstitial bone, which are most highly mineralized appear lightest. The depressions at sites of original holes within the specimen appear as nearly or completely black. Reversal lines are seen at osteonal boundaries. The osteonal lamellae are more prominent in less well mineralized bone due to the effect of polishing artefact. Field width (FW) 410 μm.

bone. Such topography created slight edge artefacts within the specimen. This resulted from an increased surface area at certain points in the scan where the size of the edge determined the amount of reflected high energy backscattered electrons detected at that point. The net effect of this phenomenon was an observable apparent increase in density in the flat surface image (Fig. 1). Hence, in the flat section images it was generally seen that alternate lamellae and the free border of the haversian canals were slightly brighter than the surrounding bone. However, the general features of varying density within the specimens were not contradicted by the topographic images.

The diagenetic effects in the alveolar bone were similar to those found previously in long bones (Bell, 1990). The alveolar bone was extensively affected and appeared to have furnished no lasting protection to the enclosed root.

Enamel

The enamel in the study appeared unaffected by diagenesis (Fig. 3). No general demineralization of the enamel surface was observed, although some apparently 'classic' carious lesions in the enamel were found. Where such an attack had taken place, the cross striations of the enamel rods could be more clearly seen than elsewhere, and the approximately weekly incremental striae of Retzius were more prominent. Micro-cavitation which resulted from partial localised removal of the enamel rods by acid dissolution, was also present. The characteristic survival of this 'white spot' type of lesion, and other more extensive enamel caries observed in this study, underlines the enamel's ability to be free of attack from body and soil flora, even when it's own microstructure had apparently been compromised by ante mortem attack. It is recognised, however, that enamel can suffer post mortem degradation to its surface as observed by Poole and Tratman (1978). In this study no widespread changes were seen and so it is concluded that the total depositional environment prevented such

changes occurring. The acidity of the soil during the time of burial is unknown.

Dentine

In striking contrast to enamel, dentine was observed to have undergone extensive diagenetic change to its microstructure (Figs. 3 and 4). The distribution of these changes tended not to affect the mantle and peripheral circumpulpal dentine of the crown (Fig. 3), but was instead concentrated quite extensively both peri-pulpally in the root and crown and peripherally in the root (Fig. 4). Whilst the underlying dentine appears to have been protected by the enamel, with attack being limited to just above the cervical location of the enamel dentine junction, no such protection was afforded by the cementum. Interglobular spaces were also maintained without change. However, where diagenetic changes to the dentine had occurred, then these tended initially to orientate along the axis of the collagen planes which run approximately at a 90°C angle to the dentine tubules. These diagenetic destruction foci were often situated preferentially around peritubular dentine, in contradistinction to in vivo carious attack, which would initially follow and locate itself within the dentine tubules. Only later, with the formation of liquefaction foci, will carious micro-organisms move outwards into the intertubular dentine. By contrast, diagenetic destruction foci were frequently clearly placed in the intertubular dentine region, often leaving intact the peritubular dentine.

Cementum

Cementum, like bone and dentine, also underwent diagenetic change (Fig. 4). The level of attack could be extensive or relatively limited. Figure 4 shows considerably attack and cracking. The cracking may have been due in total, or in-part, to the PMMA shrinking and expanding during solidification particularly if the specimen had dried previously. However, foci of diagenetically altered microstructure were clearly observed. The orientation of these foci appeared less

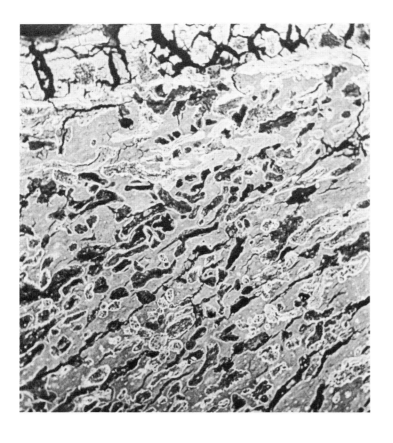

Figure 4. A longitudinal buccolingal section through a soil-buried Bronze Age mandibular permanent second premolar. The diagenetically altered dentine and cementum was distributed peri-pulpally and peripherally in the root, the enamel and most of the coronal dentine being unaffected. The orientation of the diagenetic foci in the dentine is generally parallel to the direction of the collagen fibres and so at right angles to the dentine tubules. Severe cracking is evident in the cementum alongside considerable diagenetic alteration to morphology and density. FW 455 μm.

Figure 3. SEM/BSE image. A longitudinal buccolingual section through a soil-buried Bronze Age mandibular permanent second premolar. The enamel appears unaffected by diagenetic alteration. Cracks have occurred, probably as a result of drying or during the polymerization of the PMMA, with the largest cleaving along the enamel dentine junction, usually just within the enamel. The dentine in this field beneath the enamel crown appears well maintained in morphology and density. FW 905 μm.

clearly defined than those in dentine. This may result from the complexity of the collagen order in cementum (and Sharpey fibre bone), the extrinsic fibres being approximately normal to, and the intrinsic fibres parallel to, the developing surface of the tissue.

Conclusions

BSE imaging of the dental and supporting bony tissues proved a simple and effective method for investigating the changes in mineral density that accompany the diagenetic processes. Care must be taken in interpreting the images so that the effects of density and collagen orientation on the rate of abrasion during polishing are not ignored. In the soil-buried contexts examined, all the hard tissues other than the enamel underwent diagenetic change to microstructure. The spread of the diagenetic change appeared to be strongly orientated along collagen planes. It is clear from this study that the effects of diagenesis can be extensive even in apparently well preserved dental and supporting skeletal tissues.

Acknowledgements

We thank Elaine Maconnachie for her expert assistance. The archaeological material was kindly provided by Juliet Rogers and the Trust for Wessex Archaeology. This work was supported by grants from the MRC, SERC and the Central Research Fund of the University of London.

References

Bell, L.S. (1990). Palaeopathology and diagenesis: an SEM evaluation of structural changes using backscattered electron imaging. Journal of Archaeological Science, 17, 85-102.

Boyde, A. & Jones, S.J. (1983a). Back-scattered electron imaging of skeletal tissues. Metabolic Bone Disease and Related Research, 5, 145-150.

Boyde, A. & Jones, S.J. (1983b). Backscattered electron imaging of dental tissues. *Anatomy and Embryology*, 168, 211-226.

Boyde, A. & Jones, S.J. (1987). Early scanning electron microscopic studies of hard tissue resorption: their relation to current concepts reviewed. Scanning Microscopy, 1(1), 369-381.

Boyde, A., Maconnachie, E., Reid, S.A., Delling, G. & Mundy, G.R. (1986). Scanning electron microscopy in bone pathology: review of methods, potential and applications. Scanning Electron Microscopy, IV, 1537-1554.

Dobney, K. & Brothwell, D.R. (1986). Dental calculus: its relevance to ancient diet and oral ecology. In: Teeth and Anthropology, ed. E. Cruwys & R.A. Foley, pp. 55-81. BAR International Series 291.

Falin, L.I. (1961). Histological and histochemical studies of human teeth of the Bronze and Stone Ages. Archives of Oral Biology, 5, 5-13.

Garland, A.N. (1989). Microscopical analysis of fossil bone. Applied Geochemistry, 4, 215-229.

Hackett, C.J. (1981). Microscopical focal destruction (tunnels) in excavated human bones. Medicine, Science and Law, 21(4), 243-265.

Jones, S.J. (1987). The root surface: an illustrated review of some scanning electron microscope studies. Scanning Microscopy, 1(4), 2003-2018.

Jones, S.J. & Boyde, A. (1987). Scanning microscopic observations on dental caries. Scanning Microscopy, 1(4), 1991-2002.

Martin, L.B., Boyde, A. & Grine, F.E. (1988). Enamel structure in primates: a review of scanning electron microscopic studies. Scanning Microscopy, 2(3), 1503-1526.

Meeks, N. (1988). Backscattered electron imaging of archaeological material. In:

Scanning Electron Microscopy in Archaeology, ed. S.L. Olsen, pp. 23-44. BAR International Series 452.

Olsen, S.L. (1988). Scanning Electron Microscopy in Archaeology, BAR International Series 452.

Poole, D.F.G. & Tratman, E.K. (1978). Post-mortem changes on human teeth from late upper Palaeolithic/Mesolithic occupants of an English limestone cave. Archives of Oral Biology, 23, 1115-1120.

Reid, S.A. (1986). A study of lamellar organisation in juvenile and adult human bone. Anatomy and Embryology, 174, 329-338.

Sognnaes, R.F. (1955). Postmortem microscopic defects in the teeth of ancient man. A.M.A. Archives of Pathology, 59, 559-570.

Watt, I.M. (1985). The Principles and Practice of Electron Microscopy. Cambridge: Cambridge University Press.

Wedl, C. (1864). Uber einen im Zahnbein und Knochen keimenden Pilz. Akademie der Wissenschaften in Wein. Sitzungsberichte Naturwissenschaftliche Klasse ABl. Mineralogi, Biologie Erdkunde, 50(1), 171-193.

BOG BODY LIPID TAPHONOMY

R.P. Evershed

Department of Biochemistry, University of Liverpool, P.O. Box 147, Liverpool, L69 3BX, U.K.

Introduction

While some environments result in the decomposition of animal remains in a very brief period compared to the archaeological timescale (Carbone & Keel, 1985), others ensure preservation over many millennia. The wide diversity of animal tissue types necessarily results in differential preservation on an environment dependent basis.

Although peat bogs might be regarded as somewhat specialised environments, archaeologically, the high degree of preservation of specific types of organic matter is noteworthy. The formation of peat itself is testimony to its preservative properties (Dimbleby, 1978). Preservation in peat bogs is at its most striking where bog body finds are concerned. However, while the gross physical features are remarkably well-preserved, closer microscopic examination of the tissues shows that the cellular structure is completely destroyed. The macroscopic preservation is therefore somewhat superficial with only the highly resistant protein collagen component of the skin, teeth, bones, muscle tissue, etc. remaining to preserve the characteristic physical features.

Degradation of the bog body tissues will be closely tied to the normal process of organic matter decay in peat bogs (Clymo, 1983), which include: (i) loss of organic matter, as gas or in solution following attack by animals or microorganisms, (ii) loss of physical structure, and (iii) change of chemical state, including the formation of new classes of molecules by microorganisms.

For a bog body to survive intact it must initially have been buried, or immersed in an anaerobic pool of peat bog water. Merely lying the body on the peat surface would not have ensured the classical state of preservation associated with bog bodies. Assuming the former condition existed, then cellular destruction proceeds largely under the influence of saprophytic organisms originating from the peat bog (Carbone and Keel, 1985) and those migrating from the intestines of the body itself (Evans, 1963). Enzymes present in the tissues will continue to be active for a short period after death and so exert degradative influences (Evans, 1963). Physical processes, such as osmosis and waterlogging, will also contribute to the destruction of the cellular structure of the tissues during the initial period of immersion or burial. The pH of the peat waters is also important in preservation. The bodies of mammals remain whole for many thousands of years under mildly acidic conditions, while high acidity results in rapid degradation (Coles, 1987). The aqueous environment is highly influential in the dispersion of polar components by dissolution and diffusion out of the body. A major effect of this is the complete demineralisation of bog body bone and teeth (Stead et al., 1986).

Lipids, including phospholipids and cholesterol, are major structural components of cell membranes in mammalian tissues. However, microscopic examination reveals little or nothing of their fate in bog bodies. These lipids are open to the effects of microbial and/or physico-chemical degradation during the period of burial in the peat bog. Additionally, polar lipids would be susceptible to dissolution and dispersion by diffusion in the same way as organic species. Hydrophobic lipid components, however, will be resistant to removal from the body by diffusion.

Our interests in bog bodies currently centres on investigations of the lipid constituents. The principal aims of this work are: (i) to determine the extent of preservation of endogenous lipids, and (ii) attempt to recognise processes of alteration and degradation. To date the lipid constituents of muscle and skin tissue, and adipocere of 10 bog bodies originating from the U.K., Ireland and the Netherlands have been investigated. Herein the current status of these investigations is reviewed and new data relating to the composition of the adipocere presented.

Analytical approach

The analytical techniques employed in this work have been described in detail elsewhere (Evershed & Connolly, 1988; Evershed, 1990). Briefly, the lipids are extracted from tissue samples (eg. 0.5g) by washing with organic solvent (chloroform/methanol, 2:1 v/v). Prelininary assessment of the composition of the total lipid extract is based on analytical thin layer chromatography (TLC; SiO_2) with Rf comparisons to the authentic marker compounds. Some useful compositional information, such as the presence or absence of major classes of compound, can be obtained by TLC screening. However the relatively low resolution and sensitivity limits the usefulness of TLC for detecting subtle molecular structural alterations, or compositional changes in complex lipid distributions.

Gas chromatography (GC) and combined GC/mass spectrometry (GC/MS) have become the techniques of choice in this work, owing to their ability to separate and characterise the individual molecular species comprising the complex liquid extracts. Identifications of the various compounds are based on GC retention times (t_R) compared to authentic compounds, known elution orders, computer searching of library mass spectra and the interpretation of novel mass spectra on the basis of known fragmentations of organic compounds.

Lipid Analyses

While the composition of cell membranes vary with source they contain approximately 40% of their dry weight as lipid and 60% as protein (Harrison & Lunt, 1980). Although phospholipids are the major lipid components of mammalian cell membranes, repeated TLC analyses has failed to detect intact phospholipid (1) species in any of the bog body tissues examined to date. This is perhaps not unexpected given the susceptibility of the phosphate moiety to chemical (hydrolytic) and enzymatic cleavage. Hydrolysis would also account for the low abundance of acyl lipids, eg. triacylglycerols (2), diacylglycerols (3; remaining from phospholipids following loss of the head group and phosphate moiety) and steryl fatty acyl esters (4), generally observed in the bog body tissues.

R, R' and R" = alkyl

1   2   3   4

353

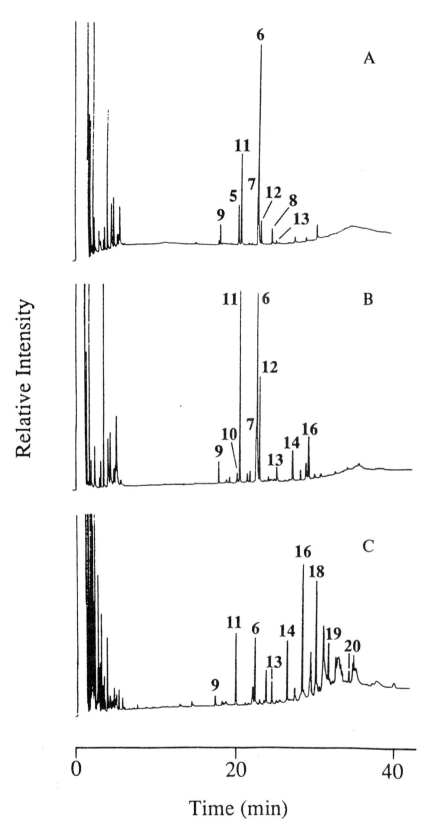

Figure 1. GC profiles of the fatty carboxylic acids (as methyl ester derivatives) of fresh human muscle tissue (A), muscle tissue from Lindow Man (Wilmslow, Cheshire, U.K.; B) and Lindow Peat (C). The numbers on the peaks refer to the structure numbers in the text. Figure redrawn from Evershed and Connolly (1988).

The profiles produced by capillary column GC are useful in 'fingerprinting' the lipid composition of various tissues. For example, Figure 1 shows the GC profiles for the fatty acid constituents (as methyl ester derivatives) of Lindow man muscle tissue, compared to those of normal human muscle and the peat surrounding the bog body. The individual components of these samples were characterised by GC/MS. The profiles show some depletion of the unsaturated species (**5** - **8**) and low level contamination by long-chain fatty acids ($>C_{20}$; **14** - **18**) migrating from the peat. Taking into account these factors and the high lipid content of peat (*ca.* 10% w/w dry weight) it is remarkable that the original distribution of fatty acid components is so well preserved. Analysis of the sterol fraction from Lindow Man muscle tissue showed large quantities of cholesterol (**21**) to be present. The fact that phytosterols (eg. sitosterol, **25**) were undetectable in this sterol fraction provided further evidence for the low level contamination of the bog body lipids by migration from the surrounding peat.

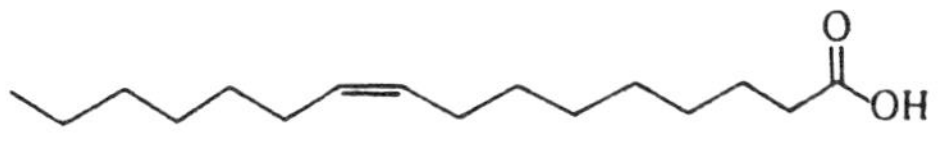

5

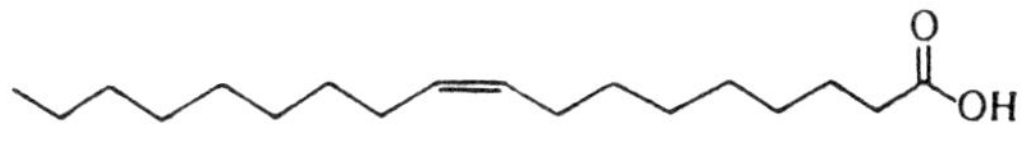

6

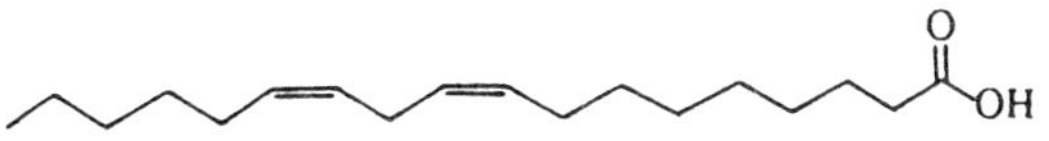

7

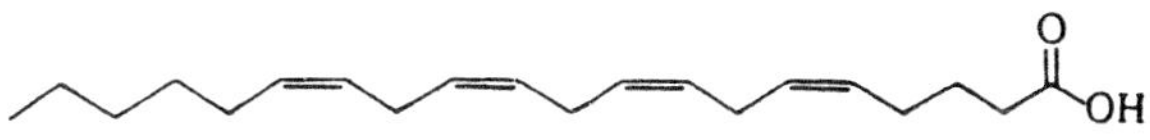

8

In contrast to muscle tissue the distributions of lipids of the skin are rather more drastically affected by decay and contamination (Evershed, 1990). This is not unexpected as skin but its very nature will be significantly more open to the processes of environmental degradation during the initial period of sub-mergence in the bog water, or burial. The intimate contact of the peat with the skin also greatly increases the possibility for contamination by migration. The summation of all these factors is represented by the GC/MS total ion chromatogram (TIC) shown in Figure 2.

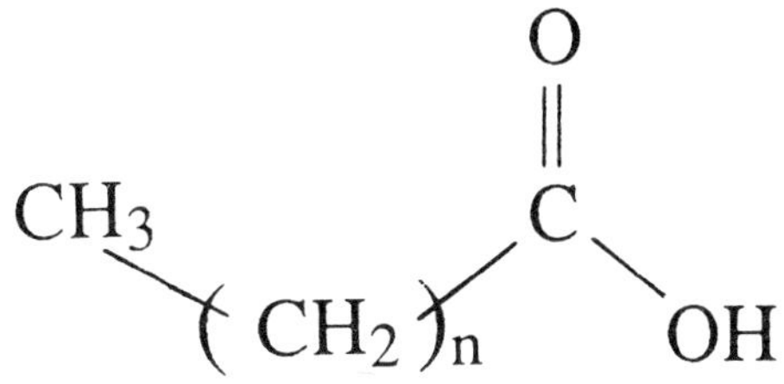

9 n = 12; **10** n = 13; **11** n = 14; **12** n = 16;
13 n = 18; **14** n = 20; **15** n = 21; **16** n = 22;
17 n = 23; **18** n = 24; **19** n = 26; **20** n = 28.

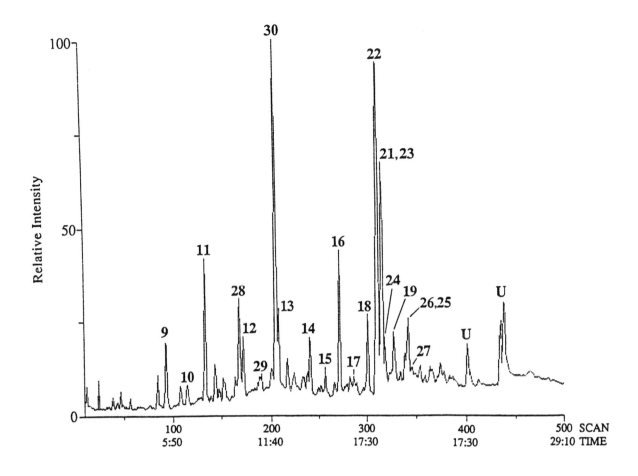

Figure 2. GC/MS total ion current profile showing fatty carboxylic acids (as methyl ester derivatives) and steroids present in the total lipid extract of a sample of bog body skin from a girl's foot (30 ± 80 BC, Yde, Netherlands). The numbers of the peaks refer to the structure numbers in the text. Redrawn from Evershed (1990).

Both fatty acids and steroidal components were determined in the same analytical run. Fatty acids are again present in high abundance. The fatty acid content of human epidermal lipids is unusual in that there exists a predominance of long-chain saturated components. For example, human sphingomyelin, one of the major epidermal phospholipids, contains nearly 20% tetracosanoic acid (C_{24}; **16**) and more than 5% hexacosanoic acid (C_{26}; **18**) (Gray & Yardley, 1975). Hence, in the total lipid extracts of bog body skins the presence of long-chain fatty acids can not be fully attributed to substances migrating from the peat. In five of the seven bog body skins examined to date the dominant fatty acid species are unusual 10-hydroxy derivatives, the major components consistently being 10-hydroxyoctadecanoic (**30**) and 10-hydroxyhexadecanoic acids (**28**), with a much lower abundance of 10-hydroxyheptadecanoic acid (**29**). These substances do not occur naturally in human epidermal lipids; they most likely arise through microbial activity. *Escherichia coli*, bacterial inhabitants of the human gut, contain large proportions of oleic (**6**) and palmitoleic acid (**5**). The spread of *E. coli* to

356

epidermal tissues would be expected during the early phase of decay of the body and would lead to a build-up of these mono-unsaturated fatty acids in this region. Subsequent stereospecific microbial hydration of these alkenoic acids, for which there is a biochemical president (Schnoepper, 1966), would yield these 10-hydroxy fatty acids in the observed proportions similar to those observed in bog body skin (Evershed, 1990).

Monohydroxy fatty acids have been characterised in the adipocere of bodies immersed for 3-6 months in the sea (Takatori & Yamaoka, 1976, 1977a,b), and tissues of a 4 000 year old Nubian mummy (Gulacar et al., 1989). In the latter instance a complex mixture of positional isomers was observed, suggesting a somewhat different mode of formation to that observed in bog bodies.

21 R = H
25 R = C$_2$H$_5$

22

23 R = H
26 R = C$_2$H$_5$

24 R = H
27 R = C$_2$H$_5$

The sterol composition of the bog body skins is of special interest. Although intact cholesterol (**21**) is readily detectable, its degradation products, coprostanol (5ß-cholestanol; **22**), 5α-cholestanol (**23**) and 5α-cholestan-3-one (**24**) have also been found in high abundance. This distribution results from microbial transformations of cholesterol postmortem. The appearance of coprostanol is distinctive, as this is an established product of the microbial decomposition of cholesterol in the human intestines (Bjorkhem & Gustafson, 1971). In bog body skin this conversion is probably mediated by intestinal flora that migrated into the tissued following death (Evans, 1963). 5α-Cholestanol, present in substantially higher abundance than in healthy tissue, would arise through an

alternative microbial reduction with 5α-cholestan-3-one as one of the intermediates (Evershed, 1990). The presence of the phytosterol sitosterol (**25**) and its reduced counterpart, 5α-stigmastanol (**26**) represents contamination through migration from the surrounding peat (Ives & O'Neill, 1958). By analogy with cholesterol

degradation, the 5α-stigmastan-3-one (**27**), present at low abundance, represents an intermediate in the microbial reduction of sitosterol. The abundance of sitosterol and stigmastanol generally increases with increased abundance of long-chain fatty carboxylic acids (C$_{22}$-C$_{30}$; **14** - **20**) which are also abundant constituents of peat.

357

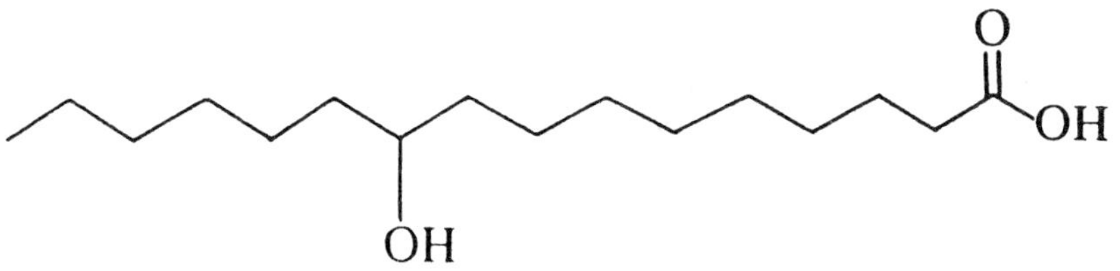

28

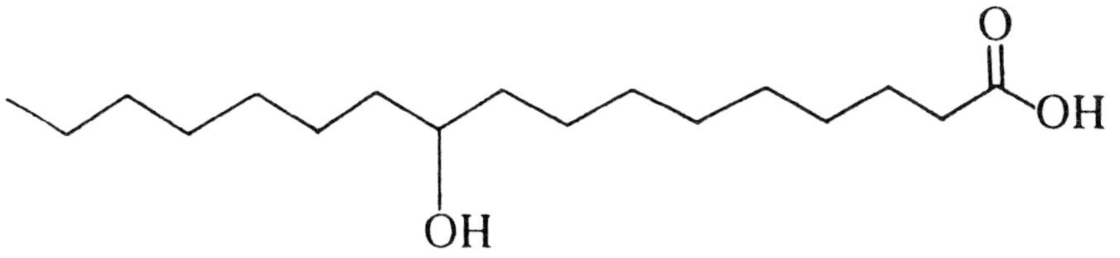

29

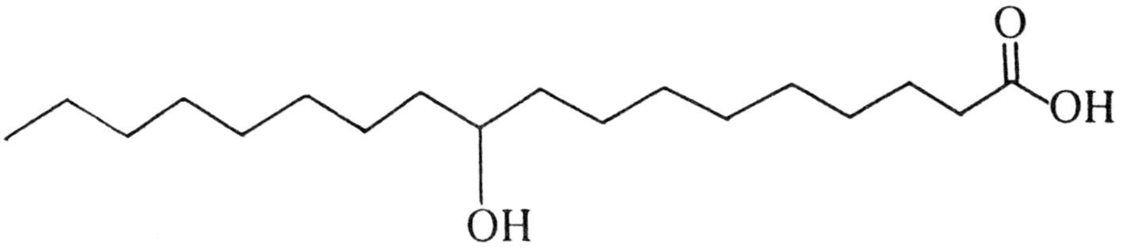

30

Large quantities of adipocere were recovered from a 16th century bog body (female) found at Meenybradden, Co. Donegal, Ireland. TLC analysis showed complete conversion of the original adipose triacylglycerols, to free fatty carboxylic acids. Subsequent GC and GC/MS analyses confirmed this and revealed a similar distribution of fatty acids to those observed previously for adipocere formed microbially from fats under anaerobic conditions (den Dooren De Jong, 1961; Morgan *et al.*, 1973). Palmitic acid (**11**) was the major component (60%) with stearic (15%; **12**) and myristic (10%; **9**) acid also present in high abundance. Although unsaturated components, octadecenoic (11.5%) and hexadecenoic (<2.5%) acids, were much reduced in abundance compared to healthy tissue, they were still clearly evident.

Di(methylthio)ethers (Francis & Veland, 1981) of the methyl esters of the unsaturated components were prepared in order to test whether or not they were structurally congruent with the monounsaturated fatty acids normally associated with human fat. GC/MS of these derivatives showed that rather than being predominantly the expected Δ^9-octadecenoic (**6**) and Δ^9-hexadecenoic (**5**) acids, mixtures of positional isomers were present. The mass spectrum of the bog body component showed quite clearly the presence of a mixture of Δ^6, Δ^7, Δ^8, Δ^9, Δ^{10}, Δ^{11}, and Δ^{12}-octadecenoic acids. In the case of the less abundant hexadecenoic acids the mixture comprised Δ^6, Δ^7, Δ^8, Δ^9 and Δ^{10} isomers. The unexpected results would appear to suggest an additional mechanism of conversion of animal fat to adipocere in an aqueous anaerobic peat environment, to the ß-oxidation and reduction of double-bonds observed by den Dooren De Jong (1961).

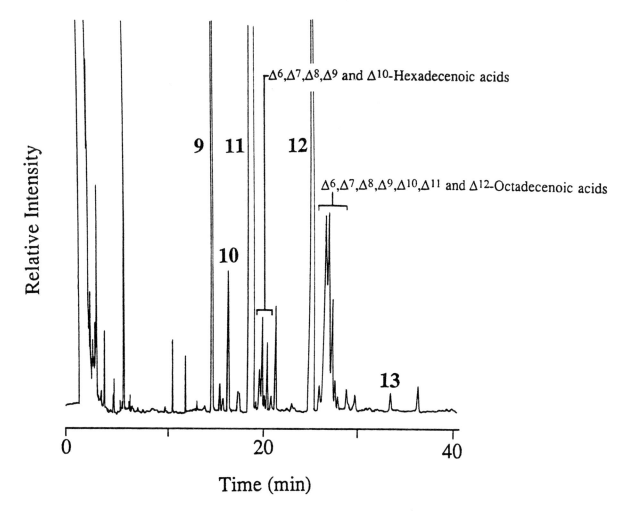

Figure 3. GC profile of fatty carboxylic acids (as methyl ester derivatives) obtained from the adipocere of the Meenybradden body (female, 16th Century, Co. Donegal, Ireland), showing the mixtures of positional isomers of hexadecenoic acid octadecenoic acids. The numbers of the peaks refers to the structure numbers in the text.

A possible explanation for the presence of a mixture of positional isomers would be lipid contributed through microbial activity. GC analysis of the adipocere fatty acid methyl ester derivatives on a high performance SP-2330 coated capillary column afforded partial resolution of the hexadecenoic and octadecenoic acid isomers (Figure 3). Vaccenic acid, (Z-11-octadecenoic acid) an established component of bacterial lipids, co-chromatographed with one of the minor, later eluting, components. While this result indicated that bacterial input of this nature was not a significant factor, it does not completely rule out bacteria as the source of the isomeric mixture. The possibility that the mixture is generated chemically (March, 1977) under the physico-chemical influence of the peat bog palaeoenvironment is currently being explored. Tests have confirmed that the mixture is not an artefact of sample work-up and derivatisation. Whatever the explanation it would appear that we have revealed a novel process for the transformation of unsaturated fatty acids in the archaeological context. A full report of these investigations of the Meenybradden body adipocere is in preparation and will be presented elsewhere.

Conclusions

It is hoped that the results presented herein serve to emphasise the importance of detailed chemical analysis in revealing the extent of preservation/degradation of organic residues of archaeological interest. Through GC and GC/MS analyses individual molecular species can be separated and characterised to reveal the subtle effects of microbial and/or chemical alteration and environmental contamination. While the study of bog bodies might be regarded as rather specialised owing to their uniqueness archaeologically, the practice of applying biochemical and biosystematic relationships, as demonstrated herein, is fundamental to the study of organic residues.

Acknowledgements

Mr M.C. Prescott is thanked for GC/MS analyses, as are Drs L.J. Goad, C. Rolph, G.A. Wolff and C. Heron for valuable discussions. Drs. R.C. Connolly, R. O'Floinn (National Museum of Ireland) and W.A.B. van der Sanden (Drents Museum, The Netherlands) are thanked for kind provision of samples.

References

Bjorkhem, I. & Gustafson, J.-A. (1971). *Eur. J. Biochem.* 21, 428-432.

Carbone, V.A. & Keel, B.C. (1985). Preservation of plant and animal remains. In: *The Analysis of Prehistoric Diets*, eds. R.I. Gilbert, Jr & J.H. Mielke, pp. 1-19, Academic Press, London.

Cylmo, R.S. (1983). Peat. In: *Mires: Swamp, Bog, Fen and Moor, A. General Studies*, ed. A.J.P. Gore, pp. 159-224, Elsevier, Amsterdam.

Coles, B. (1987). Archaeology follows a wet track. *New Scientist*, October 15th, 42-46.

den Dooren de Jong, L.E. (1961). On the formation of adipocere from fats. *Antonie van Leewenhoek J. Microbiol. Serol.* 10, 337-361.

Dimbleby, G. (1978). *Plants and Archaeology*, John Baker, London.

Evans, W.E.D. (1963). *The Chemistry of Death*. Charles Thomas, Illinois.

Evershed, R.P. & Connolly, R.C. (1988). Lipid preservation in Lindow Man. *Naturwissenschaften* 75, 143-145.

Evershed, R.P. (1990). Lipids from samples of skin from seven Dutch bog bodies: Preliminary Report. *Archaeometry* 32, 139-153.

Francis, G.W. & Veland, K. (1981). Alkylthoilation for the determination of double-bond positions in linear alkenes. *J. Chromatogr.* 219, 379-384.

Gray, G.M. & Yardley, H.J. (1975). Lipid compositions of cells isolated from pig, human and rat epidermis. *J. Lipid Res.* 16, 434-440.

Gulacar, F.O., Buchs, A. & Susini, A. (1989). Capillary gas chromatography-mass spectrometry and identification of substituted carboxylic acids in lipids extracted from a 4000-year-old Nubian burial. *J. Chromatogr.* 479, 61-73.

Harrison, R. & Lunt, G.G. (1980). *Biological Membranes: Their Structure and Function*. Blackie, Glasgow and London.

Ives, A.J. & O'Neill, A.N. (1958). The chemistry of peat. Part 1. The Sterols of peat moss (sphagnum). *Can. J. Chem.* 36, 434-439.

March, J. (1977). *Advanced Organic Chemistry: Reactions, Mechanisms and Structure*, 2nd edn., McGraw-Hill Kogakusha Ltd., Tokyo.

Morgan, E.D., Cornford, C., Pollock, D.R.J. & Isaacson, P. (1973). The composition of fatty material buried in soil. *Science and Archaeology*, 10, 9-25.

Schnoepper, G.J. (1966). Stereospecific conversion of oleic acid to 10-hydroxystearic acid. *J. Biol. Chem.* 241, 5441-5447.

Stead, I.M., Bourke, J.B. & Brothwell, D. (1986). *Lindow Man. The Body in the Bog*, British Museum Publications, London.

Takatori, T. & Yamaoka, A. (1976). Separation of some close positional isomers of hydroxy fatty acids by two dimensional thin layer chromatography. *Nippon Hoigatu Zasshi* 30, 368-370.

Takatori, T. & Yamaoka, A. (1977a). The mechanism of adipocere formation. I: Identification and chemical properties of hydroxy fatty acids in adipocere. *Forensic Sci.* 9, 63-73.

Takatori, T. & Yamaoka, A. (1977b). The mechanism of adipocere formation II: Separation and Identification of oxo fatty acids in adipocere. *Forensic Sci.* 10, 117-125.

DENTINE: AGE AND D-ASPARTIC ACID CONTENT

R.D. Gillard[1], A.M. Pollard[1], P.A. Sutton[1] and D.K. Whittaker[2]

[1]School of Chemistry and Applied Chemistry, University of Wales College of Cardiff, Cardiff, CF1 3TB, UK.

[2]Department of Basic Dental Science, Dental School, University of Wales College of Medicine, Cardiff, CF4 4XY, UK.

Abstract

The time dependence of D-aspartic acid accumulation in those metabolically stable proteins contained in tooth enamel, dentine and human eye lens has been demonstrated by several independent researchers for modern specimens. This relationship has since been assumed to hold for these same proteins recovered from either archaeological or forensic specimens, thus giving an estimate of the age at death with a reasonable degree of certainty.

Our research involves further refining the method for human dentine. It is hoped that this will enable the determination of the age at death of specimens obtained during the excavation of the 18th Century Vaults at Spitalfields, London. We are also investigating the possible existence of a D-aspartic acid gradient within teeth, this being the consequence of the tooth's morphological development. Such a gradient, should it exist, could serve as an internal standard for testing the reliability of the D-aspartic acid results obtained for dentine samples. For detailed presentation of this material, see Gillard et al. (1990).

References

Gillard, R.D., Pollard, A.M., Sutton, P.A. & Whittaker, D.K. (1990). An improved method for age at death determination from the measurement of D-aspartic acid in dental collagen. *Archaeometry* 32(1), 61-70.

METHODOLOGICAL STUDY OF THE ANALYSIS OF BONE

A.M. Pollard[1], S.E. Antoine[1], P.Q. Dresser[2] and A.W.R. Whittle[3]

[1]School of Chemistry and Applied Chemistry, [2]Geology Department and [3]School of History and Archaeology, University of Wales, College of Cardiff, Cardiff.

Introduction

Our present work addresses problems relating to the reconstruction of the palaeodiet of Neolithic communities in Southern England via chemical and isotopic analyses of human bone. This paper concentrates on a methodological study, in view of the present debate on the applicability of such an approach when applied to degraded bone (Hancock et al., 1989; Price, 1989).

Materials and methods

The bone mineral fraction is extracted by a modification of the method of Szpunar et al. (1978). Calcium, strontium and magnesium are measured by atomic emission spectroscopy, and zinc by atomic absorption. Phosphorus is measured colorimetrically by the molybdenum blue method. Collagen is isolated by the method of DeNiro and Epstein (1978). Further details are given in Antoine et al. (1988).

Results and discussion

We are convinced of the importance of two factors in the measurement of the chemical composition of archaeological bone. The first is the total organic content of the dried bone, measured as weight loss after ignition at 700°C; in modern bone this is variable but is approximately 30% (Price, 1989). The second is the calcium and phosphate content of the bone ash. In modern fired bone, this is again variable, but is normally quoted as 37% Ca and 19% P (Hancock et al., 1989; Price, 1989). This corresponds approximately to the mineral calcium hydroxyapatite, with a stoichiometry of $Ca_5(PO_4)_3.OH$; variations are due to the poor crystallinity of the bone mineral, and the incorporation of some carbonate into the structure. The theoretical value for Ca/P ratio is therefore 2.16, although in modern bone a figure of 2.0 is normally accepted. We concur with the statements of Hancock et al., (1989) and Price (1989) that deviations from these figures in the analysis of fired archaeological bone signifies diagenesis, perhaps due to the recrystallisation of the hydroxyapatite to other calcium phosphates of differing solubilities (Williams and Elliott, 1989), or the deposition of calcite crystals into the bone. It is generally expected that diagenesis will raise the Ca/P ratio (Sillen, 1989). While it is not true to say that bone which fulfils the above criteria of acceptability is not affected by diagenesis, it would be safe to assume that bone which does not has suffered from such processes. It is not yet clear what effect these processes will have on the accepted 'dietary indicators' - Sr, Zn, Mg, etc., but it is likely that recrystallisation will reduce the trace element concentrations, and that external contamination may increase them, depending on the groundwater composition, the pH, and the temperature. These principles are illustrated in the discussion of some results from Neolithic bones from Southern England, below.

It is generally accepted that the isotopic ratios $\delta^{13}C$ and $\delta^{15}N$ of bone collagen are relatively free of diagenetic effects, providing structurally-intact collagen remains in the bone (Chisholm, 1989).

DeNiro (quoted in Chisholm, 1989) expressed concern about the combustion procedure employed in the preparation of collagen for stable isotope determinations. It is easier and cheaper to combust the sample (together with Cu(II)O and Ag wires to provide oxygen and remove sulfur, respectively) in a pyrex tube at 550°C, than DeNiro's preferred method of combustion at 900°C in silica tubes. In

Table 1. Comparison of stable isotope ratios of human bone 'collagen' combusted in pyrex tubes at 550°C and in silica tubes at 800°C

Sample No.	delta ^{13}C		$\Delta\delta^{13}$C	delta ^{15}N		$\Delta\delta^{15}$N
	Silica	Pyrex		Silica	Pyrex	
1	-21.2	-21.6	0.4	12.7	14.6	1.9
2	-20.9	-21.4	0.5	13.2	15.7	2.5
3	-20.5	-21.3	0.8	10.9	13.0	2.1
4	-21.9	-22.6	0.7	16.9	16.1	-0.8
5	-21.4	-21.8	0.4	11.2	10.0	-1.2
6	-20.9	-21.5	0.6	17.3	19.5	2.2
7	-21.3	-21.9	0.6	11.7	10.9	-0.8
8	-20.9	-22.2	1.3	11.9	12.9	1.0
9	-21.1	-21.1	0.0	12.2	11.9	-0.3
10	-21.4	-21.3	-0.1	12.4	13.0	0.6
11	-21.1	-21.1	0.0	8.9	8.4	-0.5
12	-21.6	-21.9	0.3	11.4	12.2	0.8
13	-22.3	-22.2	-0.1	9.1	9.5	0.4
14	-22.3	-21.6	-0.7	12.0	12.3	0.3
15	-22.2	-22.1	-0.1	13.1	14.8	1.7
17	-22.7	-23.3	0.6	8.8	8.3	-0.5
18	-22.1	-22.6	0.5	8.2	8.3	0.1
19	-22.2	-22.5	0.3	9.0	9.4	0.4
$\bar{x}$			0.33			0.55
s			0.45			1.15
n			18			18

Table 2. Comparison of results of isotope ratios of 'collagen' duplicate samples, one of which was washed with NaOH solution before extraction with HCl

Sample No.	Washed $\delta^{13}C$	Unwashed $\delta^{13}C$	$\Delta\delta^{13}C$	Washed $\delta^{15}N$	Unwashed $\delta^{15}N$	$\Delta\delta^{15}N$
136	-21.8	-21.5	-0.3	12.2	14.3	2.2
137	-21.9	-21.9	0.0	9.3	13.7	4.4
138	-21.8	-21.8	0.0	10.8	13.6	2.8
139	-21.9	-21.6	-0.3	10.3	12.8	2.5
140	-21.8	-21.2	-0.6	9.2	12.9	3.7
141	-21.2	-21.0	-0.2	4.8	7.9	3.1
142	-22.2	-22.9	0.7	22.5	20.5	2.0
143	-22.3	-23.0	0.7	24.5	27.5	3.0
144	-22.0	-22.0	0.0	12.3	10.4	-1.9
145	-21.9	-22.8	0.9	12.3	10.0	-2.3
146	-22.0	-22.9	0.9	7.3	6.9	-0.4
147	-21.8	-22.0	0.2			
148	-22.1	-22.4	0.3			
149	-22.3	-22.9	0.6			
150	-21.8	-22.9	1.1			
$\bar{x}$			0.27			1.37
s			0.52			2.51
n			15			11

365

Table 3. Results of C/N(CHEM) ratios of 'collagen' samples measured in CHN analyser, C/N ratios measured in mass spectrometer (MS), ΔC/N and corresponding stable isotope ratios in modern human bone, pig bone and archaeological human bone.

Sample No.	C/N-CHEM	C/N-MS	ΔC/N	$\delta^{13}C$	$\delta^{15}N$
Modern Bone	2.8	3.3	0.5	-20.3	17.5
Pig	3.2	3.6	0.4	-23.5	13.3
6	5.4	2.3	-3.1	-21.5	19.5
7	3.3	1.3	-2.0	-21.9	10.5
11	3.0	2.5	-0.5	-21.9	8.4
13	5.2	2.4	-2.8	-22.2	9.5
37	3.6	1.5	-2.1	-23.0	14.2
44	2.9	1.6	-1.3	-25.8	9.6
57	4.1	2.7	-1.4	-22.7	10.2
62	3.4	1.2	-2.2	-23.9	10.7
65	2.7	0.4	-2.3	-23.4	9.8
66	3.0	0.7	-2.3	-22.4	8.7
94	5.4	1.7	-3.7	-24.5	9.5
138	2.8	1.9	-0.9	-21.8	10.8
142	3.3	1.3	-2.0	-22.2	20.2
150	3.5	1.9	-1.6	-24.3	11.3
165	4.3	2.0	-2.3	-23.2	12.8
183	3.4	2.1	-1.3	-22.0	9.5
$\bar{x}$			-1.72		
s			1.10		
n			18		

common with Chisholm *et al.* (1982, 1983) and others we have compared the two methods using duplicate samples combusted using each technique (although we carried out the silica tube combustion at 800°C - higher temperatures resulted in significant losses due to explosions in the furnace). The results are shown in Table 1. When tested using the paired Student's t-test there is no significant difference between silica and pyrex combustion for either $\delta^{13}C$ and $\delta^{15}N$. It is noticeable, however, that the reproducibility of the $\delta^{15}N$ measurements is somewhat poorer than that for $\delta^{13}C$. A second experiment consisted of pre-treating the collagen with NaOH prior to extraction with dilute HCl, to remove any contaminating humic and fulvic acids (Table 2). Again, a comparison of duplicate pairs with a paired t-test shows no significant difference between washed and unwashed samples. As before, however, the $\delta^{15}N$ reproducibility is poorer. In modern collagen, the theoretical value for the ratio C/N is 3.2. DeNiro (1985) recommended rejection of extracted collagen samples if the C/N ratio fell outside the range 2.9-3.6. We analysed a number of collagen extracts for C and N using a standard CHN analyser. Duplicate samples were combusted, and the mass spectrometer gas pressures were also used to estimate the C/N ratio during measurements of the isotopic ratios (Table 3). Clearly, the estimates obtained during the mass spectrometric measurements are unreliable when compared to the conventional results. Taking the conventional data, we note that five the sixteen archaeological samples would be rejected if the DeNiro criteria were applied - all are too high. There is, however, no discernible evidence of bias in the $\delta^{13}C$ results for these samples; the $\delta^{15}N$ are too scattered to draw any conclusions.

We conclude that the $\delta^{13}C$ measurements are acceptable after combustion at 550°C in pyrex, and that a pre-wash in NaOH has no effect on these samples from Southern England. The same may be true of the $\delta^{15}N$ results, but the reproducibility is much poorer; this is undoubtedly why less $\delta^{15}N$ measurements are reported in the literature.

Results from Southern England

A total of 185 human bone samples from three Neolithic sites in Southern England have been analysed; details are given in Table 4. Summaries of the results are given in Tables 5 and 6, and displayed graphically in Figures 1 and 2.

Table 4. Sites, types and number of bone samples and sources

Sites	Femurs	Humeri	Tibias	Fibulae	Radii	Ulnas	Ribs	Verte-brae	Scapu-lae	Digits	Skulls	Toes	Pelvis	Feet	Mixed
Fussell's Lodge[1]	16	11	13	10	6	6	15	10	8	-	-	-	-	-	-
Irthlingborough[2]	18	15	-	-	6	-	8	5	-	4	5	2	2	4	1
Haddenham[3]	7	-	-	-	-	1	-	-	-	-	11	-	-	-	-

Source: [1]BMNH, [2]English Heritage and [3]Dept. of Archaeological Sciences, University of Bradford.

Overall, the ash content of the bones from Fussell's Lodge and Irthlingborough ranged from 65.2% to 89.4%, whilst those from Haddenham were considerably higher (76.5%-99.8%). Most of the ash samples from Haddenham contained appreciable amounts of a brownish insoluble residue when dissolved in HNO_3; the insoluble residue was found to comprise between 10.9% and 28.0% of the ash weight in ten of the samples.

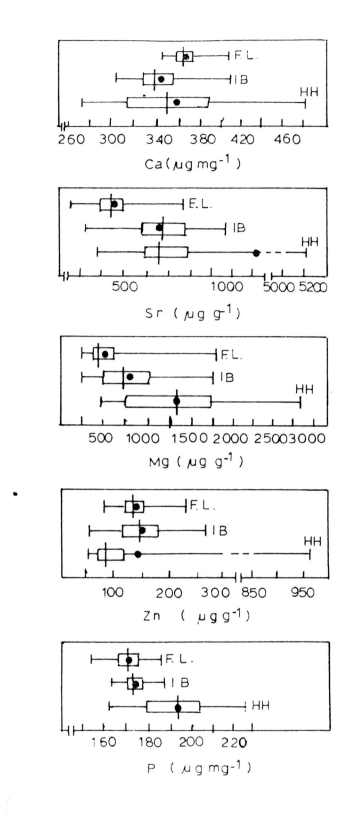

Figure 1. Range, median (bar), mean (closed circle) and upper and lower percentiles of major and trace element content.

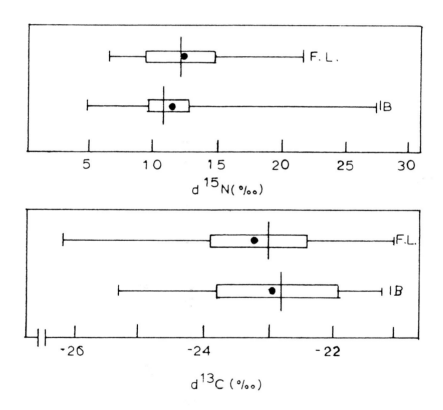

Figure 2. Range, median (bar), mean (closed circle) and upper and lower percentiles of stable isotope ratios.

The Ca and P contents were generally within the range reported in the literature for fired archaeological bones (summarised as 33.8% to 41.3% for Ca by Price (1983) - a noticeably wider range than expected from modern bone). A number of bones, however, were significantly outside this range - principally the samples from Haddenham. This confirms the doubts cast upon these data by the extremely low organic content of these bones. It is noticeable that the ranges for the 'dietary indicators' - in this case, Sr, Mg and Zn - are much wider for the Haddenham bone than the other two sites, which we must therefore suppose is due to severe diagenesis of the bone at this site.

Conclusions

1.   We emphasise the need for the following when reporting the analysis of archaeological bone:

A.   Total organic content

B.   Calcium and phosphorus content of the bone ash.

2.   For isotopic analysis, it is important to quote the C/N ratio of the collagen being analysed. We have found no variation due to combustion temperature and NaOH pre-treatment in the measurement of $\delta^{13}C$ in bones from Southern England.

3.   The reproducibility of $\delta^{15}N$ measurements by the normal method of combustion in sealed tubes gives much poorer reproducibility than the determination of $\delta^{13}C$. We are currently investigating experimental techniques for improving this procedure.

Table 5. Range, mean, standard deviation, upper and lower quartiles, skewness (SK) and coefficients of variation (cv) of the mineral contents of bone samples. Calcium and phosphorus expressed as $\mu g\ mg^{-1}$ and other minerals as $\mu g\ g^{-1}$ of ash weight.

Site		Range	Mean ± SD	Median	Quartiles		cv %	SK
					1st	3rd		
Fussell's Lodge (n=95)	Ca	344-408	365 ± 13	364	357	373	3.5	0.33
	Sr	257-765	458 ± 93	453	396	514	20.3	0.16
	Mg	266-1787	518 ± 256	449	354	573	49.3	0.81
	Zn	85-230	138 ± 27	184	121	154	19.3	-5.18
	P	154-186	171 ± 6	171	167	176	3.6	0.10
Irthling-borough (n=70)	Ca	305-409	344 ± 24	338	328	355	7.1	1.75
	Sr	327-969	673 ± 132	674	577	775	19.6	-0.03
	Mg	266-1752	795 ± 357	740	515	1022	44.8	0.46
	Zn	57-268	147 ± 49	145	115	178	33.1	0.17
	P	164-187	174 ± 5	173	171	177	2.7	0.64
Haddenham (n=20)	Ca	247-482	357 ± 57	349	315	388	15.9	0.44
	Sr	383-5208	1125 ± 1180	655	588	785	104.9	11.95
	Mg	474-2847	1305 ± 628	1318	755	172	48.1	-0.06
	Zn	54-962	138 ± 197	89	632	121	64.7	0.74
	P	163-227	194 ± 18	194	179	206	9.1	0.00

Table 6. Range, mean, median, upper and lower quartiles, skewness coefficient (SK) and coefficients of variation of $\delta^{13}C$ (upper row) and $\delta^{15}N$ (lower row) of analysed bone samples

Site	Range	Mean ± SD	Median	Quartiles		CV %	SK
				1st	3rd		
Fussell's Lodge (n=95)	-26.2 to -21.0	-23.2±1.3	-23.0	-23.9	-22.4	5.6	-0.46
	6.6 to 21.6	12.4±3.4	12.2	9.5	14.8	27.4	0.18
Irthlingborough (n=70)	-25.3 to -21.2	-22.9±1.1	-22.8	-21.9	-23.8	4.8	-0.28
	4.8 to 27.5	11.6±3.3	10.9	9.8	12.8	28.4	0.64

Table 7. Means and standard deviations of the percentage composition of organic matrix, calcium, phosphorus and Ca/P ratios and number (n) of the bone samples analysed

	Fussell's Lodge (n = 95)	Irthlingborough (n = 70)	Haddenham (n = 20)
Organic matrix	20.8±3.6	18.9±6.5	4.1±6.7
Calcium	36.5±12.7	34.4±24.4	35.7±57.1
Phosphorus	17.1±6.1	17.4±4.7	19.4±17.6
Ca/P	2.1±0.1	2.0±0.1	1.9±0.5

References

Antoine, S.E., Dresser, P.Q., Pollard, A.M. & Whittle, A.W.R. (1988). Bone Chemistry and Dietary Reconstruction: Examples from Wessex. *BAR British Series* 196, 369-380.

Chisholm, B.S. (1989). Variation in Diet Reconstruction Based on Stable Isotopic Evidence. In: *The Chemistry of Prehistoric Human Bone*, ed. T.D. Price, pp. 10-37. Cambridge University Press, Cambridge.

Chisholm, B.S., Nelson, D.E. & Schwarz, H.P. (1982). Stable isotope ratios as a measure of marine versus terrestrial protein in ancient diets. *Science* 216, 1131-32.

Chisholm, B.S., Nelson, D.E., Hobson, K.A., Schwarz, H.P. & Knyf, M. (1983). Carbon isotope measurement technique for bone collagen: notes for the archaeologist. *J. Arch. Sci.* 10, 335-60.

DeNiro, M.J. (1985). Postmortem preservation and alteration of in vivo bone collagen isotope ratios in relation to paleodietary reconstruction. *Nature* 317, 806-809.

DeNiro, M.J. & Epstein, S. (1978). Influence of diet on the distribution of carbon isotopes in animals. *Geochim. Cosmochim. Acta* 42, 495-506.

Hancock, R.G.V., Grynpas, M.D. & Pritzker, K.H.P. (1989). The abuse of bone analysis for archaeological dietary studies. *Archaeometry* 31, 169-179.

Price, T.D. (1989). Multi-element studies of diagenesis in prehistoric bone. In: *The Chemistry of Prehistoric Human Bone*, ed. T.D. Price, pp. 126-154. Cambridge University Press, Cambridge.

Sillen, A. (1989). Diagenesis of the inorganic phase of cortical bone. In: *The Chemistry of Prehistoric Human Bone*, ed. T.D. Price, pp. 211-219.. Cambridge University Press, Cambridge.

Szpunar, C.B., Lambert, J.B. & Buikstra, J.E. (1978). Analysis of excavated bone by atomic absorption. *Am. J. Phys. Anthropol.* 48, 199-202.

Williams, R.A.D. & Elliott, J.C. (1989). *Basic and Applied Dental Biochemistry*. Churchill Livingstone, London. 2nd Edition.

SCIENTIFIC METHODS IN PALAEOPATHOLOGY: PAST, PRESENT AND FUTURE

C.A. Roberts

Calvin Wells Laboratory, Department of Archaeological Sciences, University of Bradford, Bradford BD7 1DP

Introduction

Palaeopathology is the study of ancient disease using many different types of data - skeletal and soft tissues, contemporary literary and iconographical and archaeological data (eg. pollen, animal bones, seeds, coprolites, parasites etc.) in varying forms with pertinent information from modern primitive societies. This is an ideal situation in which to work but is often not the case, and certainly not for prehistoric populations.

The application of scientific methods in palaeopathology have had a long history of study but these studies have been variable in quality and quantity; they can be taken to be defined, for the purposes of this paper, as the study of disease by methods other than macroscopic observation.

Most scholars working on human remains use macroscopic analysis leading to detailed descriptions and, often, classifications of pathology (eg. Bell, 1987; Hackett, 1976; Moller-Christensen, 1961; Ortner, 1976a; Rogers et al., 1986). Description in palaeopathology is an absolute pre-requisite for further analysis whether this is to be undertaken or not. It should be emphasised that it is often not possible to assign a 'diagnosis' or cause to pathological lesions observed in human remains so description is essential. Usually, the next stage of analysis is radiography for problematical bone pathology. Beyond this stage most people do not have the time, expertise or the resources to take analysis further and this is an unfortunate limitation. As Garland has recently said (1988:328), " To examine human skeletal remains is a privelege not a right". Surely we should be making the most of the resource.

However, there have been many purely scientific studies of palaeopathology in the past. Studies are currently being undertaken and will, no doubt, continue well into the future. These studies have tended to be carried out by research personnel based within University departments of archaeology, physics, chemistry, medicine, dentistry and anatomy, working on specific scientific research projects over a finite amount of time. Scientific work in palaeopathology is usually funded by the S.E.R.C., M.R.C. and the N.E.R.C. with occasional grants from bodies such as the Wellcome or Leverhulme Trusts.

The point is that the potential for scientific applications in palaeopathology is vast and has not yet been realised fully. This paper will assess the state of the art, past, present and future.

Studies of palaeopathology have developed from the 19th century when early scholars described cases of pathology in archaic human remains. Surveys and analyses of disease prevalence were not common and palaeopathological studies were limited to case studies. There was no concerted effort to relate pathology to cultural behaviour and development of mankind; emphasis was on description only. In the early 20th century there developed a more analytical approach to palaeopathological problems, Sir Marc Armand Ruffer being instrumental in palaeopathological studies (Ruffer, 1921). From 1930-1970 syntheses and specialisms developed and from 1970 to the present day interdisciplinary approaches have become more common (Angel, 1981). The founding of the Palaeopathology Association in the United States has helped palaeopathology greatly in being recognised as a discipline.

This survey of scientific methods in palaeopathology will deal with the different types of analysis that have been and are being carried out, with suggestions for the future. A survey over the past 16 years of two of the major scientific archaeological journals (Archaeometry and Journal of Archaeological Science) reveals that the most common subject areas covered in human bone analysis are dating (particularly of early man), palaeopathology (leprosy, animal palaeopathology, dental caries, neoplastic disease and Harris' Lines), diagenesis (especially in relation to trace element analysis) and trace element analysis itself with particular emphasis on ancient diet. Table 1 shows these subjects by year of occurrence:

Table 1: Papers on human bone analysis in Journal of Archaeological Science and Archaeometry

	Dating	Palaeopathology	Trace Elements	Diagenesis
1974	(1)	2		
1975	1		1	
1976		1	1 (1)	
1977	(1)		1 (1)	
1978			1	
1979	1		1 (1)	
1980	1		1	1
1981		1	1	(1)
1982	1 (1)			
1983	1	2	2	1
1984	1		1 (1)	1 (1)
1985		1	2	1
1986			1 (1)	1
1987		3	(1)	
1988	(1)	1	2	1 (1)
1989	1	2	7	(1)

Note: Numbers in brackets refer to Archaeometry papers
        Numbers without brackets refer to Journal of Archaeological Science papers

Obviously papers on scientific methods in palaeopathology may appear in many different journals and books but the trend seems to indicate an increase in papers devoted to trace element analysis and palaeodiet from the early 1980's to the present day, with a concomitant increase in numbers of papers devoted to diagenesis, a problem well recognised and being studied by some scholars (see Bell, this volume). In addition to the most common subject areas there are other research projects which have been published in these two journals ranging from non-metrical and metrical studies of human remains to radiography, coprolite analysis, early man, ageing techniques, cremations, blood residues and DNA extraction. Strictly speaking any process involving the study of palaeopathology could be termed scientific. If medicine is, "the science of preventing, diagnosing, alleviating, or curing disease", and pathology is, "the branch of medicine concerned with the cause, origin and nature of disease", then the study of macroscopic palaeopathology is a science. However, for the purposes of this paper the use of physical and chemical techniques for human bone analysis are considered. This analysis can be divided into:

1. Histology
2. Radiography
3. Trace element analysis and diagenesis
4. Dating
5. Molecular analysis

The range of techniques now available to palaeopathologists is extensive but not exploited fully. Regrettably, the availability of funds and personnel are limiting exploitation.

Histology

Gross morphological features and measurements by physical anthropologists have been, and always will be, essential but evidence that other skeletal parameters may offer valuable information is becoming increasingly accepted. Palaeohistology has been used in the study of soft tissue more commonly than in the study of bone until the last ten years. However, microscopic analysis, in its many forms, has been used in palaeopathology since late last century (Martin, 1988; Ortner, 1976b). During this time researchers established that histological structure was preserved in human fossil bone (Stout & Simmons, 1979). Moodie (1923) often used histological sections in his work, and Ruffer (1921), in his study of Egyptian mummified material, made contributions to histological research in human remains. Of course, the study of mummified material does give much more scope for histological examination not only for the investigation of preservation of individual components of tissue, but also for palaeopathology (Tapp, 1986). The histological study of skeletal remains has developed perhaps over the last ten years as workers have realised the possibilities of obtaining more information about diet (Martin, Goodman & Armelagos, 1985), disease, characteristics of ageing (Ubelaker, 1986) and taphonomic processes (Garland, 1987) from histological examination of bone sections. Methods of sectioning, embedding and examination of bone have developed in parallel with these ideas (eg.

Stout & Teitelbaum, 1976) and methods of quantification of histological observations taken from modern clinical medicine (eg. Manaka & Malluche, 1981; Revell, 1983) are being considered. In terms of palaeopathology, conditions such as metabolic disturbances, systemic diseases and nutritional deficiencies may be identified in bone section histology. Apart from human bone and teeth, the study of coprolite and gut parasite eggs from bodies and latrine, cess and rubbish pits from waterlogged and dry contexts is developing. This has potential for reconstructing past hygiene and disease.

Another method of histological analysis is the use of the scanning electron microscope (SEM). Light microscopy may be limiting in resolution of fine detail. Using a fine probe of electrons to directly examine the microtopography of specimens, the depth of focus on an SEM is 300 times greater than for a conventional transmission light microscope and resolution is higher (Brothwell, 1969). SEM has been used in the study of archaeological bone and teeth, with Don Brothwell being one of the first to advocate its use in the study of archaeological materials generally; it has since been used intermittently by workers (Olsen, 1988).

In the case of human bone and teeth many studies have been completed (eg. Bromage, 1987 on the biological state of human bone; Dobney & Brothwell, 1986 on dental calculus; Hillson, 1986 on ancient teeth; Shipman, 1981 on taphonomic problems). Little published work has concentrated on palaeopathological aspects of human bone although the characteristic features of disease, that is remodelling of the bone tissue by osteoblasts and osteoclasts (bone forming and destroying cells) can be identified very clearly in some cases. However, this analytical method is becoming recognised by palaeopathological researchers (eg. Wenham, 1987; Wakely, Manchester & Roberts, 1989) for studying current medical problems, one of which is osteoporosis, in archaic material.

In the consideration of microscopic sections of ancient human bone and their relevance to palaeopathology it cannot be ignored that there are many problems posed by the alteration of bone structure whilst it is buried in the ground. It has already been seen that workers are becoming increasingly concerned with diagenesis (as seen in the proliferation of papers on diagenesis in the archaeological science press). Changes to bone structure will vary between bones and individuals on one site and between geographic regions (Hanson & Buikstra, 1987) and these changes are determined by, for example, bone size, temperature, water, micro-organisms, human and animal intervention and pH of the soil. This fact is particularly relevant to all scientific methods in palaeopathology and needs much more work. It is apparent that taphonomic changes in bone will need to be assessed for each cemetery site in the future before health and dietary patterns, for example, can be assessed. Postmortem change and its relevance to the study of archaeological human bone will be discussed further below.

Radiography

Radiography, or the examination of specimens using high energy electromagnetic radiation, of human skeletal remains has been undertaken since the discovery of radium in 1895 by Wilhelm Conrad Roentgen in Germany (Mould, 1980:1). The method can reveal details of composition and structure not visible macroscopically; it is therefore invaluable in palaeopathological studies. The different types of X-ray analysis used in the study of human skeletal remains can be summarised as follows:

1. Macroscopic radiography
2. Microradiography
3. Tomography
4. Electron microprobe analysis
5. X-ray diffraction

Categories 4 and 5 are both related to X-ray analysis. Electron microprobe analysis involves the combined techniques of X-ray Diffraction and Scanning Electron Microscopy. It is possible to analyse the variations in concentration of selected elements over a specimen surface (Tite, 1972:279). X-ray diffraction is usually used more often to identify mineral phases and chemical compounds present in archaeological artefacts (Tite, 1972:285). In the case of the study of archaic human bone it has been used in analysis of diagenetic problems by Sillen and Smith (1984) before more extensive analysis by atomic absorption spectrometry of bone for dietary information, for characterising normal human bone (Ortner & Von Endt, 1971, 1973) and in the discovery of the fraud of Piltdown Man (Brothwell et al., 1969).

The most commonly used of the radiographic techniques in palaeopathology is macroscopic radiography. Radiography has been used by anthropologists throughout this century for examining skeletal material from all periods (eg. Price, 1975), examination of particular aspects of palaeopathology (eg. Mays, 1985; Roberts, 1988), diagenesis of bone (Decker & Bohrod, 1939) and the study of mummified remains. Radiography has, in some instances, helped to estimate ages for skeletons from archaeological sites (eg. Ericksen, 1982) and diagnosis of palaeopathology has as its pre-requisite, in most cases, radiography. In the study of mummified material Brothwell et al. (1969) describe the importance of radiography for determining, without destruction of mummy wrappings, whether a skeleton exists, its sex, age and manifest pathological features.

Problems with postmortem change of the bone structure is, of course, a matter of great concern and importance when using this method of analysis, and suitable controls must be established.

Another method of X-ray is the microradiograph which can resolve details of specimen microstructure to better then 1 micron (Tite, 1972:253). Flat parallel sided sections of bone or tooth, 10-15 microns thick, are used to make contact X-rays which can then be observed under an optical microscope to show the microscopic structure of bone. Microradiographs can provide information about the distribution of mineral in the bone section and, in osteoarchaeology, microradiography has been used to examine ageing effects on the remodelling of bone

(Ortner, 1975), the differences in the rate of bone loss between males and females in ancient populations (Martin & Armelagos (1979) (by examining mineral distribution within bone section microradiographs which is not possible with conventional bone histological sections), investigation of certain pathological processes (Manchester and Roberts, in progress) and assessment of taphonomic processes acting on human bone (Garland, 1985). Perhaps the problem which has not been addressed to any great extent in microradiography is, again, diagenesis.

The relationship of bone loss, particularly to diet and general health of populations has much to offer in the study of human remains if the problem of diagenesis can be solved. The study of bone formation and destruction, the basis of any disease process, is helped by the examination of microradiographs of bone. There are now the hardware and techniques necessary to investigate palaeopathology at the microscopic level to obtain more detailed and more accurate diagnoses.

The final method of radiography, that of tomography, has not yet received much attention in the study of human remains other than whole bodies. A cross-sectional image of the subject is reconstructed from multiple X-ray projections transmitted through the body at 1.5-10mm intervals. This method of examination has been used since the late 1970's particularly in the examination of mummified bodies. More recently, the technique has been used to investigate the body of Lindow Man (Reznek et al., 1986). The advantage of this method over standard radiography is that small differences in tissue density and structure within bodies are well demonstrated (Notmann, 1986).

Unfortunately, the main limitation to the use of radiographic techniques in palaeopathology is funding and time. In the case of microradiography, particularly, this method is very time consuming because bone sections have to be embedded in resin before a section thin enough can be cut for contact microradiography. Tomography of mummified bodies is expensive and should perhaps be limited to especially important specimens.

Trace element analysis

Trace element analysis of human bone and teeth has been used since the early 1960's particularly in attempts to reconstruct palaeodiet of populations (Grupe & Herrman, 1988; Klepinger, 1984; Lambert et al., 1979; Mays, 1989; Price, 1989; Price et al., 1985; Runia, 1987), occupational health, for example, lead exposure (eg. Rogers & Waldron, 1985; Whittaker & Stack, 1984) and, more unique and exciting in terms of future potential, research into specific diseases and detecting the presence of elusive bodies in graves (Bethell, 1989). In the past ten years trace element analysis has seen an increasing popularity, particularly in North America as witnessed by the proliferation of published material.

Methods of analysis used in determining the trace element content of bone and teeth can be divided into two groups, destructive and non-destructive. The former group includes atomic absorption spectroscopy and optical emission spectroscopy. The latter group includes methods such as X-ray fluorescence, electron microprobe and neutron activation analysis. The assessment of palaeodietary information from human remains relies on the accumulation of certain trace elements in bone and teeth. With knowledge of elemental content of specific foods in the modern context, and with assumptions on a similar content in the past, it is believed to be possible to infer the quality and quantity of food and water ingested in the past. It has been found that some elements occur in greater abundance in meat rather than in vegetable foods. For example, zinc, copper and molybdenum are associated with animal protein and strontium, magnesium, manganese, cobalt and nickel are more prevalent in vegetable foods (Gilbert, 1985:347), and resarchers have used this data to assess aspects such as changes in amount and type of food eaten. Two approaches to trace element analysis and dietary reconstruction are the assessment of the strontium/calcium ratio and the analysis of a spectrum of elements. Thereby the relationship of specific elements to dietary characteristics may be determined and changes in trace element concentrations resulting from metabolic or physiological effects on skeletal material may be observed (Gilbert, 1985:353). The latter approach is the one recommended by most workers (eg. Price et al., 1985).

In addition to trace element analysis, the study of the abundance and ratios of stable isotopes, particularly of carbon and nitrogen, in bone collagen is becoming increasingly common (see Lee-Thorp et al., 1989; Lovell et al., 1986; Schwarz et al., 1985) in order to determine the contribution of certain isotopically distinctive nutrients to the diet. Collagen is relatively susceptible to diagenetic influences of soil and therefore should be ideal for the study of carbon and nitrogen isotopes (Runia, 1987:19). In addition to representing certain nutrients of the diet of past peoples, carbon isotope analysis has been used to distinguish marine from terrestrial diets (eg. Lazenby and McCormack, 1985).

The study of occupational exposure to hazardous substances in past periods of time has received particular attention especially with respect to lead concentrations (Rogers & Waldron, 1985). It is accepted that in the Roman period, for example, methods of food preparation and distribution of water supplies involved lead cooking vessels, lead glazed pottery and lead piping with the result that many individuals would have suffered high intakes of lead into the bloodstream. Again the problems of diagenesis have been recognised (Waldron, 1981, 1982) and recommendations for the analysis of parallel soil samples as controls have been made.

The analysis of trace elements in archaeological human bone is not without its pitfalls. Many workers are now beginning to address the problem of diagenesis, (eg. see Hancock et al., 1989; Hanson & Buikstra, 1987; Lambert et al., 1984; Pate & Hutton, 1988; Tuross et al., 1989). Methods such as neutron activation and microprobe analysis, atomic absorption spectroscopy, X-ray fluorescence, inductively coupled plasma atomic emission spectrometry and microradiography have been used to address

the problem of diagenesis and suggestions for the removal of inorganic contamination have been made recently by Lambert et al. (1989). It is obvious that diagenesis can affect the true distribution of elements in bone and teeth due to uptake from the soil. For example, an acidic soil may cause removal of zinc, copper and manganese from organic material (Gilbert, 1985:352). The analysis of soil samples and animal bones from the same site (assuming the latter derive from the same area as the human bone) is advocated. Caution must also be expressed with regard to the assessment of the potential food resources available to past communities; and further analysis is necessary on a wider range of foodstuffs for control samples. However, the problem of assessing the range of food exploitation and whether it is reflected in the human material is a difficult one to solve. There are also other factors to be taken into consideration in trace element analysis and other scientific methods. Micro-organisms may also become incorporated into the bone matrix. The choice of anatomical element to be analysed can affect results because of different bone turnover rates (Gilbert, 1985:350; Grupe, 1988) and work has been done in many cases on the variability of trace element content of bone and teeth at various depths in the actual bone and tooth. Comparable procedures for trace element analysis, inter-laboratory reference materials and addressing the problem of diagenesis is advocated urgently if meaningful results are to be obtained. There is, however, great potential for trace element analysis in the study of palaeodiet in relation to social status, age and sex of the individual. The consequence of different types of diet on the skeleton, particularly leading to subsequent disease, needs further work as does work on sources of variation within bones of different age, sex and metabolism and between individuals from different sites. In fact many of these issues have been addressed recently by Price (1989).

Dating

Scientific methods in palaeopathology also include dating of human remains, perhaps to establish a gap in the history of development of a particular disease or to provide a date for a specific burial.

The main methods used for dating human remains currently in use include radiocarbon, amino acid racemisation and electron spin resonance (ESR) (Zimmerman & Angel, 1986). Other methods include the relative dating method of nitrogen, fluorine and uranium series dating (eg. see Haddy & Hanson, 1982; Tiemei & Sixun, 1988). These methods rely on the accumulation or depletion of certain elements after burial in the soil.

For time periods older than 10,000 years BP amino acid racemisation dating provides a cross check for radiocarbon dating, and for periods older than 70,000 years amino acid dating is the only direct dating technique available (Masters, 1986:3). For dates below 10,000 BP radiocarbon dating is the only physicochemical method of absolute dating of the organic or inorganic portion of bone. Until the late 1970's conventional radiocarbon dating required a considerable quantity of bone (100mg plus). The development of the radiocarbon accelerator approach made it possible to date much smaller amounts of organic material (see Gowlett & Hedges, 1986). The accelerator method is viewed with enthusiasm because of the small quantity of bone needed for the process. Museum curators are often reluctant to allow large samples of what may be very incomplete skeletons to be sent for dating, particularly in the case of prehistoric material.

Amino acid racemisation dating is based on the fact that L-amino acids found in protein of living organisms undergo a spontaneous chemical racemisation following burial. The rate is dependent on the average temperature to which the fossil has been exposed since deposition (Masters, 1986:56). Water is necessary for racemisation to take place but copper and iron in the soil may accelerate collagen collapse. Small samples of bone can be processed (5-10g) which makes it a comparable technique to radiocarbon dating by the accelerator approach, in terms of destruction of the bone. However, the method often faces the problem of incorrect

calibrations, but dates can be obtained from several thousand to several hundred thousand years ago.

Electron spin resonance (ESR) has recently been developed (Ikeya, 1986:59) as a method for dating biological materials but has yet to develop in terms of methodology and application to bone dating.

Fluorine and uranium dating relies on the accumulation of these elements after burial of the bone in the ground and nitrogen dating is based on nitrogen loss in bone through the disappearance of protein during burial. Uptake of fluorine and uranium depends on the structure of bone; the spongy areas, internally, absorb more. This dating technique is good for determining whether a bone or burial is really the age suggested by its position in a deposit or whether it is a later intrusion as was seen with the Galley Hill and Piltdown Man bone around the turn of the century.

Molecular analysis

Molecular analysis, for the purposes of this paper, includes blood grouping and DNA extraction, both scientific techniques applicable to palaeopathology and the reconstruction of past human behaviour. The former has been well used to date but the latter is still developing.

The earlier years of this century saw the first attempt at blood grouping archaic human remains (Katsunama & Katsunama, 1929). Most early attempts were made on mummified material and it was not until the late 1930's that attempts to determine blood groups from archaic human bones developed (Candela, 1940). Lengyel (1975) established the antibody fluorescent method for blood group determination which has not been improved on to date. The majority of work on blood grouping has, as mentioned above, been carried out on mummies (eg. see Flaherty & Haigh, 1986) but promising work has been undertaken on bone material (Gruspier, 1985) with recommendations for bacterial cultures and biochemical tests on the bone material to ascertain the presence of organic substances and stability of the elements which make up the blood substance before blood grouping is attempted. Whilst blood grouping of mummified material can be undertaken now with a large degree of accuracy, the methods used on bone need further work.

Fascinating work on analysing blood residues on stone tools is currently being done (see summary by Bahn, 1987). By comparing the electrophoretic pattern of blood residues with controls from different animals, the species from which the blood came can be identified. This method has considerable potential for the reconstruction of past human behaviour including disease, genetics, human evolution, diet and environment.

Blood residues are also being investigated at the site of a Medieval hospital in Scotland (Moffat, 1988:27) in pits which are believed to contain the remnants of blood letting, a common practice in the Medieval period for relieving the afflicted of disease.

Clearly the analysis of blood groups has much to offer in reconstructing the past history of disease. For example, certain diseases are apparently associated with specific blood groups (Manchester, 1983:15) and promising work has been done on the diagnosis of the blood disease of thalassaemia in ancient bone (Ascenzi et al., 1988).

A recent development in the scientific analysis of archaeological human bone is molecular analysis (Hauswirth et al., 1988; Tuross, 1988). Hedges (1989) notes that the potential for molecular analysis of bone is vast and includes assessment of blood disorders, infections, parasitic disease, genetic associations and even allergies and colour blindness. Whilst the study of the molecular structure of bone is still in its infancy, the extraction of DNA from mummified material is relatively advanced (see Paabo, 1986).

Conclusion

In conclusion, this review of scientific methods in palaeopathology cannot be entirely complete but the major areas of past and present developments and research have been covered. It was pleasing to see more papers offered on scientific applications in human remains at this Bradford conference compared with the Glasgow proceedings of two years ago. Perhaps this is a reflection of the realisation of the potential for such scientific applications or maybe the increase in funding for such studies. Clearly, using only macroscopic and radiography in palaeopathology is often not the most accurate method of investigating pathology. However, in most cases, the hardware and knowledge of techniques are not available to the majority of people working in the field.

What is needed are opportunities for training students to carry out scientific analyses on human material, and there are some centres equipped in this respect, with adequate funding for scientific applications in palaeopathology. At present, it is only the fortunate, usually in University departments, who have facilities and finance to carry out such work. In addition, it is essential that if scientific analyses of human remains are to be a pre-requisite, pre-excavation planning with archaeologists must be carried out, the presence of a human skeletal biologist being present on site being the least requirement. In the future there will probably be further advances made in the study of palaeodiet by trace element and isotopic analysis and in molecular work and its relation to disease processes in human remains. Finally more research must be undertaken on diagenesis.

Bibliography

Angel, J.L. (1981). History and development of palaeopathology. *Amer. J. Phys. Anthrop.*, 56, 509-515.

Ascenzi, A., Bellelli, A., Brunori, M., Citro, G., Ippoliti, R., Lendaro, E. and Zito, R. (1988). in press. *The diagnosis of thalassemia in ancient bones: problems and new perspectives in pathology*. Paper presented at the International Congress of Anthropological and Ethnological Sciences symposium on Human Palaeopathology: current syntheses and future options. Zagreb, Yugoslavia, July, 1988.

Bahn, P.G. (1987). Getting blood from stones. *Nature* 330, 14.

Bell, L. (1987). *Periostitis and problems of classification*. Unpublished final year dissertation, University of Bradford.

Bethell, P. (1989). Chemical analysis of shadow burials. In: *Burial archaeology, current research, methods and developments*. British Archaeological Reports. British Series 211, ed. C. Roberts, F. Lee and J. Bintliff, pp. 205-214. Oxford.

Bromage, T. (1987). The scanning electron microscope/replica technique and recent applications to the study of fossil bone. *Scanning Electron Microscopy* 1, 607-613.

Brothwell, D. (1969). The study of archaeological materials by means of the scanning electron microscope, an important new field. In: *Science in archaeology. A survey of progress and research*, ed. D. Brothwell & E.S. Higgs, pp. 564-566. Thames and Hudson, London.

Brothwell, D., Molleson, T., Gray, P. and Harcourt, R. (1969). The application of X-rays to the study of archaeological materials. In: *Science in archaeology. A survey of progress and research*, ed. D. Brothwell & E.S. Higgs, pp. 513-525. Thames and Hudson, London.

Candela, P.B. (1940). Reliability of blood group tests on human bones. *Amer. J. Phys. Anthrop.* 27, 365-381.

Decker, F.H. & Bohrod, M.G. (1939). Medullary artefacts in prehistoric bones. *Amer. J. Roentgenol* 42, 374-5.

Dobney, K. & Brothwell, D. (1986). A scanning electron microscope study of archaeological dental calculus. In: *Scanning electron microscopy in archaeology*. British Archaeological Reports. International Series 452, ed. S.L. Olsen, pp. 372-385. Oxford.

Ericksen, M.F. (1982). Ageing changes in thickness of the proximal cortex. *Amer. J. Phys. Anthrop* 59, 121-130.

Flaherty, T. & Haigh, T.J. (1986). Blood groups in mummies. In: *Science in Egyptology*, ed. A.R. David, pp. 379-382. Manchester University Press, Manchester.

Garland, A.N. (1985) A palaeohistological study of bone decomposition. Unpublished MA Thesis, Department of Archaeology and Prehistory, University of Sheffield.

Garland, A.N. (1987). A histological study of archaeological bone decomposition. In : *Death, decay and reconstruction. Approaches to archaeology and forensic science*, ed. A. Boddington, A.N. Garland & R.C. Janaway, pp. 109-126. Manchester University Press, Manchester.

Garland, A.N. (1988). Contributions to palaeohistology. In: *Science and Archaeology, Glasgow, 1987*. British Archaeological Reports. British Series 196 (ii), ed. E. Slater & J. Tate, pp. 321-329. Oxford.

Gilbert, R.I. (1985). Stress, paleonutrition and trace elements. In: *Analysis of prehistoric diets*, ed. R.I. Gilbert & J.H. Mielke, pp. 339-358. Academic Press, London.

Gowlett, J.A.J. & Hedges, R.E.M. (eds.) (1986). *Archaeological results from accelerator dating*. Oxford University Committee for Archaeology, Oxford.

Grupe, G. (1988). Impact of the choice of bone samples on trace element data in excavated human skeletons. *J. Archaeological Sci.*, 15, 123-129.

Grupe, G. & Herrmann, B. (eds) (1988). *Trace elements in environmental history*. Springer Verlag, London.

Gruspier, K. (1985). *Paleoserology: history and new applications to the Casal San Vincenzo skeletal material*. Unpublished M.A.

thesis, Department of Archaeology and Prehistory, University of Sheffield.

Hackett, C. (1976). *Diagnostic criteria of syphylis, yaws and treponarid (treponematoses) and of some other diseases in dry bone*. Springer Verlag, New York.

Haddy, A. & Hanson, A. (1982). Research notes and application reports on nitrogen and fluorine dating of Moundville skeletal samples. *Archaeometry*, 24(1), 37-44.

Hancock, R.G.V., Grynpas, M.D. & Pritzker, K.P.H. (1989). The abuse of bone analysis for archaeological dietary studies. *Archaeometry*, 31(2), 169-179.

Hanson, D.B. & Buikstra, J.E. (1987). Histomorphological alterations in buried human bone from the Lower Illinois Valley: implications for palaeodietary research. *J. Archaeological Science*, 14, 549-563.

Hauswirth, W.W., Dickel, C.D., Doran, G.H., Laipis, P.J. & Dickel, D.N. (1988) in press. 8,000 year old brain tissue from the Windover site: anatomical, cellular and molecular analysis. Paper presented at the International Congress of Anthropological and Ethnological Sciences symposium on Human Palaeopathology: current syntheses and future options. Zagreb, Yugoslavia. July, 1988.

Hedges, R. (1989). Molecular biological information from bone. Unpublished paper presented at the 10th Anniversary Conference of the Association for Environmental Archaeology, Institute of Archaeology, London.

Hillson, S. (1986). The scanning electron microscope and the study of ancient teeth. In: *Scanning electron microscopy in archaeology*. British Archaeological Reports. International Series 452, ed. S.L. Olsen, pp. 249-260. Oxford.

Ikeya, M. (1986). Electron spin resonance. In: *Dating and age determination of biological materials*, ed. M.R. Zimmerman & J.L. Angel, pp. 59-125. Croom Helm, London.

Katsunama, R. & Katsunama, S. (1929). On the bone marrow-cells of man and animal in the Stone Age of Japan. *Proc. Imperial Academy of Japan* 5, 388-389.

Klepinger, L.L. (1984). Nutritional assessment from bone. *Annual Review Anthropol.* 13, 75-96.

Lambert, J.B., Szpunar, C.B. & Buikstra, J.E. (1979). Chemical analysis of excavated human bone from Middle and Late Woodland sites. *Archaeometry* 21(2), 115-129.

Lambert, J.B., Simpson, S.V., Szpunar, C.B. & Buikstra, J.E. (1984). Copper and barium as dietary discriminants: the effects of diagenesis. *Archaeometry* 26(2), 131-8.

Lambert, J.B., Xue, L. & Buikstra, J.E. (1989). Physical removal of contaminative inorganic material from buried human bone. *J. Archaeological Sci.* 16, 427-436.

Lazenby, R.A. & McCormack, P. (1985). Salmon and malnutrition on the north-west coast. *Current Anthropology* 26(3), 379-384.

Lee-Thorp, J.A., Sealy, J.C. & Van Der Merwe, N.J. (1989). Stable carbon isotope ratio differences between bone collagen and bone apatite and their relationship to diet. *J. Archaeological Sci.* 16, 585-599.

Lengyel, I.A. (1975). *Palaeoserology. Blood typing with the fluorescent antibody method*. Akademiai Kiado, Budapest.

Lovell, N., Nelson, D.E. & Schwarz, H.P. (1986). Carbon isotope ratios in palaeodiet: lack of age or sex effect. *Archaeometry* 28(1), 51-55.

Manaka, R.C. & Malluche, H.H. (1981). A program package for quantitative analysis of histologic structure and remodeling dynamics of bone. *Computer Programs in Biomedicine* 13, 191-202.

Manchester, K. (1983). *Archaeology of disease*. Bradford University Press, Bradford.

Manchester, K. & Roberts, C.A. (in progress) - A palaeopathological investigation of the diagnostic criteria for tuberculosis.

Martin, D.L. (1988) in press. Bone histology and palaeopathology: methodological considerations. Paper presented at the International Congress of Anthropological and Ethnological Sciences Symposium on Human Palaeopathology, Zagreb, Yugoslavia, July, 1988.

Martin, D.L. & Armelagos, G.J. (1979). Morphometrics of compact bone: an example from Sudanese Nubia. *Amer. J. Phys. Anthrop.* 51(4), 57-78.

Martin, D.L., Goodman, A.H. & Armelagos, G.J. (1985). Skeletal pathologies as indicators of quality and quantity of diet. In: *The analysis of prehistoric diets*, ed. J. Mielke & R.I. Gilbert, pp. 227-279. Academic Press, New York.

Masters, P.M. (1986). Amino acid racemisation - a review. In: *Dating and age determination of biological materials*, ed. M.R. Zimmerman & J.L. Angel, pp. 39-58. Croom Helm, London.

Mays, S. (1985). The relationship between Harris line formation and bone growth and development. *J. Archaeological Sci.* 12, 207-220.

Mays, S. (1989). Human bone strontium analysis in the investigation of palaeodiets: a case study from a British Anglo-Saxon site. In: *Burial Archaeology: current research, methods and developments.* British Archaeological Reports. British Series 211, ed. C. Roberts, F. Lee & J. Bintliff, pp. 215-233. Oxford.

Moffat, B. (1988). *Sharp practice 2. 2nd Report on researches into the Medieval hospital at Soutra, Lothian Region, Scotland.* SHARP.

Moller-Christensen, V. (1961). *Bone changes of leprosy.* Munksgaard, Copenhagen.

Moodie, R.L. (1923). *Palaeopathology.* University of Illinois Press, Urbana.

Mould, R.F. (1980). *History of X rays and radium.* London.

Notman, D.N.H. (1986). Ancient scannings: computed tomography of Egyptian mummies. In: *Science in Egyptology.*, ed. A.R. David, pp. 251-320. Manchester University Press, Manchester.

Olsen, S.L. (ed) (1988). *Scanning electron microscopy in archaeology.* British Archaeological Reports. International Series 452. Oxford.

Ortner, D.J. (1975). Ageing effects on osteon remodelling. *Calcified Tissue Research* 18, 27-36.

Ortner, D.J. (1976a). Description and classification of degenerative change in the distal joint surfaces of the humerus. *Amer. J. Phys. Anthrop.* 28(2), 139-156.

Ortner, D.J. (1976b). Microscopic and molecular biology of human compact bone: an anthropological perspective. *Yearbook of Phys. Anthrop.* 20, 35-44.

Ortner, D.J. & Von Endt, D.W. (1971). Microscopic and electron microprobe characterisation of the sclerotic lamellae in human osteons. *Israel J. Med. Sci.* 7(3), 480-482.

Ortner, D.J. & Von Endt, D.W. (1973). Electron microprobe analysis of the primary spongiosa in human foetal bone. *Proceedings of the 9th European Symposium on calcified tissues, Vienna, 1972.* Facta Publication.

Paabo, S. (1986). DNA is preserved in ancient Egyptian mummies. In: *Science in Egyptology*, ed. A.R. David, pp. 383-388. Manchester University Press, Manchester.

Pate, F.D. & Hutton, J.T. (1988). Use of soil chemistry data to address postmortem diagnosis in bone mineral. *J. Archaeological Sci.* 15, 729-739.

Price, J.L. (1975). Radiology of excavated Saxon and Medieval human remains from Winchester. *Clin. Radiol.*, 26, 363-370.

Price, T.D., Schoeninger, M.J. & Armelagos, G.J. (1985). Bone chemistry and past behaviour: an overview. *J. Human Evolution* 14, 419-447.

Price, T.D. (ed) (1989). *The chemistry of prehistoric human bone.* Cambridge University Press, Cambridge.

Revell, P. (1983). Histomorphometry of bone. *J. Clin. Pathol.* 36, 1323-1331.

Reznek, R.H., Hallett, M.G. & Charlesworth, M. (1986). Computed tomography of Lindow Man. In: *Lindow Man. The body in the bog.,* ed. I.M. Stead, J. Bourke & D. Brothwell, pp. 63-5. Guild Publishing, London.

Roberts, C.A. (1988). Trauma and treatment in British antiquity: a radiographic study. In: *Science and Archaeology, Glasgow, 1987.* British Archaeological Reports. British Series 196(ii), ed. E. Slater & J. Tate, pp. 339-359. Oxford.

Rogers, J. & Waldron, T. (1985). Lead concentrations in bones from a Neolithic long barrow. *J. Archaeological Sci.,* 12, 93-96.

Rogers, J., Waldron, T., Dieppe, P. & Watt, I. (1986). Arthropathies in palaeopathology: basis of classification according to most probable cause. *J. Archaeological Sci.,* 14, 179-193.

Ruffer, M.A. (1921). *Studies in the palaeopathology of Egypt.* University of Chicago Press, Chicago.

Runia, L.T. (1987). *Chemical analysis of prehistoric bones. A palaeodietary and ecoarchaeological study of Bronze Age West Friesland.* British Archaeological Reports. International Series. Vol. 363. Oxford.

Schwarz, H.P., Melbye, J., Katzenberg, M.A. & Knyf, M. (1985). Stable isotopes in human skeletons of southern Ontario: reconstructing palaeodiet. *J. Archaeological Sci.* 12, 187-206.

Shipman, P.L. (1981). Application of scanning electron microscopy to taphonomic problems. *Annals of the New York Academy of Science* 276, 357-385.

Sillen, A. & Smith, P. (1984). Weaning patterns are reflected in strontium-calcium ratios of juvenile skeletons. *J. Archaeological Sci.* 11, 237-245.

Stout, S.D. & Simmons, I. (1979). Use of histology in ancient bone research. *Yearbook of Physical Anthropology* 22, 228-249.

Stout, S.D. & Teitelbaum, S.L. (1976). Histological analysis of undecalcified thin sections of archaeological bone. *Amer. J. Phys. Anthrop.* 44, 263-270.

Tapp, E. (1986). Histology and histopathology of Manchester mummies. In: *Science in Egyptology.,* ed. A.R. David, pp.347-350. Manchester University Press, Manchester.

Tiemei, C. & Sixun, Y. (1988). Uranium-series dating of bones and teeth from Chinese palaeolithic sites. *Archaeometry* 30(1), 59-76.

Tite, M.S. (1972). *Methods of physical examination in archaeology.* Seminar Press, London.

Tuross, N. (1988) in press. Recovery of bone and serum proteins from human skeletal tissue: IgG, osteonectin and albumin. Paper presented at the International Congress of Anthropological Ethnological Sciences symposium on Human Palaeopathology: current syntheses and future options. Zagreb, Yugoslavia. July 1988.

Tuross, N., Behrensmeyer, A.K. & Eanes, E.D. (1989). Strontium increases and crystallinity changes in taphonomic and archaeological bone. *J. Archaeological Sci.* 16, 661-672.

Ubelaker, D. (1986). Estimation of age at death from histology of human bone. In: *Dating and age determination of biological materials,* ed. M.R. Zimmerman & J.L. Angel, pp. 240-247. Croom Helm, London.

Wakely, J., Manchester, K. & Roberts, C.A. (1989). Scanning electron microscope study of normal vertebrae and ribs from early Medieval human skeletons. *J. Archaeological Science* 16.

Waldron, T. (1981). Postmortem absorption of lead by the skeleton. *Amer. J. Phys. Anthrop.* 55, 395-398.

Waldron, T. (1982). Lead in bones: a cautionary tale. *Ecology of Disease* 1(2/3), 191-196.

Wenham, S. (1987). *Anatomical and microscopic interpretations of ancient weapon injuries.* Unpublished BSc thesis, Department of Anatomy, University of Leicester.

Whittaker, D.K. & Stack, M.V. (1984). The lead, cadmium and zinc content of some Romano-British teeth. *Archaeometry* 26(1), 37-42.

Zimmerman, M.R. & Angel, J.L. (1986). *Dating and age determination of biological materials.* Croom Helm, London.

KITCHEN CHEMISTRY? THE USE OF MICROWAVES IN THE ANALYSIS OF HUMAN BONE

L.C. Winter and C.A. Marlow

Department of Anthropology, University of Durham, 43 Old Elvet, Durham, DH1 3HN.

Elemental analysis of human bone recovered from archaeological contexts may provide valuable information about the diet and health of past populations. Certain analytical techniques, notably atomic absorption and the technique currently under investigation in Durham, inductively coupled plasma mass spectrometry (ICP-MS), require that the bone is in solution. It is obviously vital that there is complete dissolution of the bone, and that no other elements are permitted to contaminate the sample so that the investigator can be confident that the analysis relates to the elements actually present in the bone at the time of excavation. Whether it also relates to the elements present during the individual's lifetime is another question, and one which is not addressed here. The ideal dissolution method to meet these criteria would be one using the smallest possible number of vessels, to limit the possibility of contamination from inadequately cleaned equipment, and taking the shortest time, to limit the possibility of leaching occurring. In addition the method must be reproducible so that the investigator may be confident that all samples receive identical treatment and valid comparisons between samples may be made. This paper presents some preliminary work comparing the traditional method of dissolving bone samples (Szpunar et al., 1978) with a proposed method using microwaves rather than direct heating to speed the dissolution process.

Techniques

In Szpunar et al.'s (1978) technique, 1 g of ashed bone is placed in a 125 ml glass beaker, 25 ml of concentrated nitric acid is added, the beaker is covered with a watch glass and placed on a hotplate at 90°C. The sample is left overnight in a fume cupboard and evaporated to dryness, resulting in a pink, white or brownish solid sample. 5 ml concentrated nitric acid is added and the watch glass and beaker sides are washed down into the sample with deionised water. The beaker is then boiled gently for 5 minutes. The small silica residue remaining after this process is filtered out by vacuum aspiration through a scintered glass funnel, and any material adhering to the sides of the beaker is dislodged with a rubber policeman. The beaker is thoroughly washed with deionised water and the washings also filtered. The resultant solution is reduced by gentle heating to fit a 50 ml volumetric flask. The sample is then ready for dilution, if necessary, and analysis.

The technique we used to microwave bone samples is much simpler. 0.5 g of bone ash is placed in a 60 ml teflon pressure vessel and 5 ml of concentrated nitric acid is added. The vessel is left open in the fume cupboard for 1 hour, to allow any build up of gas to dissipate. The vessel is then sealed and placed in the microwave oven. The oven is set to give 1.5 minutes heating at a power level of 500 watts, and after the heating has finished the sample is left undisturbed for a period of 10 minutes to allow the pressure to reduce. This process is repeated 5 times to give a total heating period of 7.5 minutes. The vessel is not heated continuously for 7.5 minutes because of the danger of allowing too high a pressure to build up in a sealed container. The sample is then ready for dilution, if necessary, and analysis.

Precautions

There are some precautions that are necessary when heating acids to the very high level achievable in a microwave oven. Free circulating acid would cause corrosion to the inside of an oven and may also destroy the door seals which prevent microwaves escaping (Emsley, pers. comm) so it is important to carry out the heating in a sealed pressure vessel. Manufacturers of microwave ovens advise strongly against using sealed containers in their products, as there is a serious risk of explosion if pressure builds beyond the capacity of the vessel to contain it; it is therefore necessary to avoid over-filling the vessel and to heat it gently and for short periods of time. If the sample is likely to give off gasses, it is necessary to allow these to dissipate before the vessel is sealed, as these may also cause an explosion. It is also advisable to leave a beaker of water in the microwave when it is in use, as the small size of the sample may be insufficient for the magnetron to work effectively. Further information concerning the use of microwaves for chemical reactions may be obtained from Emsley (1988).

Methods

In order to compare the two techniques described above, a sample of IAEA H5 animal bone and a sample of bone recovered from a Romano-British site were prepared. The H5 bone was supplied as a fine powder and this was then dried, weighed, ashed at 500°C for 12 hours to remove the organic content and then re-weighed so that the actual organic proportion could be assessed. The archaeological bone was first cleaned and dried and then ground to a fine powder before being treated in the same way as the H5 animal bone. All the glassware and teflon vessels were cleaned in concentrated nitric acid diluted 1:1 with pure water (obtained from an Elgastat water purifier).

The techniques described above were followed closely, using two samples each of 0.5 g of the archaeological bone for the microwave technique and a further 2 for Szpunar et al.'s. (1978) technique. As only a small quantity of H5 animal bone was available, only 1 sample of this bone could be tried with each technique. Some difficulty was encountered in following Szpunar et al.'s (1978) technique as the samples did not evaporate to dryness overnight in the fume cupboard. The 3 samples and the blank prepared at the same time took between 30

and 72 hours to dry - a problem which may increase the possibility of leaching occurring during preparation. Following preparation the samples were diluted to the required level and analysed by ICP-MS.

Results

A total of 63 elements in the bone were analysed. Of these 63 elements, 54 showed a much higher concentration in the microwaved samples. 9 elements showed a higher concentration in the samples prepared on the hotplate. 8 of these 9 elements were present in concentrations of less than 4 mg/kg. The exception to this, bromine, was present in one of the hotplate samples at a level of over 3 000 mg/kg, when one would have expected a level of approximately 30 mg/kg. It seems likely that this sample had been contaminated during preparation.

Figures 1 and 2 compare the results achieved for 9 elements analysed. All were present at levels in excess of 50 mg/kg, with the exception of boron, which was the most abundant element for which the hotplate produced higher results than the microwave. The graphs have been constructed using logarithmic graph paper so that calcium, present at over 200 g/kg and boron, present at less than 5 mg/kg may be shown. The results for bromine, as suggested above, are likely to be the result of contamination, rather than to show actual levels in bone, but are included as they appear to demonstrate the potential for contamination of samples during preparation.

Conclusions

This very limited experiment would suggest that it may be worthwhile considering using a microwave oven to dissolve bone samples. At the very least it is quicker, cleaner, easier and uses less equipment; it would also appear to be a more efficient method. Further work is needed to prove this conclusion on a much larger sample size. In addition further work is needed to establish the optimum amount of

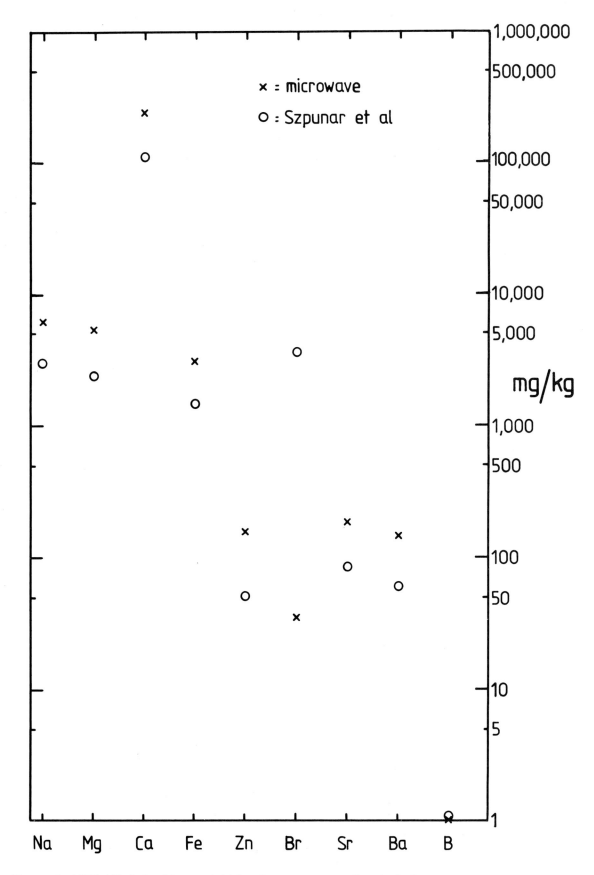

Figure 1. IAEA H5 Animal bone yield for the two preparation techniques.

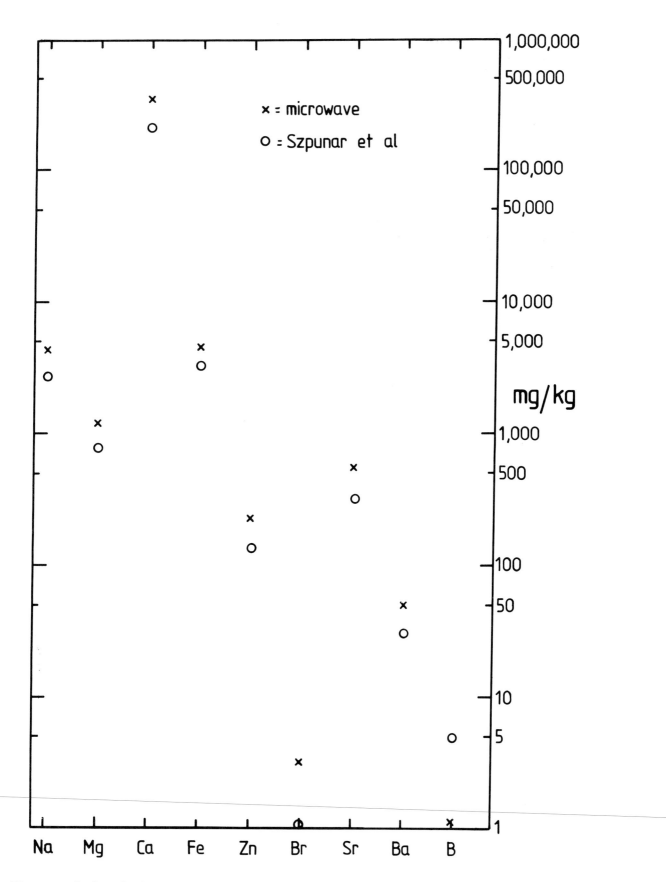

Figure 2. Archaeological bone yield for the two preparation techniques.

microwave heating required to achieve maximum dissolution of bone.

References

Emsley, J. (1988). The chemist's quick cookbook. *New Scientist* 12 November 1988, 56-60.

Szpunar, C.B., Lambert, J.B. & Buikstra, J.E. (1978). Analysis of excavated bone by atomic absorption. *American Journal of Physical Anthropology* 48, 199-202.